Metakaolin-Based Geopolymers

Dongming Yan · Shikun Chen · Yi Liu

Metakaolin-Based Geopolymers

Design, Mechanisms and Performance

Springer

Dongming Yan
Zhejiang University
Hangzhou, China

Shikun Chen
Zhejiang University
Hangzhou, China

Yi Liu
Zhejiang University
Hangzhou, China

ISBN 978-981-97-0651-8 ISBN 978-981-97-0652-5 (eBook)
https://doi.org/10.1007/978-981-97-0652-5

This Springer imprint is published by the registered company Springer Nature Singapore Pte Ltd.
The registered company address is: 152 Beach Road, #21-01/04 Gateway East, Singapore 189721, Singapore

Preface

Ordinary Portland cement (OPC) is the most widely used and essential binder material for civil engineering constructions. However, the intensively consumption of raw materials (e.g., limestone) and energy (during grinding and calcination) as well as the resulted massive CO_2 emission led to a series of environmental problems in the OPC industry. These problems drive researchers worldwide to look for more environment-friendly and high-performance binders. Geopolymer, as a new type of inorganic binder, has opened up a potential route to solve those problems and has attracted great interest of worldwide researchers and engineers in the last three decades. As one of the mainstream types of geopolymer, metakaolin-based geopolymer (MKG) is often regarded as the model system for the fundamental researches and applications of geopolymers.

This book summarizes the researches and findings about the design, mechanisms and performances of MKG conducted at Zhejiang University. It aims to reveal the underlying mechanisms behind the differences in performances of MKG and OPC. This book includes a wide range of topics around MKG. From the early-stage geopolymerization of MKG to the final mechanical performance of MKG concrete, this book tries to establish a fundamental knowledge about the performances of MKG-based materials and their correlating mechanisms with the composition and microstructure of MKG. The outcomes of this book may provide basic directions for the design and application of MKG in civil engineering constructions.

This book contains nine chapters. In Chap. 1, a brief overview of geopolymer and MKG is given. In Chap. 2, the early-stage geopolymerization process of MKG is discussed. In Chap. 3, the mechanical performance of MKG and its correlations to the composition and microstructure are discussed. In Chap. 4, the drying shrinkage mechanism of MKG is discussed. In Chap. 5, the sulfate corrosion performance of MKG and the underlying mechanism are discussed. In Chap. 6, the high temperature deterioration mechanism of MKG is discussed. In Chap. 7, the freeze-thaw deterioration mechanism of MKG is discussed. In Chap. 8, the influence of aggregate on mechanical performance of MKG concrete is discussed. In Chap. 9, the reinforcement bonding behaviors of MKG concrete and the influence factors are discussed.

This book is intended for scientists and engineers in material science and civil engineering who are interested and devoted to using geopolymer (especially MKG) as a new technology to make a green future.

Hangzhou, China

Dongming Yan
Shikun Chen
Yi Liu

Acknowledgments

The authors gratefully acknowledge the National Key R&D Program of China (Nos. 2022YFE0109200 and 2022YFB4102100), the National Natural Science Foundation of China (Nos. 52171277, 52101328 and 51978608), the Natural Science Foundation of Zhejiang Province (No. LTGS23E090005), and Shanxi-Zheda Institute of Advanced Materials and Chemical Engineering (No. 2022SZ-TD006) for the financial supports.

The authors would also thank the professors, postdoctoral researchers, and postgraduate students who worked together on the projects. They are Qiang Zeng, Chenglin Wu, Hedong Li, Hailong Ye, Jiyang Wang, Tian Ye, Xiaoqian Qian, Shaoqin Ruan, Hongyuan Fang, Yu Peng, Satoru Fujitsu, Tian-Nan Ye, Yunjin Hu, Wenbin Zheng, Xingliang Sun, Hao Lin, Shilang Xu, Fan Yang, Xiuyu Zhu, Hamed Fazile, Yajun Zhang, Jiangchuxiong Jin, Yu Ao, Nv Han, Lingjun Xie, Yaogu He, Shengqian Ruan, Jiaxi Mao, Yilu Qiu, and Piaoxue Shiyang. Their hardwork, dedication, and intelligence have contributed considerably to the work presented in this book.

Any opinions, findings, and conclusions or recommendations expressed in this book are those of the authors and do not necessarily reflect the views of the sponsors, any institutions, or organizations.

Contents

Abbreviations

AAM	Alkali-activated material
AE%	Alkali equivalent
APSD	Accumulative pore size distribution
BET	Brunauer–Emmett–Teller
BSE	Backscattered electron
CFRP	Carbon fiber-reinforced polymer
CH	Calcium hydroxide
CPSD	Cumulative pore surface area distribution
CPVD	Cumulative pore volume distribution
C-S-H	Calcium silicate hydrate
DB	Deformed bar
DIF	Dynamic increasing factor
DPSD	Differential pore size distribution
DSC	Differential scanning calorimetry
EDS	Energy-dispersive spectroscopy
FAG	Fly ash-based geopolymers
FRP	Fiber-reinforced polymer
F-T	Freezing-thawing
FTIR	Fourier transform infrared
GPC	Geopolymer concrete
H_2O%	Mass content of water
ITZ	Interfacial transition zone
LOI	Loss on ignition
LVDT	Linear variable displacement transducer
MAS	Magic angle spinning
MIP	Mercury intrusion porosimetry
MK	Metakaolin
MKG	Metakaolin-based geopolymer
MS	Magnesium sulfate
Ms	Modulus of silicate
MSD	Mean square of deviation

M-S-H	Magnesium silicate hydrate
N-A-S-H	Sodium aluminosilicate hydrate
NMR	Nuclear magnetic resonance
OPC	Ordinary Portland cement
PB	Plain bar
PSD	Pore size distribution
RH	Relative humidity
SD	Standard deviation
SEM	Scanning electronic microscopy
SS	Sodium sulfate
SSD	Sum of squares of deviations
TGA	Thermogravimetric analysis
w/b	Water-to-binder ratio
WG	Water-glass
WSR	Water to solid ratio
XRD	X-ray diffraction
XRF	X-ray fluorescence analysis
μCT	Computed microtomography

List of Figures

List of Tables

Chapter 1
Introduction

Abstract This chapter provides an introduction to the fundamental background of geopolymer and metakaolin-based geopolymer (MKG). Firstly, an overview of geopolymer and geopolymerization is presented. Subsequently, two crucial constituents of MKG, namely metakaolin and alkali activator, are introduced. Furthermore, notable distinctions between geopolymer and ordinary Portland cement (OPC) are highlighted. Lastly, this chapter discusses the various applications of geopolymer. The primary objective of this chapter is to establish a brief foundation that enables readers to comprehend the concepts of geopolymer and MKG more effectively.

1.1 Overview of Geopolymer and Geopolymerization

Geopolymer is a class of synthetic aluminosilicate materials. It was named by the French scientist Prof. Joseph Davidovits in 1970s (Davidovits 1991). In different research fields, these materials are also known as 'inorganic polymer', 'mineral polymer', 'alkali-bonded ceramics', etc. (Provis and Van Deventer 2009). The term 'geopolymer' usually stands for the aluminosilicate material obtained from low-calcium-contented pozzolanic natural minerals or industrial wastes by alkaline activation. So, it is also regarded as a sub-set of alkali-activated materials (AAMs). It usually has low calcium content and high aluminate and alkali (Na, K) contents (Provis and Van Deventer 2009).

The major component of geopolymer is amorphous gel formed by the spatial polymerization of SiO_4 tetrahedra and AlO_4 tetrahedra, which commonly named as the geopolymerization process (Duxson et al. 2007). Usually, the chemical composition of geopolymer can be expressed as $M_n[-(SiO_2)_z-AlO_2-]n\ wH_2O$, where M represents the alkali metal cations (such as K^+, Na^+, etc.); z and n represent the monomer type and the degree of polymerization, respectively; w is the amount of water bound to the gel (Provis 2014).

The geopolymerization process is an alkali-activation reaction. It can be roughly divided into four stages: (1) the aluminosilicate raw materials powders dissolves in an

D. Yan et al., *Metakaolin-Based Geopolymers*,
https://doi.org/10.1007/978-981-97-0652-5_1

alkaline solution environment; (2) the dissolved aluminosilicate monomers diffuse from solid surface into the solution and polymerized into oligomers; (3) aluminosilicate oligomers undergo further polycondensation to form spatial networked polymers (4) The polymer network further polymerization and rearrangement, and finally forms a hardened geopolymer solid.

Davidovits proposed that the oligomer products generated by geopolymerization can be divided into three forms:

1. Poly(sialate)

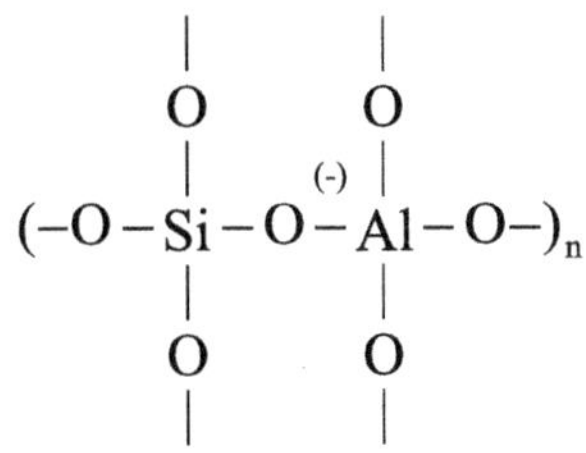

2. Poly(sialate-siloxo)

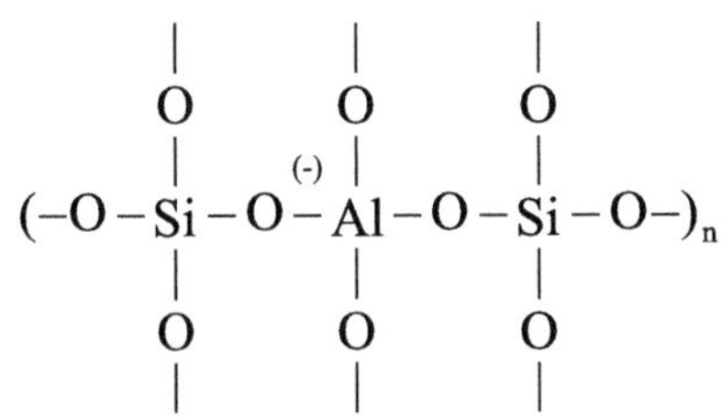

3. Poly(sialate-disiloxo)

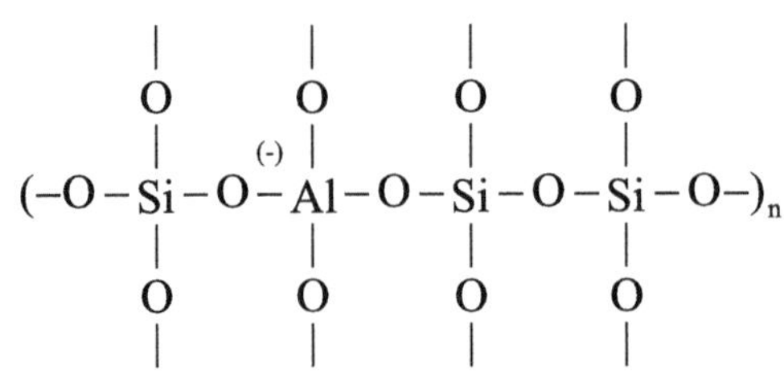

The three types of oligomers form a spatial network structure through polymerization, which constitutes the matrix of the geopolymer. Among these three types of oligomers, silicon element is in a stable + 4 valence state and the silicon-oxygen tetrahedron is electrically neutral. On the other hand, the aluminum element exists in the form of IV coordination. The aluminum-oxygen tetrahedron is electrically negative need the cations to balance its electronegativity. The following chemical reactions may occur during geopolymerization (Fig. 1.1).

The geopolymerization process is significantly different from the hydration process of OPC. The water only serves as a reaction environment. At the end of

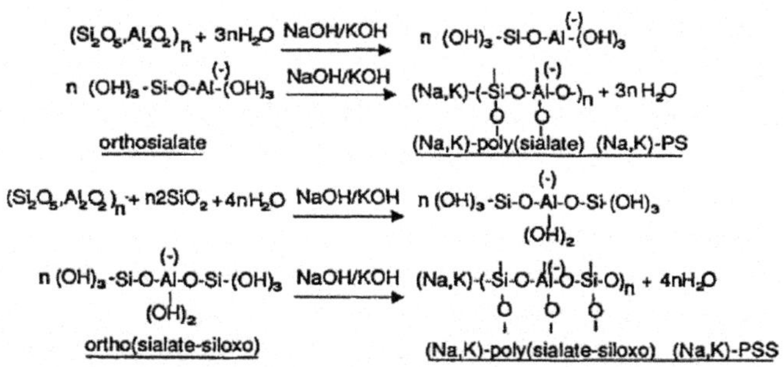

Fig. 1.1 Reactions during geopolymerization proposed by Davidovits

geopolymerization, most of the water is removed from the gel network. Provis and van Deventer proposed a more detailed conceptual model of geopolymerization (Fig. 1.2) (Provis and Van Deventer 2009). This conceptual model depicts the product changes during geopolymerization and lays the foundation for a better understanding of the mechanism of geopolymer formation. During the geopolymerization, the glassy aluminosilicate raw materials (such as metakaolin and fly ash) dissolves into silicate and aluminate monomers. At the same time, the polymerized silicate species, which provided in the alkaline solution, depolymerize into monomers as well. Then, the silicate and aluminate monomers polymerize into aluminosilicate oligomers. These oligomers further polymerize into amorphous polymers or quasi-/nano-crystalline 'nuclei'. The amorphous phase further polymerizes, rearranges and forms the geopolymer gel, while the quasi-/nano-crystalline phase further forms the 'zeolitic' nano-crystalline.

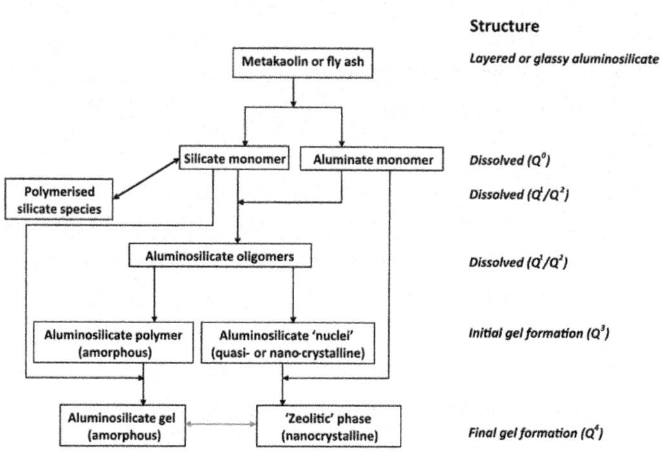

Fig. 1.2 Conceptual geopolymerization model proposed by Provis and van Deventer

1.2 Metakaolin and Alkali Activators

Geopolymers can be prepared from various raw materials, such as metakaolin, fly ash, coal gangue, granulated blast furnace slag, etc. Among them, metakaolin and fly ash are the two main raw materials. The geopolymer prepared by metakaolin and fly ash usually has a low calcium content.

Metakaolin is an amorphous material with pozzolanic activity. Its main chemical component is silicon oxide and aluminum oxide ($Al_2O_3 \cdot 2SiO_2$). To obtain metakaolin, the natural kaolin is calcined at 650–800 °C. Metakaolin has high reactivity in alkaline environments, so MKG is often used as a model material for studying geopolymers.

Alkali activator is another important chemical component in the preparation of geopolymers. Differences in ions and concentrations in alkali activators can significantly influence pore structure of geopolymer. In a geopolymer, the silicate species present in the activator solution undergoes transformation and forms –Si–O–Al– networks that constitute geopolymer gels. Increasing the amount of soluble silicate content in the activator leads to an increase in the Si/Al ratio in geopolymers. This, in turn, reduces the volume of pores exceeding 100 nm in size and contributes towards achieving a more homogeneous gel structure.

Reducing large pores is beneficial in reducing the risk of microcrack cracking. It could also increase the strength of the material and reduce the permeability. At the same time, in geopolymer slurries with low soluble silica content, there are fewer gel phases around unreacted particles. Duxson et al. (2007) found that an increase in the Si/Al ratio in geopolymer led to an increase in small-sized pores. Duxson et al. (2007) also noted that the presence of small molecular weight, unstable species in alkali activators with lower SiO_2/Na_2O ratios. Geopolymers containing higher concentrations of soluble silica may be hindered from structural reorganization during the gelation process due to the slower rate of exchange between cyclic or caged oligomeric species. The density and porosity of geopolymers can also be affected by the silicate content. The type of alkali metal ions also affects the pore structure, as a certain proportion of alkali metal ions (Na^+, K^+) enters the network to balance the negative charge at the position of the aluminum atom.

As reported by Kriven and Bell (2003), the replacement of sodium by potassium in geopolymers results in a decrease in mean pore size. The addition of KOH, NaOH, and $Ca(OH)_2$ influences the porosity of geopolymers, potentially due to differences in the size and hydrophilicity of alkali ions.

1.3 Geopolymer Versus Ordinary Portland Cement

Ordinary Portland cement (OPC) is the major cementitious material for concrete today. After more than 100 years of development, the production and application of OPC have laid a solid foundation for the modern construction industry. As of

2015, the global cement production has exceeded 4.1 billion tons. However, every ton of OPC produced will produce 0.55–0.95 tons of carbon dioxide emissions. With the increasing demand for concrete and OPC, the environmental problems caused by the production process of OPC also increasingly prominent. Green cementitious materials, such as geopolymer, with excellent performance and few environmental burdens have become important direction for the civil engineering.

Geopolymers and OPC are fundamentally different in formation, chemical composition and microstructure. OPC mainly relies on calcium silicate hydrate (CSH) gel formed from the hydration of cement clinker. On the other hand, geopolymer is majorly composited of aluminosilicate gel, which generated from the geopolymerization process and with alkali-cation (such as K^+, Na^+) and hydroxyls chemically bonded. Unlike CSH gel, the water molecules are not chemically integrated into geopolymer gel. Sometimes, these aluminosilicate gels are called sodium (or potassium) aluminosilicate hydrate (NASH or KASH). However, the formation of these gel is far from the hydration reaction of OPC (Provis and Van Deventer 2009).

Due to the chemical difference between geopolymer and OPC, there are also great differences in their macroscopic properties. Compared with OPC, geopolymer has the characteristics of high early strength, corrosion resistance, high temperature resistance, etc. At the same time, geopolymer is brittle, significant in drying shrinkage and efflorescence. The difference in the properties of these cementitious materials is a problem that must be considered in the engineering application of geopolymer concrete, and the understanding of its in-depth mechanism is also an important basis for realizing the rational design and utilization of geopolymer concrete.

At the same time, the carbon emissions and energy consumption of geopolymers are also significantly lower than those of Portland cement. According to estimation, the production energy consumption and CO_2 emissions required to produce 1 ton of geopolymer are 41% and 20% of that of Portland cement, respectively (Fig. 1.3). If geopolymers are produced from industrial wastes (such as fly ash, slag, and coal gangue), the energy consumption and carbon emissions can be further reduced. Therefore, geopolymer concrete is a new green and low-carbon building material, which is considered to be a new path to promote the sustainable development of civil engineering.

1.4 Applications of Geopolymer

Geopolymers possess exceptional acid and fire resistance properties, thus rendering them a promising multifunctional material. The versatility of this class of materials has been shown in various applications, such as protective coatings for diverse surfaces, including metals, sustainable concrete, and fireproof building materials. Furthermore, their outstanding mechanical, chemical, and heat resistance characteristics make them an eco-friendly choice for constructions (Jindal et al. 2022).

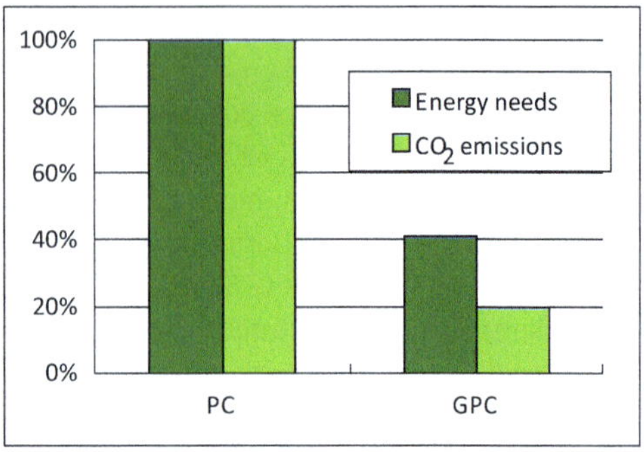

Fig. 1.3 The CO_2 emissions and energy needs comparison between geopolymer and ordinary Portland cement

This section offers a comprehensive overview of the potential applications of geopolymers.

(1) Structural components

Geopolymers possess several advantages, including rapid polymerization and high early strength development. Previous studies have indicated that geopolymers demonstrate comparable or superior mechanical properties to conventional cement-based concrete materials, and can serve as a partial replacement for traditional cement-based concrete materials. Additionally, the incorporation of metakaolin in concrete has been shown to enhance its durability and decrease the environmental impact of the cement industry.

Refaie et al. (2020) conducted compression tests on insulation sandwich panels utilizing geopolymer concrete as structural boards. The results demonstrated that sandwich wall panels containing geopolymer concrete boards displayed superior stiffness, axial capacity, flexural strength, and less deflection during the initial formation of cracks in comparison to conventional concrete panels. These outcomes suggest excellent constructability and competent construction costs. Correspondingly, Neupane (2016) compared geopolymer concrete and ordinary Portland cement (OPC) concrete of similar grades and reported that geopolymer concrete exhibited comparable drying shrinkage and elastic modulus but higher tensile and flexural strength. Nazeer and Kumar (2014) proposed that partially substituting cement with a combination of class F fly ash and metakaolin enhances the strength properties of concrete, including durability, impact resistance, and reduced environmental impact. However, the addition of metakaolin lowers the compressive and tensile strength.

(2) Anti-corrosion Products

Geopolymer gels possess a unique three-dimensional network structure and extremely low calcium hydroxide content, resulting in superior corrosion resistance when compared to traditional cement-based materials (Zhang et al. 2010). Moreover, geopolymers can be tailored with primers or textured paints for different material surfaces. Geopolymer exhibits relatively high bond strength with steel, aluminum, and borosilicate glass (Bell et al. 2009). The bond strength of geopolymer coating to steel surface is significantly influenced by the composition of geopolymer coating. To investigate the tidal impact resistance of an in-situ geopolymer coating applied to concrete surfaces in coastal areas, Zhang et al. (2010) tested and found that the coating displayed a strong bond to the concrete and good resistance to tidal waves for up to four hours. Additionally, the incorporation of PP fibers and some magnesium-based expander proved effective at reducing the shrinkage rate to acceptable levels. As such, geopolymers can serve as a coating for corrosion protection of marine concrete and metals. It could also improve the appearance, wear resistance, scratch resistance, and heat resistance.

(3) Fireproof Materials

Geopolymer researchers have conducted extensive investigations into the potential of geopolymers as heat and refractory materials. Kong et al. (2007) reported a reduction in strength of metakaolin-based geopolymers when subjected to high temperatures, which may be attributed to excessive water loss when compared to fly ash-based geopolymers (FAG). Temuujin et al. (2009) demonstrated that geopolymer based on metakaolin produced coating materials that are both highly resistant to water and fireproof. The produced coating material forms a fireproof protective layer with low thermal conductivity. The endothermic process of MKG leads to a further reduction in surface temperature. Endothermic processes of MKG lead to the expansion of porous structures, enabling it to behaves as a barrier between the fire and the metal.

In a study by Temuujin et al. (2010), it was found that geopolymers with a high silica content exhibit the best bond strength, and increasing the coating thickness enhances the fire insulation capability. Hung et al. (2013) made a geopolymer composing with metakaolin and blast furnace slag. They produced specimens using the mechanical foaming process, and when the relative density was less than 0.2, the thermal conductivity was less than 0.1 W/(m K). Wu et al. (2014) investigated the fireproof geopolymer coatings using MKG with rice husk ash. They measured the thermal conductivity of the coatings and studied the effect of the concrete removal, loading rate, and thickness of the coating on the fire resistance of the composite panels. The results obtained by the researchers demonstrate that the fireproof coating exhibits exceptional thermal insulation properties, notably when the binder is activated with sodium silicate. Sakkas et al. (2014) made a metakaolin geopolymer containing solid SiO_2 and activated by potassium hydroxide solution and found that the material exhibited outstanding fire-resistant properties and had a lower thermal conductivity, making it suitable for use as an effective heat flux barrier and for fire protection in concrete tunnel linings. Duan et al. (2015) reported that the microstructures of fly ash/metakaolin composite geopolymer become denser with increasing temperature up to 400 °C due to geopolymerization and sintering. Additionally, Rashad (2019) discovered that MKG mortars display good fire resistance, with 10.25–11.58% higher relative strength than FAG mortars. In contrast, FAG mortars exhibited better thermal insulation (lower thermal conductivity) than MKG counterpart. Furthermore, substituting sand with expanded perlite can provide greater thermal insulation and better fire resistance. However, Rovnanik and Šafránková (2016) observed a significant decline in the mechanical strengths of MKG when exposed to 1200 °C, which was mitigated by adding chamotte aggregate. Their results showed that MKG mortar with chamotte aggregate had significantly high strength. In summary, MKG hold great promise for high-temperature applications.

(4) Soil treatment

The use of soil stabilizers, such as lime and cement, has been widespread for improving the strength of weak subgrade soils and expansive soils. However, concerns have been raised regarding the negative impact of their production and use on the environment, prompting a search for more sustainable alternatives. Geopolymer treated soils have demonstrated comparable strength and stiffness properties to those treated with lime and cement, and have emerged as a more sustainable alternative in numerous studies. Samuel et al. (2019) reported that metakaolin-based geopolymers effectively reduced the swell-shrink potential of high plasticity index soils in North Texas. Nevertheless, further research is needed to assess the durability and sustainability of geopolymer stabilization, as well as other factors that may impact its performance. Khadka et al. (2018) also showed the feasibility of using geopolymers, specifically MKG and FA, to enhance the strength and shrinkage/expansion behavior of high plasticity index natural clays. The study revealed that clay stabilized with MK geopolymer had higher strength than that stabilized with FA geopolymer, while the swelling behavior of the latter was considerably diminished. Further research is necessary to evaluate the potential of geopolymers compared to traditional stabilizers for treating high plasticity index clay soil containing sulfate. Zhang et al. (2015) investigated the viability of calcium-free geopolymers as a soil stabilizer for sulfate-rich soils. The study found that higher MKG concentrations led to better ductility and less expansive potential. Nevertheless, additional research is required to determine the optimal chemical composition and curing conditions for effective and economical geopolymer stabilization. Besides, Samuel et al. (2020) evaluated the sustainability benefits of metakaolin-based geopolymers versus conventional lime treatment for stabilizing high plasticity expansive soil. Their weighted multi-criterial evaluation framework indicated that the sustainability index of geopolymer treatment was approximately 10% lower than that of conventional treatment, reflecting its greater potential for sustainable development.

(5) Organic pollution removal

The microstructure of metakaolin particles exhibits irregular shapes due to their porous structure, specific surface area, and pore size. Therefore, geopolymer is an effective adsorbent for removing organic pollutants from water media. To enhance its adsorption capacity, Barbosa et al. (2018) proposed a novel method for synthesizing mesoporous geopolymers using metakaolin and rice husk ash as raw materials and soybean oil as a mesoporous structure guiding agent. The maximum adsorption capacity for methyl violet 10B dye removal was found to be 276.9 mg/g when the concentration of the geopolymer synthesized using this approach was 1.5 g/L, which is higher than that of the conventional method without oil. El Alouani et al. (2019) optimized the adsorption conditions for methylene blue using geopolymer synthesized from metakaolin and alkaline activators. They discovered that in an alkaline environment, with an initial concentration of 40 mg/L 100 ml methylene blue, 1 g of the geopolymer adsorbent can spontaneously adsorb 43.48 mg/g of the maximum monolayer adsorption capacity. Sanguanpak et al. (2021) investigated the feasibility of porous metakaolin-based polymers as adsorbents for removing ammonium ions from wastewater. They demonstrated that at 15 wt% air content in the porous geopolymer, the maximum adsorption capacity for an ammonium concentration of 200 mg/L wastewater was 47.17 mg/g. These newly developed geopolymer particles possess adsorption properties over 35% higher than traditional adsorbents and can be reused in combination with sodium chloride and sodium hydroxide solutions. Further research is necessary to explore the potential of these materials as sustainable and efficient adsorbents for various pollutants in diverse water media.

(6) Heavy metal immobilization

Geopolymers, being inorganic polymers, possess a porous structure due to the formation of pores during condensation, rendering them suitable for heavy metal adsorption and industrial wastewater treatment. Previous studies (Wang et al. 2007; Mužek et al. 2014) have demonstrated the efficacy of using fly ash-based geopolymers for Cu(II) removal. Cheng et al. (2012) systematically investigated the adsorption performance of metakaolin-based polymers for various heavy metal ions (Pb^{2+}, Cu^{2+}, Cr^{3+}, and Cd^{2+}), finding that the adsorption capacity was particularly remarkable for Pb^{2+} ions. Kara et al. (2018) studied the adsorption capacity of metakaolin-based polymers for Mn(II) and Co(II) in aqueous solution, obtaining a maximum adsorption capacity of 72.34 mg/g and 69.23 mg/g, respectively. Metakaolin-based polymers exhibit excellent adsorption properties in both batch and continuous systems. Sun et al. (2014) reported the feasibility and mechanism of using metakaolin-based geopolymer added with sodium sulfide as an adsorbent to detoxify and immobilize Cr(VI) from chromite ore processing residue. Reduction products exist in the form of Cr(III), and the samples have considerable compressive strength, which can be used as potential building materials. Fu et al. (2020) innovatively combined hydrothermal treatment with the preparation process of metakaolin-based polymers to efficiently extract Cs in synthesizing pollucite. The optimal fixation performance of Cs occurred at 230 °C with an initial Na/Cs ratio of 1:4 after six hours of treatment, and the immobilization capacity and compressive strength of the hydrothermal products reached 94.95% and 18.9 MPa, respectively.

1.5 Content of This Book

MKG plays a crucial role in the current field of geopolymer materials research, encompassing a broad range of applications such as green environmental protection, fire resistance, heat resistance, corrosion resistance, fast hardening, and early strength. This study aims to enhance our comprehension of the outstanding properties and diverse applications of geopolymers. In Chap. 2, the mechanism of geopolymerization in MKG is explored. In Chap. 3, the mechanical performance of MKG and its corresponding correlation mechanism with composition are discussed. In Chap. 4, the drying shrinkage mechanism of MKG is examined. In Chap. 5, the sulfate corrosion mechanism and performance of MKG are investigated. In Chap. 6, the high-temperature deterioration mechanism of MKG is analyzed. In Chap. 7, the freeze–thaw deterioration mechanism of MKG is scrutinized. In Chap. 8, the effect of aggregate on MKG concrete is evaluated. Lastly, in Chap. 9, the reinforcement bonding behaviors of MKG are studied.

References

Barbosa TR, Foletto EL, Dotto GL, et al (2018) Preparation of mesoporous geopolymer using metakaolin and rice husk ash as synthesis precursors and its use as potential adsorbent to remove organic dye from aqueous solutions [J]. Ceramics Int 44(1):416–423

Bell JL, Driemeyer PE, Kriven WM (2009) Formation of ceramics from metakaolin-based geopolymers. Part II: K-based geopolymer. J Am Ceram Soc 92(3):607–615

Cheng TW, Lee ML, Ko MS, Ueng TH, Yang SF (2012) The heavy metal adsorption characteristics on metakaolin-based geopolymer. Appl Clay Sci 56:90–96

Davidovits J (1991) Geopolymers. J Therm Anal Calorim 37(8):1633–1656

Duxson P, Mallicoat SW, Lukey GC et al (2007) The effect of alkali and Si/Al ratio on the development of mechanical properties of metakaolin-based geopolymers. Colloids Surf A 292(1):8–20

Duan P, Yan C, Zhou W, Luo W, Shen C (2015) An investigation of the microstructure and durability of a fluidized bed fly ash–metakaolin geopolymer after heat and acid exposure. Mat Design 74:125–137

El Alouani M, Alehyen S, El Achouri M (2019) Preparation, characterization, and application of metakaolin-based geopolymer for removal of methylene blue from aqueous solution. J Chem 2019

Fu S, He P, Wang M, Cui J, Wang M, Duan X et al (2020) Hydrothermal synthesis of pollucite from metakaolin-based geopolymer for hazardous wastes storage. J Cleaner Prod 248:119240

Hung TC, Huang JS, Wang YW, Fan YC (2013) Microstructure and properties of metakaolin-based inorganic polymer foams. J Mat Sci 48(21):7446–7455. https://doi.org/10.1007/s10853-013-7559-3

Jindal BB, Alomayri T, Hasan A, Kaze CR (2022) Geopolymer concrete with metakaolin for sustainability: a comprehensive review on raw material's properties, synthesis, performance, and potential application. Environ Sci Pollut Res 1–26

Kara İ, Yilmazer D, Akar ST (2017) Metakaolin based geopolymer as an effective adsorbent for adsorption of zinc (II) and nickel (II) ions from aqueous solutions. Appl Clay Sci 139

Kara I, Tunc D, Sayin F, Akar ST (2018) Study on the performance of metakaolin based geopolymer for Mn (II) and Co (II) removal. Appl Clay Sci 161:184–193

Khadka SD, Jayawickrama PW, Senadheera S (2018) Strength and shrink/swell behavior of highly plastic claytreated with geopolymer. Transp Res Rec 2672:174–184

Kriven WM, Bell JL, Gordon M (2003) Microstructure and microchemistry of fully reacted geopolymers and geopolymer matrix composites, Ceram Trans 153:227–250. https://doi.org/10.1002/9781118406892.ch15

Kong DLY, Sanjayan JG, Sagoe-Crentsil K (2007a) Comparative performance of geopolymers made with metakaolin and fly ash after exposure to elevated temperatures Cem Concr Res 37:1583–1589.https://doi.org/10.1016/j.cemconres.2007.08.021

Luukkonen T, Sarkkinen M, Kemppainen K, Rämö J, Lassi U (2016) Metakaolin geopolymer characterization and application for ammonium removal from model solutions and landfill leachate. Appl Clay Sci 119:266–276

Muzek MN, Svilovic S, Zelic J (2014) Fly ash-based geopolymeric adsorbent for copper ion removal from wastewater. Desalin Water Treat 52:2519–2526

Nazeer M, Kumar AR (2014) Strength studies on metakaolin blended high-volume fly ash concrete. ISSN. 2249-8958

Neupane K (2016) Fly ash and GGBFS based powder-activated geopolymer binders: a viable sustainable alternative of portland cement in concrete industry. Mech Mater 103:110–122. ISSN 0167-6636

Provis JL (2014) Geopolymers and other alkali activated materials: why, how, and what? Mater Struct/Mater Constr 47(1–2):11–25

Provis JL, Van Deventer JSJ (2009) Geopolymers, Structures, processing, properties and industrial applications, Geopolymers Struct Process Prop Ind Appl 1–454. https://doi.org/10.1533/9781845696382

Rashad AM (2019) Insulating and fire-resistant behaviour of metakaolin and fly ash geopolymer mortars. Proceed Institution of Civil Engineers-Const Mat 172(1):37–44

Refaie FAZ, Abbas R, Fouad FH (2020) Sustainable construction system with Egyptian metakaolin based geopolymer concrete sandwich panels. Case Stud Constr Mater 13:e00436

Rovnaník P, Šafránková K (2016) Thermal behaviour of metakaolin/fly ash geopolymers with chamotte aggregate. Materials 9(7):535

Sakkas K, Panias D, Nomikos PP, Sofianos AI (2014) Potassium based geopolymer for passive fire protection of concrete tunnels linings. Tunnelling underground space Tech. 43:148–156

Samuel R, Huang O, Banerjee A, Puppala A, Das J, Radovic M (2019) Case study: use of geopolymers to evaluate the swell-shrink behavior of native clay in North Texas. In: Eighth international conference on case histories in geotechnical engineering. pp 167–178

Samuel R, Puppala AJ, Radovic M (2020) Sustainability benefits assessment of metakaolin-based geopolymer treatment of high plasticity clay. Sustainability 12(24):10495

Sanguanpak S, Wannagon A, Saengam C, Chiemchaisri W, Chiemchaisri C (2021) Porous metakaolin-based geopolymer granules for removal of ammonium in aqueous solution and anaerobically pretreated piggery wastewater. J Cleaner Prod 297:126643

Siyal AA, Shamsuddin MR, Khan MI (2018) A review on geopolymers as emerging materials for the adsorption of heavy metals and dyes. J Environ Manage

Sun T, Chen J, Lei X, Zhou C (2014) Detoxification and immobilization of chromite ore processing residue with metakaolin-based geopolymer. J Environ Chem Eng 2(1):304–309

Temuujin J, Minjigmaa A, Rickard W, Lee M, Williams I, Van Riessen A (2009) Preparation of metakaolin based geopolymer coatings on metal substrates as thermal barriers. Appl Clay Sci 46(3):265–270

Temuujin J et al (2010) Fly ash based geopolymer thin coatings on metal substrates and its thermal evaluation. J Hazard Mat 180(1-3):748–752

Wang SB, Li L, Zhu ZH (2007) Solid-state conversion of fly ash to effective adsorbents for Cu removal from wastewater. J Hazard Mater 139:254–259

Wu Bo et al (2014) Development of metakaolin–fly ash based geopolymers for fire resistance applications. Const Building Mat 55:38-45

Zhang Z, Yao X, Zhu H (2010) Potential application of geopolymers as protection coatings for marine concrete I. Basic properties Appl Clay Sci 49:1–6. https://doi.org/10.1016/j.clay.2010.01.014

Zhang M, Zhao M, Zhang G, Nowak P, Coen A, Tao M (2015) Calcium-free geopolymer as a stabilizer for sulfate-rich soils. Appl Clay Sci 108:199–207

Chapter 2
Geopolymerization of MKG

Abstract The geopolymerization process is a key influencing factor of the mechanical properties of materials and chemical durability of geopolymers. In this chapter, vacuum dehydration and nuclear magnetic resonance analysis were conducted to investigate the geopolymerization process of fresh MKG paste. Four distinct stages of geopolymerization were identified, including dissolution, short-ranged polymerization, structural rearrangement, and long-ranged condensation. Every stage was characterized by its reaction duration and $Q_m(nAl)$ evolution. The findings from this study offer new insights into the mechanism through which the loss of free water affects the structural rearrangement of geopolymers.

2.1 Introduction

According to Li et al. (2019), geopolymer cements are tending to replace the traditional ordinary Portland cement (OPC) as a promising construction material. Geopolymer, prepared from reactive aluminum silicate precursors and optimized formulations offer excellent mechanical properties with low energy consumption and low greenhouse gas emissions (Provis et al. 2015). In comparison to OPC, geopolymer shows faster solidification, higher strength, and stronger adhesion to steel reinforcement, as reported by Irfan Khan et al. (2015), Bhutta et al. (2017), Thermou and Hajirasouliha (2018), respectively. Additionally, geopolymer exhibits excellent thermal resistance (Subekti et al. 2017; Chindaprasirt and Rattanasak 2018; Uddin Ahmed Shaikh et al. 2019) and possesses high chemical durability in corrosive environments such as saline or alkali land, coastal areas, and alpine regions (Zhang et al. 2010; Reddy et al. 2013). Furthermore, geopolymer manufacture requires less fuel and energy consumption as well as minimal secondary processing, resulting in lower discharge of flue gas and smoke dust, thereby reducing environmental pollution (Scrivener and Kirkpatrick 2008; Provis 2014).

The impressive efficacy of geopolymer can be ascribed to the three-dimensional molecular configuration of alkali aluminosilicate (N–A–S–H) gels with bridged silicon and aluminum tetrahedrons by means of oxygen bond (Šmilauer et al. 2011;

D. Yan et al., *Metakaolin-Based Geopolymers*,
https://doi.org/10.1007/978-981-97-0652-5_2

Ismail et al. 2014; Nath et al. 2016). As arranged in an organized way, the atomic structure of N–A–S–H is compact, adhesive, and robust in the nanoscale (Provis 2014; Nath et al. 2016). Geopolymers exhibit a unique pore distribution with peaks primarily in the nanometer scale (Steins et al. 2014), which has a substantial impact on their mechanical features, such as strength and stiffness (Duxson et al. 2005a). The microstructure and pore network of geopolymer cement are formed during the process of geopolymerization.

The geopolymerization process first passes through the stages of dissolution, dispersion, and oligomerization, followed by further polymerization and a dehydration-hardening stage of the silica-aluminum source in an alkaline solution (Provis 2014). Provis and Van Deventer proposed a four-stage model of geopolymerization: dissolution, reorientation, gelatinization, and cross-linking (Provis and van Deventer 2007). Duxson brought the conceptual model full circle by proposing dissolution of aluminosilicate precursors in alkaline solution, producing a supersaturated solution, which leads to the gradual formation of gels (Duxson et al. 2007). In the pre-polymerization process of geopolymerization, the dissolution phase is the main factor influencing the dynamic process of gel formation and polymer structure formation. These theories examine different reaction stages on the basis of multiple detection methods and may have limitations in clearly identifying stage transitions. Therefore, in-depth validation by continuous detection techniques is necessary to improve the consistency and reliability of stage evolution characterization. Many studies have investigated the geopolymerization process through micro- and nanoscale characterization techniques. For example, Gordon et al. used x-ray diffraction to identify crystals in geopolymers cured under different conditions and clearly analyzed the mineralogical and chemical alteration processes of the different components (Gordon et al. 2014). Walkley used scanning electron microscopy (SEM) and energy spectroscopy (EDS) techniques based on backscattered electron (BSE) imaging to put reaction product compositions into elemental distribution maps. The results enabled in situ tracking of product formation and chemical elemental distributions (Walkley et al. 2016a). Provis used synchrotron radiation-based Fourier transform infrared (SR-FTIR) spectroscopic microscopy to continuously monitor the reaction product system and illustrate the processes of change. In their study, attenuated total reflection Fourier transform infrared (ATR-FTIR) spectroscopy was used to compare the degree of aggregation on a cross profile (Hajimohammadi et al. 2010, 2011). The essence of both methods is the use of image recognition and classification to identify phase changes. In Walkley's study, nuclear magnetic resonance (NMR) was used to analyze the Si, Al, Na, and H content and distribution, reflecting the distribution of Si and Al tetrahedra in the samples during the evolution of the growth of the accompanying polymer gel (Walkley et al. 2016b). Although the approach is widely applied, current spectral mining and analysis is deficient and often restricted by the accuracy of the measurements. Neutron diffraction and x-ray pair distribution function analysis can provide more information about chemical bonding changes through peak shifts on smaller nanoscale (White et al. 2013). However, few studies have used consistent methods and instrumental parameters, especially focusing on early geopolymerization within the first 14 days. Nuclear magnetic resonance (NMR) detection, a widely used assay for gelling materials, has the potential to reveal more details of chemical bonding changes at the property scale. Therefore, there is an urgent need for a simple method to prepare powder samples for NMR detection without changing the state of ground polymerization.

Geopolymerization occurs in an alkaline environment containing free water. Free water mainly serves as an ion-transport medium and is involved in the precursor dissolution process and the polymerization reaction itself (Yunsheng et al. 2010). Chen proposed a combined water-solvent extraction method for dehydrating cement. The method is based on acetone, acetate or ethanol/acetone mixtures to remove water from the geopolymerization system (Chen et al. 2014). The process of removing organic solvents is necessary to be carried out at temperatures higher than 50 °C to evaporate the organic solvents and free water from the samples more quickly. However, the geopolymerization reaction is strongly influenced by temperature, which may significantly accelerate the polymerization process within a few hours of sample preparation time, leading to erroneous test results (Mustafa Al-Bakria et al. 2011). According to North and Swaddle, NMR testing at low temperatures between − 5 and 5 °C can provide more information about monomers and dimers that is often overlooked at higher temperatures (North and Swaddle 2000).

The feasibility of using solvent extraction methods to prevent geopolymerization has yet to be proven, and current research tends to explore more robust and reliable dehydration methods. In this study, the vacuum dehydration processing method was examined by NMR and the results obtained by this method were compared with those obtained using solvent extraction methods. The vacuum dehydration method is a more efficient way to stop the geopolymerization process during geopolymer sample preparation. By this method, early geopolymers from 0 to 14 days were trace detected by ^{27}Al and ^{29}Si NMR spectroscopy. The spectra were analyzed by comparison of successive views, inverse fold accumulation with strict parameter control, and fuzzy analysis to identify reaction features that distinguish and delineate geopolymerization phases, thus describing the entire geopolymerization process. Existing modeling theories are typically correlated with each other through conclusions from different stages of several different detection methods (Duxson et al. 2007; Provis and van Deventer 2007). The synthesis process therefore establishes a baseline to validate and complement existing theories.

2.2 Experiments

2.2.1 Preparation of Samples

Geopolymer samples were prepared from metakaolin and alkali activator. The alkaline activator was prepared from water glass, sodium hydroxide and water. The modulus ($n(SiO_2)/n(Na_2O)$) of the water glass was 3.28. Tables 2.1 and 2.2 show the

Table 2.1 Chemical composition of metakaolin

Composition	Al_2O_3	SiO_2	K_2O	Na_2O	CaO	TiO_2	Fe_2O_3	LOI
Mass ratio (%)	39.68	57.26	0.21	0.27	0.04	1.78	0.43	0.34

Reproduced from [Early-stage geopolymerization revealed by ^{27}Al and ^{29}Si nuclear magnetic resonance spectroscopy based on vacuum dehydration] by [Xiuyu Zhu] with permission from [Construction and Building Materials]

Table 2.2 Chemical composition of water glass

Composition	Na_2O	SiO_2	H_2O
Mass ratio (%)	8.2	26.0	65.8

Reproduced from [Early-stage geopolymerization revealed by ^{27}Al and ^{29}Si nuclear magnetic resonance spectroscopy based on vacuum dehydration] by [Xiuyu Zhu] with permission from [Construction and Building Materials]

chemical composition of the metakaolin and water glass. Sodium hydroxide (chemical purity, purity $\geq$ 98%) was used as a formulated alkali exciter. Tap water was used as reaction medium. Ethanol with a purity greater than 99.7% (AR) and acetone with a purity beyond 99.5% (analytical purity) were also employed.

The silica-aluminum, sodium-aluminum, and water-binder ratios in the reactants were as follows: $SiO_2/Al_2O_3 = 4$, $Na_2O/Al_2O_3 = 1$, and water/binder = 0.7. The setting time at 20 °C was approximately 12–18 h, whereas at 5 °C, it was 60–70 h. The low-temperature solidification provided sufficient time to collect samples for all stages of geopolymerization. During preparation, water and sodium hydroxide solids were added to the water glass and stirred thoroughly until they were completely dissolved and homogeneous. The clarified solution was waited to cool to room temperature to obtain the alkali activators. Metakaolin was blended into the alkali activator and rapidly stirred for 90 s until well combined. The mixture was poured into 20 mm cubic silica gel molds and the upper surface was covered with plastic film. All samples need to be cured at a temperature of 5 °C for sufficient time.

2.2.2 *Dehydration Process*

Two approaches, vacuum dehydration and solvent extraction, were compared for arresting geopolymerization (Chen et al. 2014).

For vacuum dehydration, 0.5–0.8 g of the sample in the transition state (from flexible slurry to gradually losing fluidity) was removed from the reaction product and placed in a small container. If the slurry did not lose fluidity, it was transferred into an oil pump-driven liquid nitrogen condensing gas recirculation unit, and then water was removed by pumping for 40 min. The production was then ground using a mortar and pestle for 10 min until absolutely pulverized. This dehydrating and grinding process was repeated twice. If the sample had completely lost its plasticity, it was ground directly into powder, filtered through a 70-mesh sieve, and then pumped for 2.5 h.

For solvent extraction, 0.5–0.8 g of the material was put in a flask with 100 ml of an ethanol/acetone solution. The container was then placed on a rotating evaporator with a rotational speed of 90 r/min and a temperature of 50 °C. After the organic solvent had completely evaporated, the sample was moved to a mortar and finely pulverized. It took more than four hours to complete three cycles of evaporation and grinding.

The samples were cured for the required amount of time at 5 °C while being wrapped in plastic wrap. They were split into two groups, with the group that underwent vacuum dehydration receiving the label V-X-Y and the group that underwent solvent extraction receiving the label S-X-Y. Here, X stands for the day the samples underwent solvent or vacuum dehydration extraction, and Y stands for the day they were subjected to NMR analysis. As an illustration, V-1-1 signifies that the sample was subjected to vacuum dehydration after being cured for one day and analyzed by NMR after one day of curing. Similarly, V-1-7 indicates that the sample was vacuum dehydrated after one day of curing and tested by NMR after seven days of curing. The purpose of S-1-1, S-1-7, and S-7-7 is the same as what was previously stated, with the exception that the samples underwent solvent extraction as opposed to vacuum dehydration. The meaning of S-1-1, S-1-7, and S-7-7 is similar to that outlined above, except that the samples underwent solvent extraction instead of vacuum dehydration. In the G-X group, G represents geopolymer samples, and X signifies the day on which the samples were treated with vacuum dehydration and analyzed by NMR.

2.2.3 *Early Geopolymerization*

All of the geopolymer samples were cured at 5 °C and exposed to vacuum dehydration after 1, 2, 3, 5, 7, and 14 days in order to track the early phases of geopolymerization continually. Subsequently, the dehydrated powder was analyzed by NMR. Samples from days 1 and 2 had a viscous paste consistency, while those from day 3 were solids but had limited strength. The remaining samples had lost all free water and had solidified into block-like shapes.

Each of the specimens were divided into eight groups, with M standing for metakaolin as the fundamental control group and G–X standing for geopolymer samples, with X standing for the length of time the samples were cured.

2.2.4 *NMR Detection*

Solid-state ^{27}Al and ^{29}Si NMR spectra were acquired for all the samples. On a Bruker Avance III 400 solid-state spectrometer with 3.2 mm double resonance probes spinning at a frequency of 15 kHz, the magic angle spinning (MAS) tests were carried out. To overcome the poor signal sensitivity of Si, the ^{29}Si spectrum was acquired via cross-polarization/MAS NMR from ^{1}H to ^{29}Si, and a single band was produced from 2048 scans. There was a 5.5 s recycling delay between each successive scan. In contrast, the ^{27}Al spectra was produced using a one-pulse sequence, and because the Al signal was sensitive enough, a single spectrum was generated from 512 scans. The pulse duration was 1 μs (10^{-6} s) with a recycle delay of 1 s. The basal chemical shifts were referenced to tetramethylsilane for the ^{29}Si nuclei and to aqueous $AlCl_3$ for the ^{27}Al nuclei.

Spectrum peak chemical shifts provided information on Al's coordination environment and Si's linkage configurations. In the ^{27}Al MAS NMR spectrum, the peak between 50 and 70 ppm corresponded to four-coordinate aluminum (denoted as Al(IV)) in aluminosilicates. According to Davidovits (1991), the signal around 0–20 ppm suggested six-coordinate aluminum resonance from $[Al(H_2O)6]^{3+}$, while the peak for Al(V) was at about 25 ppm. The Si atoms with various connections were visible as a variety of peaks, designated as $Q_m(nAl)$, with a regular interval ranging from − 72 to − 110 ppm. The central Si atom's chemical shift demonstrated a change of approximately 8–10 ppm relative to the pure Si tetrahedron space structure when there was a deficiency of one linked Si atom. This change could be accumulated when more Si atoms disappeared. Additionally, the Al atoms were linked in the Si tetrahedron space structure through oxygen bridges. Because the connected Al atom would change the bond angle and length of the core Si tetrahedron, those Si atoms coupled with one Al atom and three Si atoms displayed a tiny interval of around 3–5 ppm in comparison to the Si atom linked with four other Si atoms. This change could be accumulated when more Al atoms were linked. The positions of these characteristic peaks could be determined based on the given reference scope and the actual spectrum interval. Furthermore, Si sites linked with two or more Al atoms were deemed challenging to form with the given Si/Al atom ratio in this chapter since they overlapped at a larger chemical shift and were not distinctive enough to identify. Based on the above principles and previous experiences, the peak of Al(IV) was identified at 58 ppm, Al(V) at 28 ppm, and Al(VI) at 0 ppm. In the ^{29}Si MAS NMR spectrum, peaks were recorded near − 76 ppm as Q_0, − 79 ppm as Q_1, − 82 ppm as Q_2(1Al), − 85 ppm as Q_2, − 88 ppm as Q_3(1Al), − 94 ppm as Q_3(0Al), 101.5 ppm as Q_4(1Al), and 108 ppm as Q_4(0Al) during the peak-fitting stage (Lippmaa et al. 1980; Tazawa et al. 1995; Duxson et al. 2005b; MacKenzie et al. 2008; Walkley et al. 2016a; Li et al. 2019) (see Fig. 2.1).

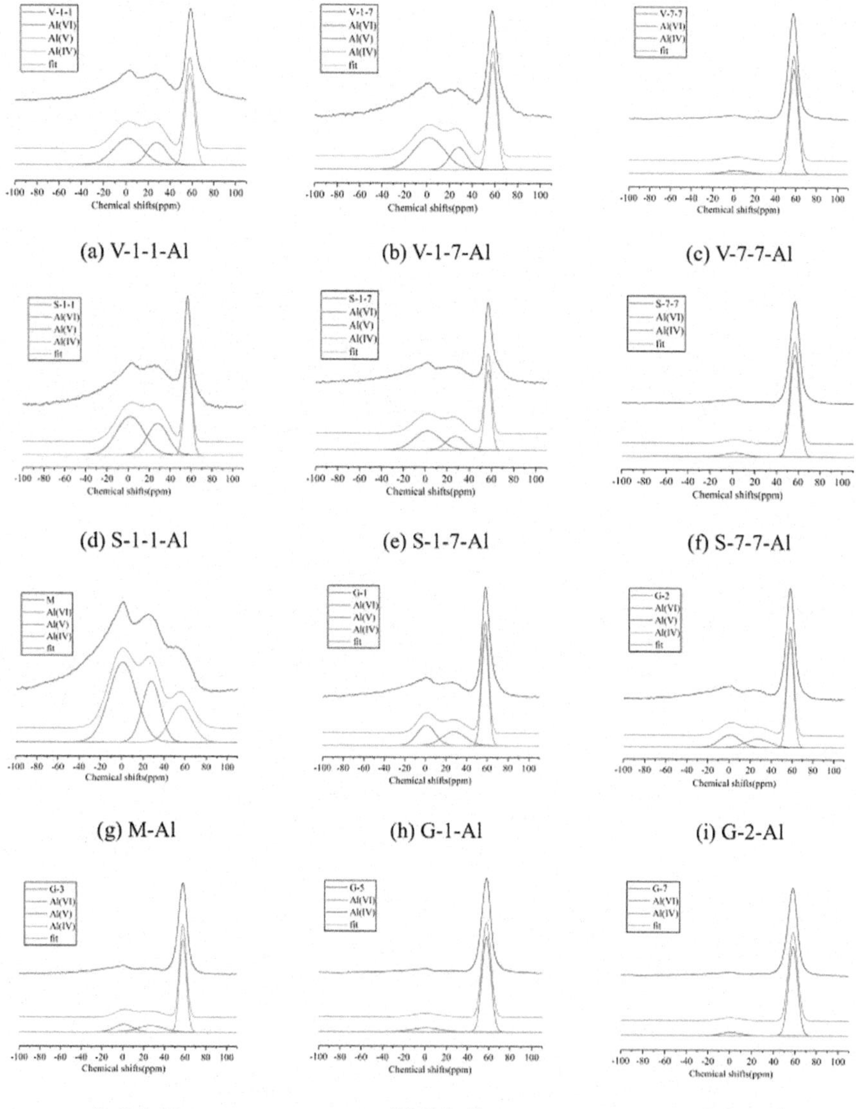

(a) V-1-1-Al (b) V-1-7-Al (c) V-7-7-Al

(d) S-1-1-Al (e) S-1-7-Al (f) S-7-7-Al

(g) M-Al (h) G-1-Al (i) G-2-Al

(j) G-3-Al (k) G-5-Al (l) G-7-Al

Fig. 2.1 Deconvolution process of all spectrums. Reproduced from [Early-stage geopolymerization revealed by ^{27}Al and ^{29}Si nuclear magnetic resonance spectroscopy based on vacuum dehydration] by [Xiuyu Zhu] with permission from [Construction and Building Materials]

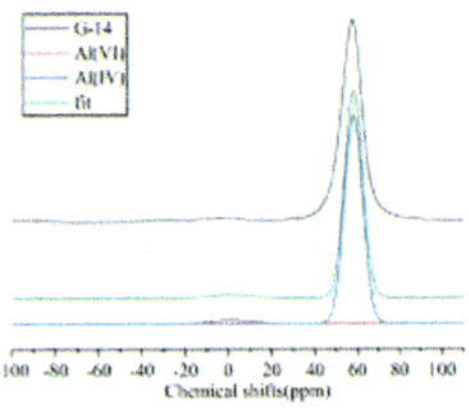

(m) G-14-Al

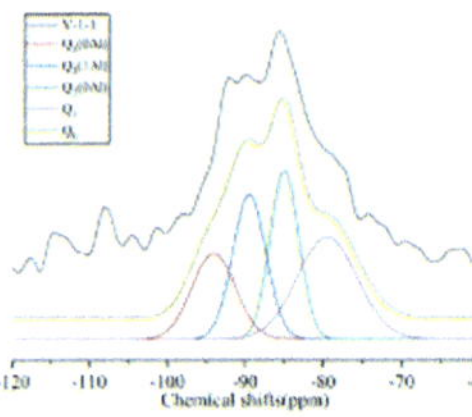

(n) V-1-1-Si

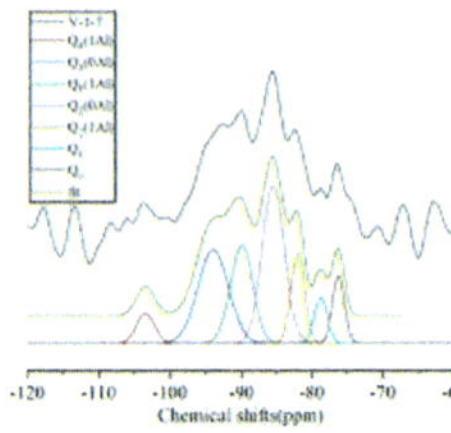

(o) V-1-7-Si

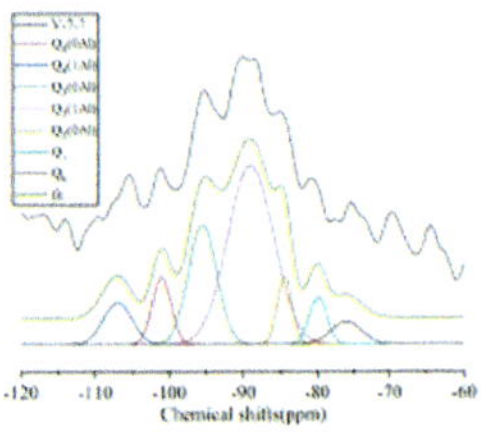

(p) V-7-7-Si

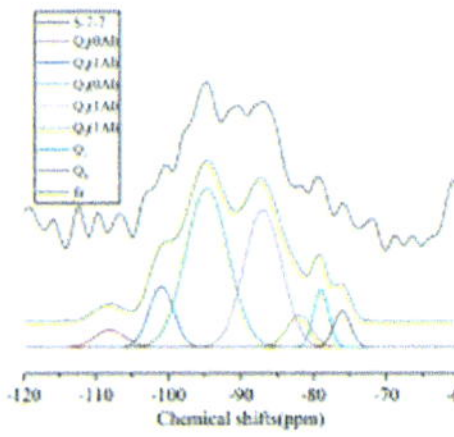

(q) S-7-7-Si

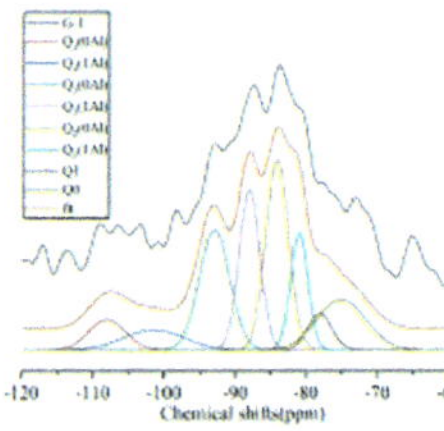

(r) G-1-Si

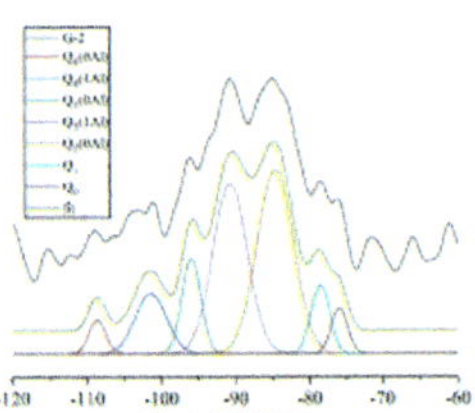

(s) G-2-Si

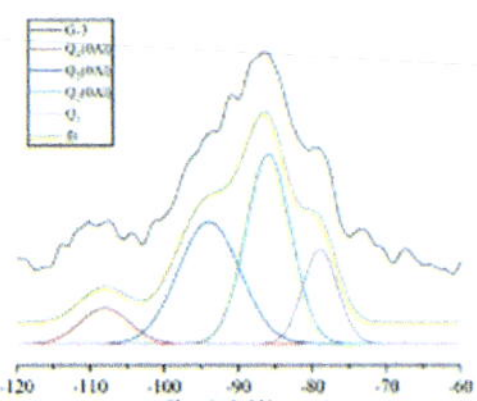

(t) G-3-Si

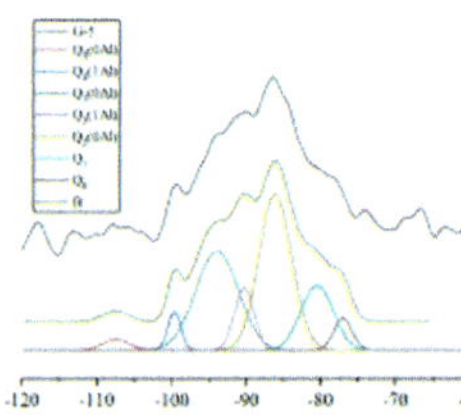

(u) G-5-Si

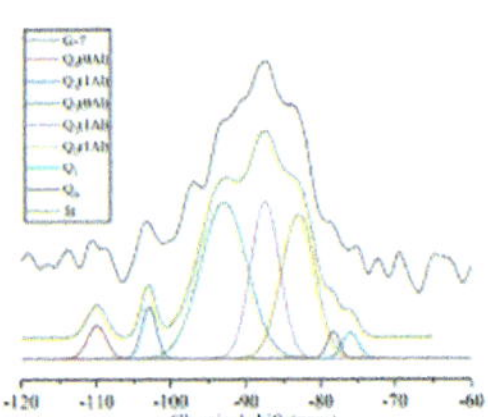

(v) G-7-Si

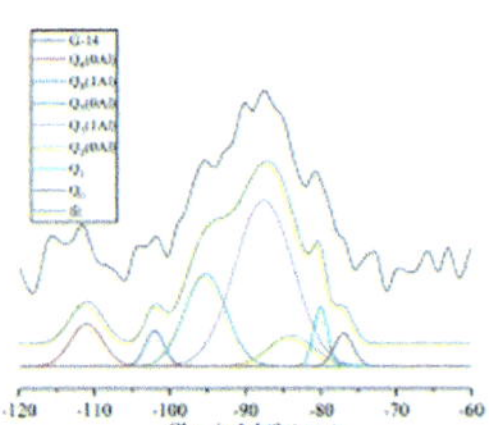

(w) G-14-Si

Fig. 2.1 (continued)

2.3 Results and Discussion

2.3.1 Dehydration Process Analysis

2.3.1.1 Fuzzy Analysis

Fuzzy analysis was used for secondary processing on all NMR spectral lines. A Gaussian curve was used for the fitting of each sub-peak. The areal convolution result of each sub-peak should typically match with the observed spectral line. This was the main fitting requirement. The ratio of homologous atomic structures to all the atoms in various coordination structures was specifically expressed by the ratio of each areal convolution of the Gaussian curve to the entire area (see Tables 2.3 and 2.4). The deconvolution process is vital in the quantitative analysis of geopolymers (Walkley et al. 2016a). To standardize and compare peak-fitting results, fixed parameters and rules were adopted from similar spectral analyses (Duxson et al. 2005a, 2007; Li et al. 2019). To guarantee the deconvolution for polymerization analysis was more accurate and consistent, new criteria were established. Taking into account the flaws in reference adjustment for each specimen, all chemical shifts of each peak permitted an adjustment range of 2 ppm. Each peak's complete width at half maximum was constrained to a range between 2 and 10, as the majority of distinctive peak values (more than 90%) were spread between 4 and 8. This criteria might prevent peaks that were less noticeable but exceedingly short and wide from being found and fitted, peaks that were not typical peaks and would contribute significantly to the area integral (Li et al. 2019). The coefficient of determination (R^2) was higher than 0.93 for the ^{27}Al MAS NMR spectrum and 0.95 for the ^{29}Si MAS NMR spectrum, indicating that the fitting and deconvolution were effective (see Fig. 2.1). These restrictions provided more reliable features for each category and avoided greater irregularity, which could be caused by more restrained deconvolution.

To enhance tolerance for unstable chemical shift migration caused by aluminum sites, Q_3(1Al) and Q_3(0Al) were introduced, as previously noted in studies conducted

Table 2.3 Deconvolution ratio results of ^{27}Al MAS NMR spectrum with two dehydration methods

Sample name	Al(IV) (%)	Al(V) (%)	Al(VI) (%)	Al(V) + Al(VI) (%)
V-1–1	41.93	21.22	36.86	58.07
V-1–7	40.71	16.84	42.46	59.29
V-7–7	93.12	–	6.88	6.88
S-1–1	30.18	26.93	42.90	69.82
S-1–7	40.69	20.53	38.77	59.31
S-7–7	92.86	–	7.14	7.14

Reproduced from [Early-stage geopolymerization revealed by ^{27}Al and ^{29}Si nuclear magnetic resonance spectroscopy based on vacuum dehydration] by [Xiuyu Zhu] with permission from [Construction and Building Materials]

Table 2.4 Deconvolution ratio results of ^{29}Si MS NMR spectrum with two dehydration methods

Sample name	Q_0 (%)	Q_1 (%)	Q_2(1Al) (%)	Q_2(0Al) (%)	Q_3(1Al) (%)	Q_3(0Al) (%)	Q_4(1Al) (%)	Q_4(0Al) (%)
V-1-1	–	31.51	–	24.66	24.66	19.18	–	–
V-1-7	6.98	5.22	10.52	29.99	18.79	24.16	4.64	–
V-7-7	4.22	5.05	–	5.96	49.34	20.56	7.69	7.17
S-1-1	–	–	–	–	–	–	–	–
S-1-7	–	–	–	–	–	–	–	–
S-7-7	3.72	5.52	4.92	–	31.60	41.63	9.26	3.35

Reproduced from [Early-stage geopolymerization revealed by ^{27}Al and ^{29}Si nuclear magnetic resonance spectroscopy based on vacuum dehydration] by [Xiuyu Zhu] with permission from [Construction and Building Materials]

Table 2.5 Deconvolution ratio retreatment results of ^{29}Si MAS NMR spectrum with two dehydration methods

Sample name	Q_0 (%)	Q_1 (%)	Q_2 (%)	Q_3 (%)	Q_4 (%)
V-1-1	–	31.51	24.66	43.84	
V-1-7	6.98	5.22	40.20	42.96	4.64
V-7-7	4.22	5.05	5.96	69.91	14.86
S-1-1	–	–	–	–	–
S-1-7	–	–	–	–	–
S-7-7	3.72	5.52	4.92	73.23	12.60

Reproduced from [Early-stage geopolymerization revealed by ^{27}Al and ^{29}Si nuclear magnetic resonance spectroscopy based on vacuum dehydration] by [Xiuyu Zhu] with permission from [Construction and Building Materials]

by Duxson et al. (2005a). Similar treatments were applied to Q_4(1Al) and Q_4(0Al), as well as Q_2(1Al) and Q_2(0Al), as listed in Table 2.5. The fuzzy analysis retreatment attenuated the quantitative impact of joint Al sites on peak-fitting and deconvolution, but intensified the variation in deconvolution values of Si sites. This method provided more insights into Si polymer structure evolution while mitigating signal noise and enhancing deconvolution precision.

2.3.1.2 Vacuum Dehydration Analysis

For the group subjected to vacuum dehydration, the ^{27}Al MAS NMR spectrum exhibited three recognized peaks (refer to Fig. 2.2a). The coexistence of these three peaks indicated that there were residual reactants present since Al(V) and Al(VI) are derived from metakaolin, as previously reported (Li et al. 2019). After a 7-day curing time, the Al(V) peak totally vanished as compared to V-1-1 and V-7-7. The other peaks changed a little at the same time, going from 58.7 ppm to 57.4 ppm and from 3.9 ppm

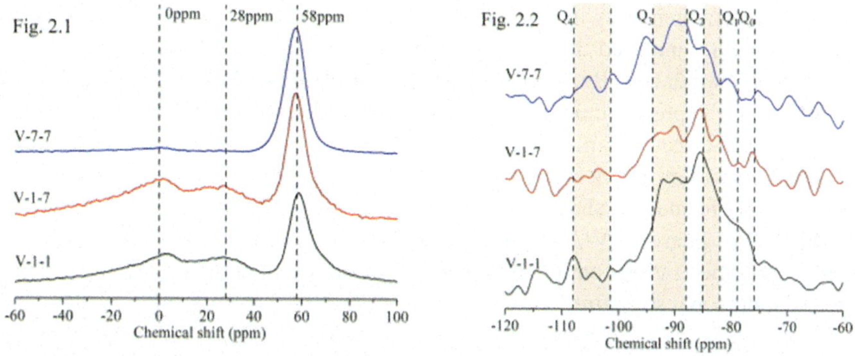

Fig. 2.2 ^{27}Al and ^{29}Si MAS NMR spectrum of specimens with vacuum dehydration. Reproduced from [Early-stage geopolymerization revealed by ^{27}Al and ^{29}Si nuclear magnetic resonance spectroscopy based on vacuum dehydration] by [Xiuyu Zhu] with permission from [Construction and Building Materials]

to 1.4 ppm, respectively. The little change in the peak showed that the aluminosilicate product's previously faultless spatial extension had become better. Referring to Fig. 2.2b, the center of the hump in the ^{29}Si MAS NMR spectra showed a clear change from around 85.0 ppm to 89.5 ppm. The lower chemical shift half of the hump became plumper, indicating an increase in Si tetrahedrons with more bridge oxygen. Newly emerging peaks below 100 ppm corresponded to Q_4 and three-dimensional structures, as noted by Duxson et al. (2005a). These clues from the ^{27}Al and ^{29}Si MAS NMR spectra were consistent and suggested that the degree of polymerization of the gel constantly increased during the normal cure of 7 days. However, there was no discernible variation in peak or form between V-1-1 and V-1-7, proving that the majority of polymer structures were stable and did not alter with additional curing. Consequently, geopolymerization had stopped.

Further details can be obtained from the deconvolution data of NMR analysis. As shown in Table 2.3, the sum of Al(V) and Al(VI) in V-1-7 was 59.29%, while in V-1-1 it was 58.07%. This little shift suggests that the unstable Al in the reactant did not further dissolve and become stable Al(IV), taking into account the slight variation of the data fit induced by the broad base when compared to the Gaussian curve. The total of Al(V) and Al(VI) was just 6.88% for V-7-7, in comparison. It is known that the formation of Al(IV) from metakaolin is the result of dehydroxylation, and therefore, it becomes evident that Al(VI) is continuously converted to Al(IV) during days 1–7, and the evolution of Al(V) indicates the mineral activity, as previously reported (Rocha and Klinowski 1990). This implies that vacuum dehydration on the first day is quite efficient in halting future geopolymerization processes and preventing additional modifications to polymeric structures.

Regarding the ^{29}Si MAS NMR spectrum, there was only a slight change observed in Q_3 and Q_4 from V-1-1 to V-1-7, but there was noticeable movement from Q_1 to Q_2, as presented in Table 2.4. This indicated that the degree of polymerization slightly changed, but more linkages were built between the end tetrahedrons, suggesting local structure rearrangement within the system. According to Duxson et al. (2007), this rearrangement connected the ends of dispersed gel clusters, leading to a larger gel structure and water loss. Because the overlap of the Si signal and base noise could not be ruled out, the presence of Q_0 could not be explained. When V-7-7 was compared to V-1-1, the large rise in Q_3 and Q_4 and drop in Q_2 indicated that the reactant system's oligomers had decreased and that a highly polymerized three-dimensional structure had formed, as shown in Table 2.3. The large proportion of the stable state of the Al(IV) phase in V-7-7 could provide verification from another perspective that geopolymerization continued to proceed towards a more integrated and ordered three-dimensional structure.

2.3.1.3 Solvent Extraction Analysis

The experimental results indicate that the group subjected to solvent extraction exhibits comparable basic peaking characteristics, as shown in Fig. 2.3a. However, there is a significant decrease in the ratio sum of Al(V) and Al(VI) from 69.82% (S-1-7) to 59.31% (S-1-1), accompanied by a considerable increase in Al(IV), as presented in Table 2.3. This implies that some unstable Al species in the structure are transformed into stable Al(IV) even after undergoing solvent extraction. These Al(IV) atoms are present in discrete monomers and oligomers or in local spatial structures through polymerization.

The proportion of unstable aluminum in S-7-7 is found to be merely 7.14%. This outcome implies that the utilization of solvent extraction on the first day partially restrains the reaction. Nevertheless, it is worth noting that the polymerization reaction cannot be fully suppressed. It could be hypothesized that the dissolution of

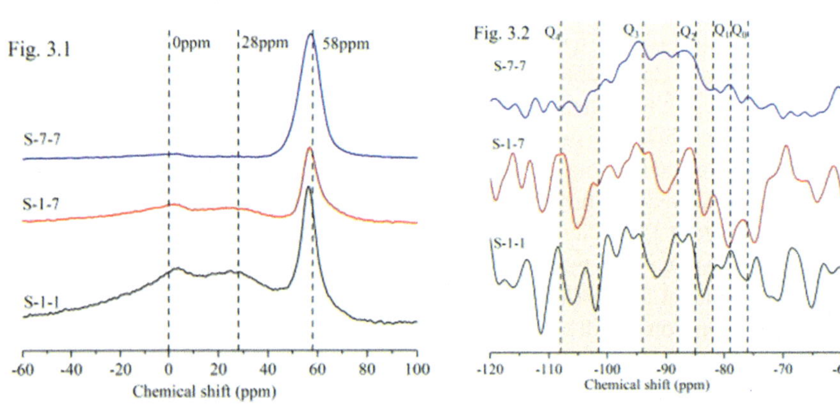

Fig. 2.3 ^{27}Al and ^{29}Si MAS NMR spectrum of specimens with solvent extraction. Reproduced from [Early-stage geopolymerization revealed by ^{27}Al and ^{29}Si nuclear magnetic resonance spectroscopy based on vacuum dehydration] by [Xiuyu Zhu] with permission from [Construction and Building Materials]

monomers may still occur post solvent extraction, leading to continued conversion from unstable Al(V) to stable Al(IV) until the complete disappearance of Al(V). The dissolution process plays a vital role since it acts as the limiting step for the entire geopolymerization reaction (Favier et al. 2015). Additionally, the increase in temperature during rotary evaporation procedure may accelerate the dissolution process. Based on the spectral analysis, it can be inferred that approximately 4 h in the rotary evaporator is adequate to bring about a prominent spectrum variation.

Upon examining the ^{29}Si MAS NMR spectrum, it was observed that the usage of the rotary evaporator disrupted the regular resonant response between H and Si in the reactant system. This was reflected by the absence of peaks in S-1-1 and S-1-7 (refer to Fig. 2.3b). The results obtained from S-7-7 were similar to V-7-7, indicating that solvent extraction is applicable for bulk and powder samples (as presented in Table 2.4). However, it should be noted that solvent extraction may not yield desirable outcomes for early pulp or paste samples containing a significant amount of free water. In such cases, more organic solvent and additional steaming are necessary to convert the sample into a dry powder suitable for NMR analysis. The organic solvent also has stringent prerequisites for effective NMR detection, which raises doubts about the reliability of solvent extraction.

2.3.1.4 Vacuum Dehydration Versus Solvent Extraction

Upon comparing the two techniques, the vacuum dehydration approach was more effective and efficient in handling the geopolymerization reactant system, particularly during the early stages where there is an abundance of free water. The temperature within the vacuum system remained stable at approximately 15 °C, thereby precluding any rise in temperature as seen in the rotary evaporator. Furthermore, the vacuum dehydration process was comparatively easier, standardized, and dependable. In conclusion, vacuum dehydration was deemed highly dependable when dealing with early pulp or paste geopolymer specimens. The deconvolution data verified that vacuum dehydration was superior to the rotary evaporation approach for NMR quantitative analysis. It offered relevant and precise data that were dependable and reliable.

2.3.2 Structure Evolution of Geopolymer

2.3.2.1 ^{27}Al NMR Analysis

Figure 2.4a depicts all the ^{27}Al MAS NMR spectral lines. Relative to M, the Al(V) and Al(VI) peaks of geopolymer samples are visibly diminished, even on the first day in G-1. Simultaneously, the resonance of Al(IV) became noticeably stronger. These observations suggest that the metakaolin dissolution process was predominantly completed within the initial 24 h post the addition of alkali activator. Among the geopolymer samples, the appearance of the Al(V) peak was only evident within the first three days, indicating that Al(V) was more vulnerable than Al(VI) and dissolved and transformed sooner. The Al(VI) peak persisted after 14 days, wherein self-dehydration occurred following the third day. This suggests that, in comparison to Al(V), the structure adjustment of Al(VI) was less dependent on the presence of free water and high alkalinity (refer to Table 2.6).

The gradual decline of Al(VI) from G-1 to G-14 indicated the transformation of the Al coordination state. It also provided insights into the chemical nature of the metakaolin dissolution process (Rocha and Klinowski 1990). By considering the sum of the proportion of Al(V) and Al(VI) in metakaolin as a new unit (per Eq. (2.1)), the dissolution ratio was determined to be 42.69% after day 1, 61.50% after day 3, and 95.80% after day 14, as illustrated in Fig. 2.5.

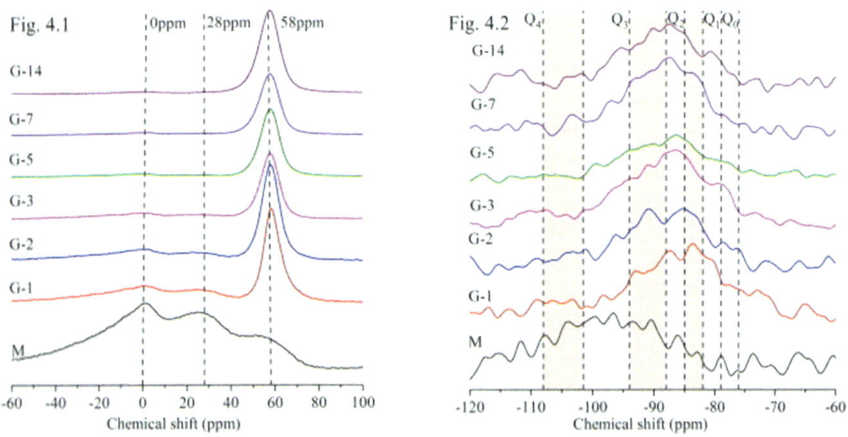

Fig. 2.4 ^{27}Al and ^{29}Si MAS NMR spectrum of metakaolin and geopolymer samples for the first 14 days. Reproduced from [Early-stage geopolymerization revealed by ^{27}Al and ^{29}Si nuclear magnetic resonance spectroscopy based on vacuum dehydration] by [Xiuyu Zhu] with permission from [Construction and Building Materials]

Table 2.6 Deconvolution ratio results of ^{27}Al MAS NMR spectrum in metakaolin and geopolymer samples for the first 14 days

Sample name	Al(IV) (%)	Al(V) (%)	Al(VI) (%)	Al(V) + Al(VI) (%)
M	20.08	27.03	52.89	79.92
G-1	54.05	21.65	24.30	45.95
G-2	63.92	16.04	20.04	36.08
G-3	69.23	16.69	14.08	30.77
G-5	89.18	–	10.82	10.82
G-7	93.32	–	6.68	6.68
G-14	96.64	–	3.36	3.36

Reproduced from [Early-stage geopolymerization revealed by ^{27}Al and ^{29}Si nuclear magnetic resonance spectroscopy based on vacuum dehydration] by [Xiuyu Zhu] with permission from [Construction and Building Materials]

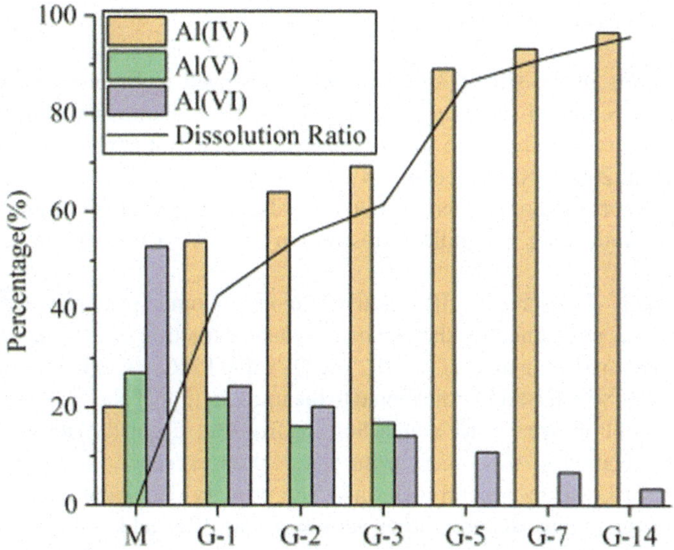

Fig. 2.5 Deconvolution re-treatment results of ^{27}Al MAS NMR spectrum of metakaolin and geopolymer samples. Reproduced from [Early-stage geopolymerization revealed by ^{27}Al and ^{29}Si nuclear magnetic resonance spectroscopy based on vacuum dehydration] by [Xiuyu Zhu] with permission from [Construction and Building Materials]

$$Dissolution\ Ratio = \frac{\mathrm{Al(IV)}_{(G-X)} - \mathrm{Al(IV)}_{(M)}}{1 - \mathrm{Al(IV)}_{(M)}} \tag{2.1}$$

These findings elucidate the rapid strength development observed in the early stages of geopolymerization. Even at lower temperatures, such as 5 °C, metakaolin dissolution was approximately 75% complete within the first three days and more than 90% complete within the first 14 days. Laboratory-produced samples are typically cured at higher temperatures (usually around 20 °C), thereby generating optimal chemical reaction kinetics. The swift release of aluminate is theoretically conducive to nucleation during geopolymerization. Undissolved residues and amorphous aluminosilicate nanoparticles serve as nucleation seeds and furnish numerous solid–liquid interfacial surfaces as nucleation sites (Serrano and Van Grieken 2001; Cundy and Cox 2003). Additionally, a high concentration of aluminate anions promotes nucleation and forms the first layer known as Al-rich gels (Provis et al. 2005; Rees et al. 2008). It is therefore evident that sufficient availability of Al is requisite for cross-linking, hardening, and strength enhancement during geopolymerization (Fernández-Jiménez et al. 2006; Provis and van Deventer 2007). This finding is a significant addition to the limited body of literature on dissolution research since it highlights the potent role of quantitative tracing of Al as a strong proponent for nucleation theoretically (Rees et al. 2008).

2.3.2.2 ^{29}Si NMR Analysis

The alterations in the silicon spectrum were observed to be more intricate than those in the aluminum spectrum (see Fig. 2.4b). Prior studies indicate that hump of metakaolin ranges from 115 to 85 ppm, with a center at approximately 97 ppm (Duxson et al. 2005a; Kuenzel et al. 2012). Nonetheless, our findings reveal that all the hump centers of geopolymers shifted towards a higher chemical shift range, i.e., 88–84 ppm, indicating a significant disruption in ordered linkages in the mineral structure.

Q_0 typically represents the dissociative silicon monomer, primarily derived from the water glass component in the activator, where a fraction of it is depolymerized from metakaolin (Lippmaa et al. 1980). The Q_0 value for G-1 was highest during the initial 14 days, which could be attributed to the introduction of Na_2SiO_3 and provides valuable insights into reaction kinetics. Starting from the second day, the consumption and polymerization of monomers from water glass exceeded the production and dissolution of Q_0 from mineral materials. The primary depolymerization occurred within two days; thus, incorporating additives that offer additional monomers to facilitate polymerization during sample curing might prove advantageous.

While Q_2 has two bridge oxygens, Q_1 only has one, often placing them at the end of the structural branch and the middle of the catenaries, respectively. Q_1 also only has one bridge oxygen connected to other tetrahedrons. This provides a clear representation of morphology evolution. The increase in Q_1 during days 1–3 followed by a decrease during days 3–14 coincided with the self-dehydration of samples on day 3 (refer to Table 2.7 and Fig. 2.6). Additionally, G-3 lacked the Q_0 peak. Based on the stabilization of the sum of Q_1 and Q_0 during days 1–5, we hypothesize that the rapid self-dehydration process resulted in the temporary transmission of Q_0 and an increase in Q_1. Due to the collapse of interlayer pores, all monomers are captured and attached at the end of the structure (Zhou et al. 2018), with water movement serving as a driving force for changes in the reaction system.

Table 2.7 Deconvolution ratio results of ^{29}Si MAS NMR spectrum in geopolymer samples for the first 14 days

Sample name	Q_0 (%)	Q_1 (%)	Q_2(1Al) (%)	Q_2(0Al) (%)	Q_3(1Al) (%)	Q_3(0Al) (%)	Q_4(1Al) (%)	Q_4(0Al) (%)
G-1	12.72	4.58	10.81	22.89	17.81	19.08	6.36	5.75
G-2	4.02	7.00	–	35.28	30.92	9.58	10.20	2.98
G-3	–	15.66	–	40.50	–	35.54	–	8.30
G-5	4.02	14.64	–	36.85	8.74	30.14	3.52	2.09
G-7	2.20	1.82	25.99	–	24.10	38.37	4.13	3.39
G-14	3.42	4.61	–	7.44	52.59	20.80	3.71	7.42

Reproduced from [Early-stage geopolymerization revealed by ^{27}Al and ^{29}Si nuclear magnetic resonance spectroscopy based on vacuum dehydration] by [Xiuyu Zhu] with permission from [Construction and Building Materials]

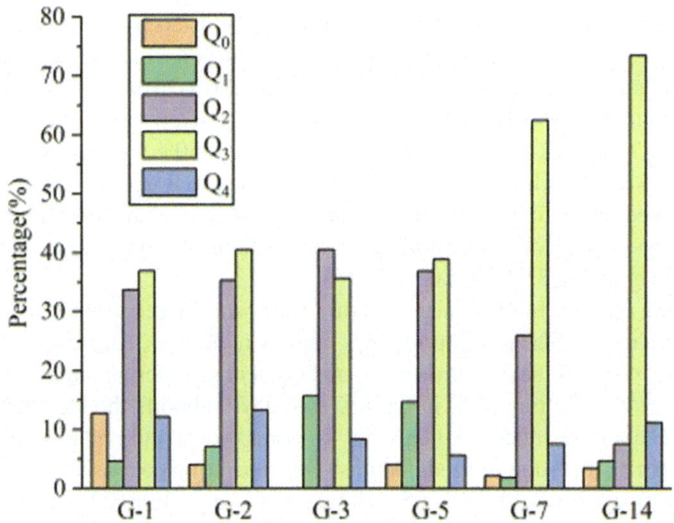

Fig. 2.6 Deconvolution re-treatment results of ^{29}Si MAS NMR spectrum of geopolymer samples. Reproduced from [Early-stage geopolymerization revealed by ^{27}Al and ^{29}Si nuclear magnetic resonance spectroscopy based on vacuum dehydration] by [Xiuyu Zhu] with permission from [Construction and Building Materials]

A critical stage in the geopolymerization process is the evolution of Q_2, which might result from serially linked monomers and partial mineral material separation. The defective hexagonal network complex ion, which is more prevalent in kaolinite and talcum, undergoes weak bond breakage and interlamination dissociation during the first stage of the geopolymerization process, which results in the production of catenarian tetrahedral compound ions of various lengths. This is consistent with the existing form of the infinite silicon-oxygen tetrahedral compound ion in mineralogy. The prior increase in the first three days can be explained as aforementioned, while the subsequent decrease is attributed to different factors. Notably, the primary reduction of Q_2 in G-5 and G-7 occurred later than the decline of Q_0 in G-1 and G-2. From the degree of linkage, both reductions suggest possible polymerization of the gel spatial structure. Indeed, the evolution of Q_0 denotes the polymerization that occurs following dissolution and depolymerization, as noted in most geopolymerization theories. This initial polymerization process, referred to as "pre-crystallization" results in catenarian compound ions of longer length and even spatial compound ions, with no more than 4-unit cells or 8–10 nm in length observed in zeolite studies (Yang et al. 2000). The reactant system is a primary amorphous gel containing colloidal-sized globular units (Van Deventer et al. 2012). After three days, there is a propensity for Q_2 to change into Q_3 and potentially Q_4 on a greater scale due to the steady loss of free water. This transformation may have distinct dynamic mechanisms compared to "pre-crystallization" given the lack of free water and limited available space (Kuenzel et al. 2012). It can be speculated that polymerization after the loss of free water primarily arises from linkage to nearby available tetrahedron sites due to the insufficient impetus of incompact gel clusters from water and confinement by existing bonds. According to this theory, drying shrinkage brought on by the loss of "structural" water exists (Kuenzel et al. 2012).

The fluctuating increase in Q_3 comprehensively reflects other forms of tetrahedrons. In G-7 and G-14, the proportion of Q_3 exceeds 50%, indicating that a three-dimensional network is the dominant form of structure in geopolymeric gels. The lack of linked bridge oxygen is the morphology-based explanation for geopolymer to encapsulate large ions or impurities in the structure-forming matrix defect sites (Lambertin et al. 2013; Rooses et al. 2013).

Q_4 is the optimal fundamental structure for powerful geopolymers since it is a perfect tetrahedron. The proportion of Q_4 initially undergoes a decrease followed by an increase, whereas the evolution of Q_1 and Q_2 demonstrates an initial increase and a quick decline. The tiny peak in the first two days supports the idea that some of Q_4 may be about residues left behind by the breakdown and collapse of defective sites. It may also be due to seeds defined beforehand that act as the prior zone of polymerization (Provis et al. 2005). However, the dip observed during days 3–7 suggests that there may be other parts affected or destroyed by the loss of free water since recovery takes more than ten days. This dip is reported for the first time in this study, with unclear reasons for its occurrence. It is not observed in analogous NMR studies. Considering the relative changes in Q_0 and Q_2, the loss of water may affect Q_4 concurrently. The decrease in Q_4 indicates that highly ordered local structures may have been destroyed, which can be observed through the FTIR showing a shift in the main Si–O–T bond towards lower wave numbers initially before increasing towards higher wave numbers. This may be explained by the fact that Al is present from the start, broadening the bond and reducing the wave numbers, whereas Si is added to the gel on the second or third curing day, leveling out the bond (Lippmaa et al. 1980). These consistent findings lend support to the nucleation theory, where initially dissolved Al and nano-residues trigger nucleation leading to the formation of the initial Al-rich gel (Van Deventer et al. 2012; Lambertin et al. 2013).

The nucleation gels serve as the source of the ordered part in the short-range due to the free water environment. The loss of free water, however, has an impact on neighboring connection and causes unequal stress within the gel, which causes cracks in the nucleus. This phenomenon could account for the release of Si and the formation of Si-rich gel in the later stage of geopolymerization (Hajimohammadi et al. 2010), which gradually enhances the final strength, as demonstrated in G-5 to G-14. The evolution process of Q_4 reflects the rearrangement of the gels, with the kinetic impetus for rearrangement potentially lying in the change of the free water environment.

The occurrence of microcracking in various lengths during the curing process has been attributed to the drying effect (Collins and Sanjayan 2001). It can be inferred that the connections between tetrahedrons are driven by different factors. Prior to dehydration and hardening, monomers and small catenarian compound ions tend to form ordered tetrahedral 3D structures surrounding core sites such as aluminum tetrahedrons (Provis et al. 2013). The gaps among these ordered 3D structures are filled with atactic and incomplete bonds, which provide ample space for water and positive ions. Dehydration generates energy from within, causing the collapse and rearrangement of atactic bonds. This results in the release of more silica from covered clusters and the formation of new connections, ultimately leading to a massive structure.

2.3.3 *Geopolymerization Analysis*

The geopolymerization process during the first 14 days can be categorized into four stages: dissolution (day 0–1), polymerization (days 2–3), collapse and rearrangement (days 3–5), and repolymerization (days 5–14). These stages are distinguishable over time, as indicated by changes in NMR deconvolution results. On day 1, a significant depolymerization or dissolving process produces a significant amount of monomers in the pore solution. These monomers, tiny residues, and compound ions easily mix in an ordered way prior to dehydration (days 1–3) to form longer and larger structures mimicking nucleation. The gel structure develops from several beginning points simultaneously as a result of this combination's tendency to produce an ordered atomic structure from a core, with multiple cores spread throughout the system. As a result, the product's general homogeneity is compromised and it cannot link completely. The product created during the first polymerization step is also organized and dense in the short term, but it is flawed and uneven in the long term. After then, structural reorganization and dehydration occur (days 3–5). Thus, it is postulated that the removal of free water causes uneven inner stress, which may lead to local structural breakage and collapse. Fragments may fill in the defective space, making the system denser and more disordered. During the curing period after dehydration (days 5–14), the low amount of water inhibits the dissociation of ions. In the repolymerization stage, these fragments can only form new bonds in adjacent zones, between layers, and inside the frame structure. The new links can improve the overall degree of polymerization and strengthen the spatial structure with flaws, as Q_3 represents the most representative molecular species in geopolymers.

2.4 Conclusions

The following conclusions can be drawn from this chapter:

(1) When compared to the traditional solvent extraction procedure, vacuum dehydration efficiently halts the geopolymerization reaction in its early stages. Since it swiftly eliminates water without raising the temperature, it is appropriate for samples with a high water content. Vacuum dehydration can test materials with high variability utilizing the same detection techniques and parameters for pulpy and paste samples before to self-dehydration as well as solid samples following water removal. The process is straightforward and prevents organic contamination and temperature fluctuations.

(2) NMR assisted by a fuzzy analysis of deconvolution data reveals an integrated process of early-stage geopolymerization that can be divided into four stages: (i) dissolution from raw materials into monomers, (ii) polymerization free from nucleation, (iii) collapse and rearrangement through the loss of free water, and (iv) repolymerization in flaw sites. The development of Qm (m = 0–4) may be used as a reflection of each stage. The genuine geopolymerization process is accurately reflected by the fuzzy analysis of NMR deconvolution data, which offers both qualitative and quantitative data. These geopolymerization findings validate the Duxson model and demonstrate the validity of the nucleation hypothesis in incipient geopolymerization, as seen by the first rapid and vigorous increase of Q_4. The two-gel-phase idea is somewhat supported by the observation of swiftly generated Al(IV) and gradually formed Q_4.

(3) Q_1's falling trend, which indicates connection, and Q_4's valley, which depicts breakdown and rebuilding, reveal the primary impact of losing free water on the gel structure. On the one hand, end tetrahedrons cling to the matrix or the ordered polymerization as a result of the loss of free water. On the other hand, this may result in the matrix collapsing and breaking, an issue that has frequently gone unnoticed in earlier research. Following the loss of water, Q_2 demonstrates the connection between surrounding catenaries to increase the degree of polymerization. The distinctive structure, which is ordered in the short-range and unordered in the long-range, may be explained by the various driving forces present throughout the two polymerization phases. The driving forces change significantly as free water is lost, resulting in different linkage categories due to different water and kinetic environments. However, more details are needed about the different kinetic factors before and after the loss of free water.

References

Bhutta A, Farooq M, Zanotti C, Banthia N (2017) Pull-out behavior of different fibers in geopolymer mortars: effects of alkaline solution concentration and curing. Mater Struct Constr 50.https://doi.org/10.1617/s11527-016-0889-2

Chen X, Meawad A, Struble LJ (2014) Method to stop geopolymer reaction. J Am Ceram Soc 97:3270–3275. https://doi.org/10.1111/jace.13071

Chindaprasirt P, Rattanasak U (2018) Fire-resistant geopolymer bricks synthesized from high-calcium fly ash with outdoor heat exposure. Clean Technol Environ Policy 20:1097–1103. https://doi.org/10.1007/s10098-018-1532-4

Collins F, Sanjayan JG (2001) Microcracking and strength development of alkali activated slag concrete. Cem Concr Compos 23:345–352. https://doi.org/10.1016/S0958-9465(01)00003-8

Cundy CS, Cox PA (2003) The hydrothermal synthesis of zeolites: history and development from the earliest days to the present time. ChemInform 34.https://doi.org/10.1002/chin.200319217

Davidovits J (1991) Geopolymers: inorganic polymeric new materials. J Therm Anal 37:1633–1656

Duxson P, Provis JL, Lukey GC et al (2005a) Understanding the relationship between geopolymer composition, microstructure and mechanical properties. Colloids Surfaces A Physicochem Eng Asp 269:47–58. https://doi.org/10.1016/j.colsurfa.2005.06.060

Duxson P, Provis JL, Lukey GC et al (2005b) ^{29}Si NMR study of structural ordering in aluminosilicate geopolymer gels. Langmuir 21:3028–3036. https://doi.org/10.1021/la047336x

Duxson P, Fernández-Jiménez A, Provis JL et al (2007) Geopolymer technology: the current state of the art. J Mater Sci 42:2917–2933. https://doi.org/10.1007/s10853-006-0637-z

Favier A, Habert G, Roussel N, D'Espinosede Lacaillerie JB (2015) A multinuclear static NMR study of geopolymerisation. Cem Concr Res 75:104–109. https://doi.org/10.1016/j.cemconres.2015.03.003

Fernández-Jiménez A, Palomo A, Sobrados I, Sanz J (2006) The role played by the reactive alumina content in the alkaline activation of fly ashes. Microporous Mesoporous Mater 91:111–119. https://doi.org/10.1016/j.micromeso.2005.11.015

Gordon LE, Nicolas RS, Provis JL (2014) Chemical characterisation of metakaolin and fly ash based geopolymers during exposure to solvents used in carbon capture. Int J Greenh Gas Control 27:255–266. https://doi.org/10.1016/j.ijggc.2014.06.005

Hajimohammadi A, Provis JL, Van Deventer JSJ (2010) Effect of alumina release rate on the mechanism of geopolymer gel formation. Chem Mater 22:5199–5208. https://doi.org/10.1021/cm101151n

Hajimohammadi A, Provis JL, Van Deventer JSJ (2011) The effect of silica availability on the mechanism of geopolymerisation. Cem Concr Res 41:210–216. https://doi.org/10.1016/j.cemconres.2011.02.001

Irfan Khan M, Azizli K, Sufian S, Man Z (2015) Sodium silicate-free geopolymers as coating materials: effects of Na/Al and water/solid ratios on adhesion strength. Ceram Int 41:2794–2805. https://doi.org/10.1016/j.ceramint.2014.10.099

Ismail I, Bernal SA, Provis JL et al (2014) Modification of phase evolution in alkali-activated blast furnace slag by the incorporation of fly ash. Cem Concr Compos 45:125–135. https://doi.org/10.1016/j.cemconcomp.2013.09.006

Kuenzel C, Vandeperre LJ, Donatello S et al (2012) Ambient temperature drying shrinkage and cracking in metakaolin-based geopolymers. J Am Ceram Soc 95:3270–3277. https://doi.org/10.1111/j.1551-2916.2012.05380.x

Lambertin D, Boher C, Dannoux-Papin A et al (2013) Influence of gamma ray irradiation on metakaolin based sodium geopolymer. J Nucl Mater 443:311–315. https://doi.org/10.1016/j.jnucmat.2013.06.044

Li Z, Zhang S, Zuo Y et al (2019) Chemical deformation of metakaolin based geopolymer. Cem Concr Res 120:108–118. https://doi.org/10.1016/j.cemconres.2019.03.017

Lippmaa E, Magi M, Samoson A et al (1980) Structural studies of silicates by solid-state high-resolution ^{29}Si NMR. J Am Chem Soc 102:4889–4893. https://doi.org/10.1021/ja00535a008

MacKenzie KJD, Komphanchai S, Vagana R (2008) Formation of inorganic polymers (geopolymers) from 2:1 layer lattice aluminosilicates. J Eur Ceram Soc 28:177–181. https://doi.org/10.1016/j.jeurceramsoc.2007.06.004

Mustafa Al Bakria AM, Kamarudin H, Bin Hussain M et al (2011) The effect of curing temperature on physical and chemical properties of geopolymers. Phys Procedia 22:286–291. https://doi.org/10.1016/j.phpro.2011.11.045

Nath SK, Maitra S, Mukherjee S, Kumar S (2016) Microstructural and morphological evolution of fly ash based geopolymers. Constr Build Mater 111:758–765. https://doi.org/10.1016/j.conbuildmat.2016.02.106

North MR, Swaddle TW (2000) Kinetics of silicate exchange in alkaline aluminosilicate solutions. Inorg Chem 39:2661–2665. https://doi.org/10.1021/ic0000707

Provis JL (2014) Geopolymers and other alkali activated materials: why, how, and what? Mater Struct Constr 47:11–25. https://doi.org/10.1617/s11527-013-0211-5

Provis JL, van Deventer JSJ (2007) Geopolymerisation kinetics. 2. Reaction kinetic modelling. Chem Eng Sci 62:2318–2329. https://doi.org/10.1016/j.ces.2007.01.028

Provis JL, Lukey GC, van Deventer JSJ (2005) Do geopolymers actually contain nanocrystalline zeolites? A reexamination of existing results. Chem Mater 17:3075–3085

Provis JL, Hajimohammadi A, White CE et al (2013) Nanostructural characterization of geopolymers by advanced beamline techniques. Cem Concr Compos 36:56–64. https://doi.org/10.1016/j.cemconcomp.2012.07.003
Provis JL, Palomo A, Shi C (2015) Advances in understanding alkali-activated materials. Cem Concr Res 78:110–125. https://doi.org/10.1016/j.cemconres.2015.04.013
Reddy DV, Edouard J-B, Sobhan K (2013) Durability of fly ash-based geopolymer structural concrete in the marine environment. J Mater Civ Eng 25:781–787. https://doi.org/10.1061/(asce)mt.1943-5533.0000632
Rees CA, Provis JL, Lukey GC, van Deventer JSJ (2008) The mechanism of geopolymer gel formation investigated through seeded nucleation. Colloids Surfaces A Physicochem Eng Asp 318:97–105. https://doi.org/10.1016/j.colsurfa.2007.12.019
Rocha J, Klinowski J (1990) ^{29}Si and ^{27}Al magic-angle-spinning NMR studies of the thermal transformation of kaolinite. Phys Chem Miner 17:179–186. https://doi.org/10.1007/BF00199671
Rooses A, Steins P, Dannoux-Papin A et al (2013) Encapsulation of Mg–Zr alloy in metakaolin-based geopolymer. Appl Clay Sci 73:86–92. https://doi.org/10.1016/j.clay.2012.09.023
Scrivener KL, Kirkpatrick RJ (2008) Innovation in use and research on cementitious material. Cem Concr Res 38:128–136. https://doi.org/10.1016/j.cemconres.2007.09.025
Serrano DP, Van Grieken R (2001) Heterogenous events in the crystallization of zeolites. J Mater Chem 11:2391–2407. https://doi.org/10.1039/b100818h
Šmilauer V, Hlaváček P, Škvára F et al (2011) Micromechanical multiscale model for alkali activation of fly ash and metakaolin. J Mater Sci 46:6545–6555. https://doi.org/10.1007/s10853-011-5601-x
Steins P, Poulesquen A, Frizon F et al (2014) Effect of aging and alkali activator on the porous structure of a geopolymer. J Appl Crystallogr 47:316–324. https://doi.org/10.1107/S160057671303197X
Subekti S, Bayuaji R, Darmawan MS et al (2017) Review: potential strength of fly ash-based geopolymer paste with substitution of local waste materials with higherature effect. IOP Conf Ser Mater Sci Eng 267.https://doi.org/10.1088/1757-899X/267/1/012001
Tazawa E, Miyazawa S, Kasai T (1995) Chemical shrinkage and autogenous shrinkage of hydrating cement paste. Cem Concr Res 25:288–292. https://doi.org/10.1016/0008-8846(95)00011-9
Thermou GE, Hajirasouliha I (2018) Compressive behaviour of concrete columns confined with steel-reinforced grout jackets. Compos Part B Eng 138:222–231. https://doi.org/10.1016/j.compositesb.2017.11.041
Uddin Ahmed Shaikh F, Haque S, Sanjayan J (2019) Behavior of fly ash geopolymer as fire resistant coating for timber. J Sustain Cem Mater 8:259–274. https://doi.org/10.1080/21650373.2018.1537015
Van Deventer JSJ, Provis JL, Duxson P (2012) Technical and commercial progress in the adoption of geopolymer cement. Miner Eng 29:89–104. https://doi.org/10.1016/j.mineng.2011.09.009
Walkley B, San Nicolas R, Sani MA et al (2016a) Phase evolution of Na_2O-Al_2O_3-SiO_2-H_2O gels in synthetic aluminosilicate binders. Dalt Trans 45:5521–5535. https://doi.org/10.1039/c5dt04878h
Walkley B, San Nicolas R, Sani MA et al (2016b) Phase evolution of C-(N)-A-S-H/N-A-S-H gel blends investigated via alkali-activation of synthetic calcium aluminosilicate precursors. Cem Concr Res 89:120–135. https://doi.org/10.1016/j.cemconres.2016.08.010
White CE, Page K, Henson NJ, Provis JL (2013) In situ synchrotron X-ray pair distribution function analysis of the early stages of gel formation in metakaolin-based geopolymers. Appl Clay Sci 73:17–25. https://doi.org/10.1016/j.clay.2012.09.009
Yang S, Navrotsky A, Phillips BL (2000) In situ calorimetric, structural, and compositional study of zeolite synthesis in the system $5.15Na_2O$-$1.00Al_2O_3$-$3.28SiO_2$–$165H_2O$. J Phys Chem B 104:6071–6080. https://doi.org/10.1021/jp9944278

Yunsheng Z, Wei S, Zongjin L (2010) Composition design and microstructural characterization of calcined kaolin-based geopolymer cement. Appl Clay Sci 47:271–275. https://doi.org/10.1016/j.clay.2009.11.002

Zhang Z, Yao X, Zhu H (2010) Potential application of geopolymers as protection coatings for marine concrete I. Basic properties. Appl Clay Sci 49:1–6. https://doi.org/10.1016/j.clay.2010.01.014

Zhou C, Ren F, Zeng Q et al (2018) Pore-size resolved water vapor adsorption kinetics of white cement mortars as viewed from proton NMR relaxation. Cem Concr Res 105:31–43. https://doi.org/10.1016/j.cemconres.2017.12.002

Chapter 3
Composition-Dependent Mechanical Performance of MKG

Abstract A quantitative assessment of the mechanical performance and the comprehension of correlations between composition and performance are vital for the rational design of MKG. This chapter investigates the elastic and creep behavior of MKG. The elastic stress–strain behaviors of MKG are measured under uniaxial compression, and the basic elastic parameters (including Young's, bulk, and shear moduli and Poisson's ratio) are evaluated. The binder phase creep behavior in MKG is also studied using nanoindentation techniques. Moreover, by combining microstructure analysis, the underlying influence of composition and microstructure on the elastic and creep behavior of MKG is further analyzed.

3.1 Introduction

For its engineering uses, geopolymer's mechanical properties are essential. Therefore, comprehensive experimental investigations have been carried out on the compressive and/or split strength of geopolymer paste, mortar, and concrete by various researchers (e.g. Rowles et al. 2003; Wang et al. 2005; Zhang et al. 2010; Posi et al. 2013; Gao et al. 2014; Islam et al. 2014; Cheng et al. 2015a, b). This study aims to investigate the elastic and creep mechanical properties of geopolymer-based materials. Specifically, the Young's moduli, Poisson's ratio, and creep modulus are examined as these are essential parameters for calculating the internal forces and deformations of structures made with geopolymer, particularly those subjected to spatial stress states such as building foundations, bridges, and dams.

Research has shown that the composition (including the type of raw materials and mixture), synthesis process, curing time, and temperature have a significant impact on the Young's modulus of geopolymer-based materials, such as paste, mortar, or concrete (Duxson et al. 2005, 2007; Wongpa et al. 2010; Lizcano et al. 2012). Measurements of the Young's modulus of MKG pastes with various Si/Al ratios and alkali types were made by Duxson et al. (2005, 2007), who found that compositional parameters have complicated impacts on the geopolymers' Young's moduli and that these effects are closely tied to changes in microstructure. Further micro-indentation

D. Yan et al., *Metakaolin-Based Geopolymers*,
https://doi.org/10.1007/978-981-97-0652-5_3

tests on Na and K-based MKG pastes conducted by Lizcano et al. (2012) further confirmed the dependence of mechanical properties on composition. Similar observations have been made for fly-ash-based geopolymer pastes (Andini et al. 2008). However, there is insufficient knowledge about the Poisson's ratio of geopolymer-type materials, despite its significant importance for engineering applications, as it measures the ratio of deformations in different directions. Other elastic properties, including bulk modulus and shear modulus, can be found directly by calculating Young's modulus and Poisson ratio values. Therefore, investigating the elastic properties of geopolymer materials and understanding their composition correlations is of great scientific and engineering interest.

Concrete constructions' serviceability and durability are substantially impacted by the creep behavior of cementitious binders (Bazant and Panula 1980; Sicard et al. 1992; Bažant 2001; Barpi and Valente 2003; Benboudjema et al. 2005). As a newly developed binder material, creep behavior of geopolymer has garnered considerable attention from researchers (Hardjito et al. 2004; Wallah and Rangan 2006; Sagoe-Crentsil et al. 2013; Islam 2015; Castel et al. 2016; Lee et al. 2018). Using bulk-scale creep tests, Hardjito et al. (2004) demonstrated that fly-ash-based geopolymer concrete exhibits low creep strain over 12 weeks of loading. Subsequent research found that the specific creep (creep strain normalized by creep stress) (Wallah and Rangan 2006; Lee 2007; Castel et al. 2016) and creep coefficient (the ratio between creep strain and elastic strain) (Sagoe-Crentsil et al. 2013; Castel et al. 2016) of fly-ash based geopolymer concrete are generally lower than those of OPC concrete. According to some experts, the unreacted fly ash particles in the binder had a micro-aggregation effect, which was the cause of the low creep (Wallah and Rangan 2006; Wallah 2010). However, because there was no direct correlation between the bulk concrete characteristics and the binder microstructure, the bulk creep experiment could only offer a general explanation for the creep behavior in geopolymer. To provide small-scale direct measurements, Lee et al. (2018) conducted minute-long nanoindentation experiments to characterize the creep behavior of alkali-activated fly ash binders. The largest geopolymerization byproduct, the N–A–S–H gel phase, had more creep compliance whereas the partially activated and non-activated phases displayed less creep, according to a deconvolution analysis of the grid indentation. This discovery backs up earlier inferences from bulk-scale findings. Additionally, They discovered that certain composition factors, such as the Si/Al ratio, liquid-to-solid ratio, sand-to-cement ratio, silica fume, and superplasticizer dose, will vary the creep behavior of specific phases. Results from bulk studies are compatible with the composition-dependent creep behavior of geopolymer at a small scale (Wallah 2010; Islam 2015). However, the correlation between composition and creep behavior of geopolymer remains unclear.

The objective of the current work is to carry out extensive experimental testing to determine the Young's, bulk, and shear moduli of MKG pastes as well as their Poisson's ratio. To explore their connections with composition, two series of MKG pastes are created by varying the dosage of alkali equivalent to aluminosilicate material within a range of 10–50% and the molar ratio of SiO_2 to Na_2O in the activator within a range of 1.25–2.25. Stress–strain curves and microstructures are utilized to perform an in-depth analysis of the fracture behaviors of MKG pastes under compression tests and elucidate the impact of composition on their elastic properties. The study also investigated the composition-dependent creep properties of geopolymer. Since the Si/Al ratio is a significant compositional factor of geopolymer (Rowles et al. 2003; Duxson et al. 2007; White et al. 2012; He et al. 2016; Wan et al. 2017), it was varied to produce geopolymers with different compositions. Metakaolin-based geopolymer (MKG) was used due to its high purity and reproducibility, making it an ideal model system for studying geopolymers (Duxson et al. 2007; White et al. 2012; Wan et al. 2017). To investigate the creep behavior of geopolymer at the binder phase length scale, indentation creep experiments were carried out, greatly avoiding the impact of large voids and flaws in bulk-scale studies. The extracted elastic modulus, hardness, and creep parameters of MKGs with different Si/Al ratios were obtained by analyzing the findings. To describe the chemical and physical structures of MKG with various Si/Al ratios, experiments using X-ray diffraction (XRD), scanning electron microscopy (SEM), and mercury intrusion porosimetry (MIP) were carried out simultaneously. The experimental results and micromechanical analyses were used to determine the underlying creep mechanism in MKG with varying Si/Al ratios and the relationship between composition and structural properties. The study found that the binder-scale creep mechanism in MKGs depends on the sizes of nanopores and micropores, which are determined by the Si/Al ratio. These findings may aid in designing and tailoring MKG materials for further engineering applications.

3.2 Elastic Behavior of MKG

3.2.1 Experimental Program

3.2.1.1 Materials

As the aluminosilicate source for the geopolymer, a commercial metakaolin powder (Metamax, Basf Co.) was used. Aluminate and silicate made up around 97% of the total mass of the MK powder, according to an X-Ray Fluorescence study (XRF-1800, Shimadzu) of its chemical makeup. Refer to Table 3.1 for detailed composition data. The particle size distribution of the MK powder was determined by laser particle size analysis (LS-230, Coulter), indicating a mean particle size of 5.91 μm and a 90%-passed particle size of 13.59 μm (Fig. 3.1a). The particle size distribution results were consistent with SEM observation (Fig. 3.1b).

To make the alkaline activator, a liquid sodium silicate and pellet sodium hydrate were used. The molar ratio of SiO_2 to Na_2O in the liquid sodium silicate, 2.87, was used to define the modulus of the substance. The sodium hydrate pellet had a 96% purity rating and was of AR grade. According to the needed silicate modulus, activator solutions were made by combining the right amount of liquid sodium silicate, pellet sodium hydrate, and water. The solutions were then made, sealed, and stored at a

Table 3.1 Chemical composition of the metakaolin powder

Component	Al_2O_3	SiO_2	K_2O	Na_2O	CaO	TiO_2	Fe_2O_3	LOI
Mass content (%)	39.68	57.26	0.21	0.27	0.04	1.78	0.43	0.34

Reproduced from [Correlating the elastic properties of metakaolin-based geopolymer with its composition] by [Dongming Yan] with permission from [Materials & Design]

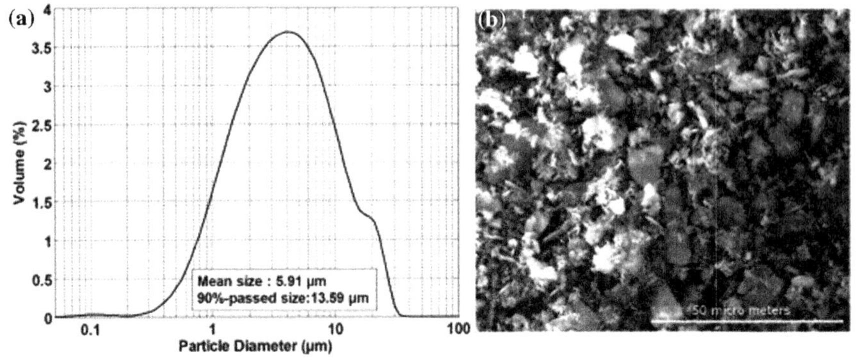

Fig. 3.1 **a** Particle size distribution and **b** SEM photograph of the metakaolin powder. Reproduced from [Correlating the elastic properties of metakaolin-based geopolymer with its composition] by [Dongming Yan] with permission from [Materials & Design]

room temperature of 20 ± 1 °C for at least 24 h before geopolymer synthesis. No specific precautions were taken to exclude CO_2 from the solutions.

3.2.1.2 Mixture Design

As previously mentioned, the polymeric character and polymerization degree of geopolymer can be affected by the Si/Al ratio and/or M/Si ratio (or M/Al). As a result, most experiments have been conducted by controlling either the Si/Al ratio or M/Si ratio or both, requiring careful measurement of raw powders and activators. However, this approach may not be ideal for engineering applications due to the dependence of the targeted Si/Al and M/Si ratios on the chemical composition of raw powders and activators. Instead, a mixture design strategy can be used to control the mass (or molar) ratios of key ingredients, with the detailed Si/Al and M/Si (or M/Al) ratios calculated accordingly. This approach has been used for the mixture design of MKG foams (Chen et al. 2011; Hung et al. 2013). Geopolymer composition can be controlled by three parameters using this strategy: molar ratio of SiO_2 to Na_2O in the activator (or modulus of silicate—Ms), mass percentage of the alkali equivalent to the aluminosilicate material (AE%), and mass percentage of water in the combination (H_2O%). These are how they are expressed:

$$\text{AE\%} = \frac{m(\text{Na}_2\text{O})_{\text{Activator}}}{m(\text{Metakaolin powder})} \times 100\% \tag{3.1a}$$

$$\text{Ms} = \frac{mol(\text{SiO}_2)_{\text{Activator}}}{mol(\text{Na}_2\text{O})_{\text{Activator}}} \tag{3.1b}$$

$$\text{H}_2\text{O\%} = \frac{m(\text{H}_2\text{O})_{\text{Mixture}}}{m(\text{Mixture})} \times 100\% \tag{3.1c}$$

The mol(·) denotes the molar content and m(·) denotes the mass content.

Two series of MKG pastes were developed by manipulating the AE dose and Ms level in order to evaluate the possible impact of composition on the elastic mechanical characteristics of MKG-based materials. With regard to the first batch of MKG pastes, a constant activator Ms level of 1.5 and a variable AE dose ranging from 10 to 50% with a stepwise increase of 10% were necessary. In the second group of MKG pastes, Ms varied from 1.25 to 2.25 with a stepwise increase of 0.25 while maintaining a constant AE dosage of 25%. The total water content was kept under control at 41% for both sets of MKG pastes. The detailed synthesis proportions of the MKG pastes are listed in Table 3.2. Note that different amounts of water were required for each mixture due to variations in the water content in activator, as presented in Table 3.2. Moreover, the detailed Si/Al ratio and Na/Si ratio for all mixtures based on mass conservation are present in the last two columns of Table 3.2. The resulting MKG pastes possessed final Si/Al and Na/Al ratios similar to those prepared by Rowles and O'Connor (Rowles et al. 2003), with Si/Al ratios ranging from 1.08 to 3.0 and Na/Al ratios between 0.51 and 2.0.

Table 3.2 Synthesis mixture of the AE- and Ms-grouped MKG pastes

Composition	MK powder (g)	Activator (g)	Water (g)	Ms (−)	AE (%)	Si/Al (−)	Na/Al (−)
AE-grouped MKG pastes (Ms = 1.5)							
AE10	677	389	378	1.5	10	1.5	0.4
AE20	500	575	211	1.5	20	1.8	0.8
AE30	430	741	123	1.5	30	2.2	1.3
AE40	375	862	56	1.5	40	2.5	1.7
AE50	330	948	5	1.5	50	2.8	2.1
Ms-grouped MKG pastes (AE = 25%)							
Ms125	480	619	187	1.25	25	1.9	1.0
Ms150	460	661	163	1.50	25	2.0	1.0
Ms175	440	697	140	1.75	25	2.1	1.0
Ms200	430	774	122	2.00	25	2.3	1.0
Ms225	410	770	102	2.25	25	2.4	1.0

3.2.1.3 Specimen Preparation

In a Hobart mixer, fresh MKG pastes were made. First, Table 3.2 was used to weigh the activators, MK powders, and more distilled water. The activators were added to the mixing bowl, followed by gradual addition of MK powders and distilled water while the activators were being stirred. After 3–5 min of mixing, a smooth slurry (fresh paste) was obtained, which was then poured into cylindrical molds with diameters of 37 mm and lengths of 85 mm. The casted molds were vibrated for 2 min to remove any bubbles, which is particularly important for high Ms level specimens where the viscosity of the activator solution is high, leading to sticky pastes and poor fluidity that may cause heterogeneity and lower strength. The cylindrical specimens and molds were then sealed with plastic sheets and put in a curing room with controls to maintain a temperature of 20 °C and a relative humidity of above 90%. After 7 days of curing, the specimens were demolded and transferred to sealed plastic bags to prevent carbonation during the hardening process. The specimens were then further cured in the same curing chamber for an additional 21 days before undergoing elastic mechanical and microstructure tests.

3.2.1.4 Mechanical Measurement

The compression test was conducted using an Instron 8802 full-functional test machine, with the moving speed of the loading head maintained at 0.48 mm/min by its electro-hydraulic servo system to maintain uniaxial loading. A spherical bearing was fitted between the specimen and lower loading head to guarantee uniaxial loading. A force sensor mounted to the upper loading head measured the load, with maximum force values recorded and used to determine the compressive strength.

To prevent stress concentration and friction between the surfaces of the specimens and loading heads during mechanical tests, all MKG specimens had their ends polished. The aspect length-to-diameter ratio of the polished specimens was retained at 2.16 and the height at 80 ± 0.5 mm. To obtain statistically significant results, three specimens were tested for each mixture.

To measure elastic deformation, four strain gauges were glued to the middle of each specimen: two along the axial direction and two along the lateral direction. In addition, three linear variable displacement transducer (LVDT) meters were positioned between the top loading head and the spherical bearing to track deformation following strain gauge failure and quantify the total axial deformation of the MKG specimens. Figure 3.2 illustrates the entire measurement system setup. The Integrated Measurement & Control system recorded data simultaneously, providing stress–strain curves of the cylindrical MKG specimens.

The relationships between stress and strain for a cylindrical elastic solid under uniaxial compression, as described by Hooke's Law, can be expressed in Eqs. (3.2a), (3.2b), (3.2c), and (3.2d).

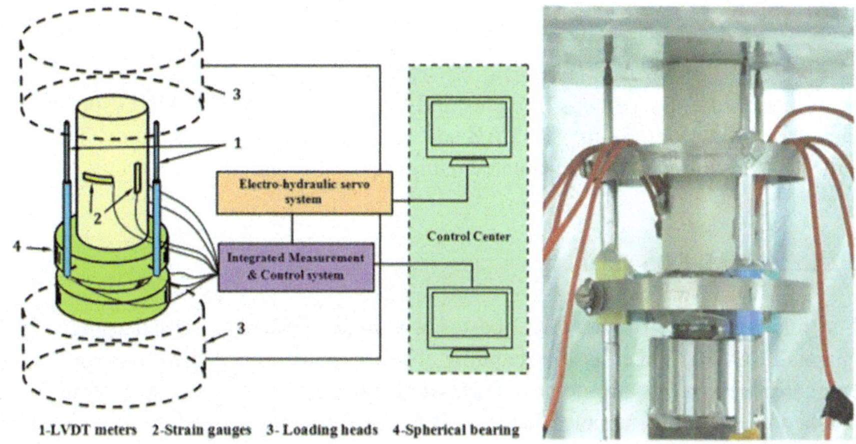

Fig. 3.2 Schematic illustration of the measurement system (left) and an in-situ picture of a specimen and apparatus under testing (right). Reproduced from [Correlating the elastic properties of metakaolin-based geopolymer with its composition] by [Dongming Yan] with permission from [Materials & Design]

$$\sigma_z = E\varepsilon_z \tag{3.2a}$$

$$\sigma_z = -\frac{E\varepsilon_\theta}{\nu} = -\frac{E\varepsilon_r}{\nu} \tag{3.2b}$$

$$\sigma_r = \sigma_\theta = \tau_{rz} = \tau_{\theta z} = \tau_{r\theta} = 0 \tag{3.2c}$$

$$\gamma_{rz} = \gamma_{\theta z} = \gamma_{r\theta} = 0 \tag{3.2d}$$

The equations are denoted by subscripts z, r, and θ, which represent the axes along the axial, radial, and lateral directions of the cylinder, respectively. E and v represent the Young's modulus and Poisson's ratio of the elastic solid.

The Young's modulus, E, and Poisson's ratio, v, of the cylindrical elastic solid under uniaxial compression in the elastic stage can be determined by fitting Eqs. 3.1.2a and 3.1.2b to the measured stress-axial strain curve and stress-lateral strain curve, respectively. The solid's bulk modulus, K, and shear modulus, G, may be computed after the Young's modulus and Poisson's ratio have been determined, using Eqs. (3.3) and (3.4), respectively.

$$K = \frac{E}{3(1-2\nu)} \tag{3.3}$$

$$G = \frac{E}{2(1+\nu)} \tag{3.4}$$

To determine the Young's modulus and Poisson's ratio of the MKG specimens, the curve fitting toolbox in MATLAB software was utilized in this study. The bulk and shear moduli were then evaluated using Eqs. 3.1.3 and 3.1.4, respectively.

3.2.1.5 SEM–EDS Analysis

The microstructure of the MKG pastes was analyzed and observed using a FEI Quanta FEG650 field emission scanning electron microscope (SEM) equipped with energy dispersive spectrometer (EDS). After compression testing, specimens were crushed to create the samples for SEM–EDS analysis. The samples were cut to an acceptable size without polishing the observation surfaces in order to maintain the natural morphology of the fracture surfaces. An accelerating voltage of 20 kV was selected to prevent damaging the microstructure of the samples while ensuring clear results. The working distance was set to 10.5 mm, and the spot was controlled at 3.5.

3.2.2 Result and Discussion

3.2.2.1 Compressive Strength

The compressive strengths of different mixtures are presented in Fig. 3.3. Under normal curing circumstances, the compressive strength values of MKG pastes often exhibit composition-dependent behavior, with their average values being close to those of OPC pastes with a w/c ratio of 0.5 (20–40 MPa) (Metha and Monterio 2006). As shown in Fig. 3.3a, the AE dosage of the MKG pastes has a dual effect on compressive strength. Augmentation of compressive strength occurred as AE% increased from 10 to 40%, from 4.2 to 36 MPa, an increase of approximately 875%. When AE% further increased from 40 to 50%, compressive strength sharply decreased from 36 to 17 MPa, by approximately 56%. The maximum compressive strength was achieved for the AE40 specimens (Si/Al = 2.42, Table 3.2). Although some studies claimed that the compressive strength of MKG pastes monotonously increased with increasing alkaline content or concentration, this dual effect of alkali content on mechanical strength in MKG pastes is consistent with previous research findings (Steveson and Sagoe-Crentsil 2005; Wang et al. 2005). It should be noted that curing conditions of 70 °C for the first 24 h were used for specimens in Rowles and O'Connor (2003), while 85 °C for the first 2 h were utilized for specimens in Steveson and Sagoe-Crentsil (2005). These findings suggest that there exists an appropriate level of alkali content within the geopolymerization process (Rashad 2013). Below this level, insufficient alkaline reactant is provided by the activator to dissolve powder

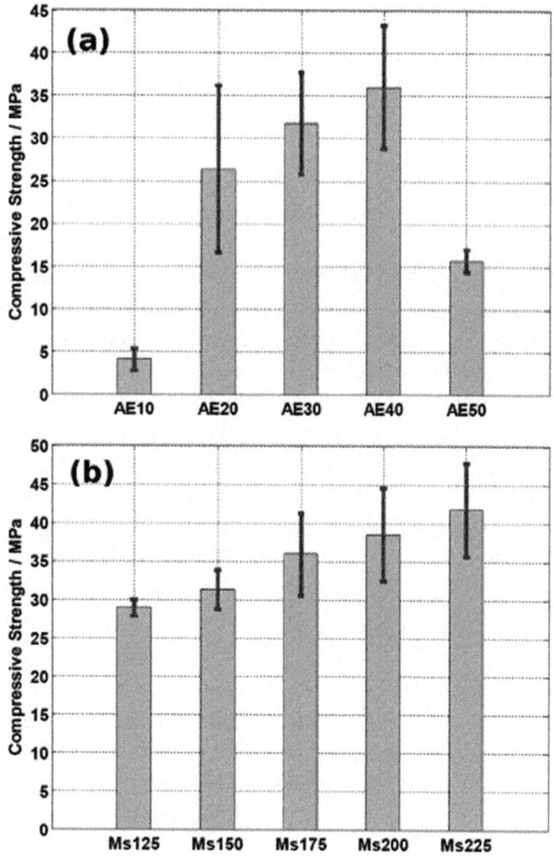

Fig. 3.3 **a** Distribution of compressive strengths for the AE-grouped MKG pastes; **b** distribution of compressive strengths for the Ms-grouped MKG pastes. Reproduced from [Correlating the elastic properties of metakaolin-based geopolymer with its composition] by [Dongming Yan] with permission from [Materials & Design]

materials and form a geopolymer matrix, while excess alkali beyond that level would appear in the geopolymer matrix and weaken the structure formed.

The compressive strengths of the Ms-grouped MKG pastes are shown in Fig. 3.3b. The compressive strengths of the MKG pastes are increased from 28 to 42 MPa as Ms rises from 1.25 to 2.25, a rise of almost 44%. The element ratio of Si/Na or Si/Al in the mixture is determined by the molar ratio of SiO_2 to Na_2O in the activator, Ms, with AE dose in this investigation being regulated at 25% (Na/Al $\approx$ 1). A strong correlation between the Si/Na ratio or Si/Al ratio (silicate content or Ms level) and mechanical strength has been widely reported by other studies, including Cheng et al. (2015a, b), Pelisser et al. (2013), and Ozer and Soyer-Uzun (2015). Some studies (e.g., Barbosa et al. 2000; Duxson et al. 2005; Lyu et al. 2013; Gao et al. 2014) have reported a negative effect of high Ms levels on mechanical strength. This may be due to poor workability of geopolymer pastes caused by the viscous nature of the sodium-silicate solution (Gao et al. 2014) and the inhibitory effect of high amounts of sodium-silicate solution on the geopolymerization process (Heah et al. 2012). However, the compressive strength of MKG pastes was not adversely affected by the Ms ratio of 2.25 (Si/Al 2.35), according to our experiment. Given that the Si/Al ratio is still below 2.42 (AE = 40% in Fig. 3.3a), where the measured compressive strength is at its highest, this seems logical. Additionally, higher strengths have been reported for MKG materials with Si/Al ratios of 2.75 by Zhang et al. (2010).

3.2.2.2 Stress–Strain Behavior

The stress–strain behavior of a solid is crucial to characterizing its elastic properties and depicting the fracture process under loading. We selected the test results of an Ms200 specimen from among the 30 axial load-strain and/or displacement tests conducted on MKG specimens to demonstrate mechanical states and corresponding superficial damages observed by eye; see Fig. 3.4. The stress–strain curves for other mixtures are shown in Fig. 3.5, with only one sample for each composition displayed for clarity. These results may assist in analyzing the failure process of MKG pastes under uniaxial compression. Based on our experiments, we can characterize the failure process of MKG pastes as occurring in four stages.

Stage I: Elastic Stage

The specimens' surfaces were undamaged at this point, which is represented by State A in Fig. 3.4b. Indicating that the MKG specimens are in a condition of linear elasticity, the stress–strain curves produced from both axial and lateral strain gauges show linear correlations beginning from the origin point. The elastic properties such as Young's modulus, bulk modulus, shear modulus, and Poisson's ratio can be evaluated using the data obtained during this stage. It is worth noting that a slight vibration was observed in the initial part of the stress–strain curves measured by the LVDTs within the range of 0.5–6 MPa, which is usually 10–20% of the ultimate strengths of the MKG specimens. However, the remaining parts of the stress–strain curves remained linear until the stresses approached about 60–100% of the ultimate strengths. The LVDTs' observed behavior of a little amount of vibration at the start of stress–strain curves might be explained by a lack of complete contact between the loading heads and the specimens, which would disturb the measured data. The stress–strain curves acquired by the LVDTs would be compatible with the findings obtained by the gauges as the compression loads rose since this impact would be lessened. The differences between the elastic moduli of all the MKG specimens obtained from the gauge data and the LVDT data were found to be less than 5%, indicating the reliability of the measurement.

Stage II: Crack Initiation and Propagation

At this stage, the specimens may exhibit visible cracks along the axial direction on their side surfaces, as observed in State B shown in Fig. 3.4b. The density of these

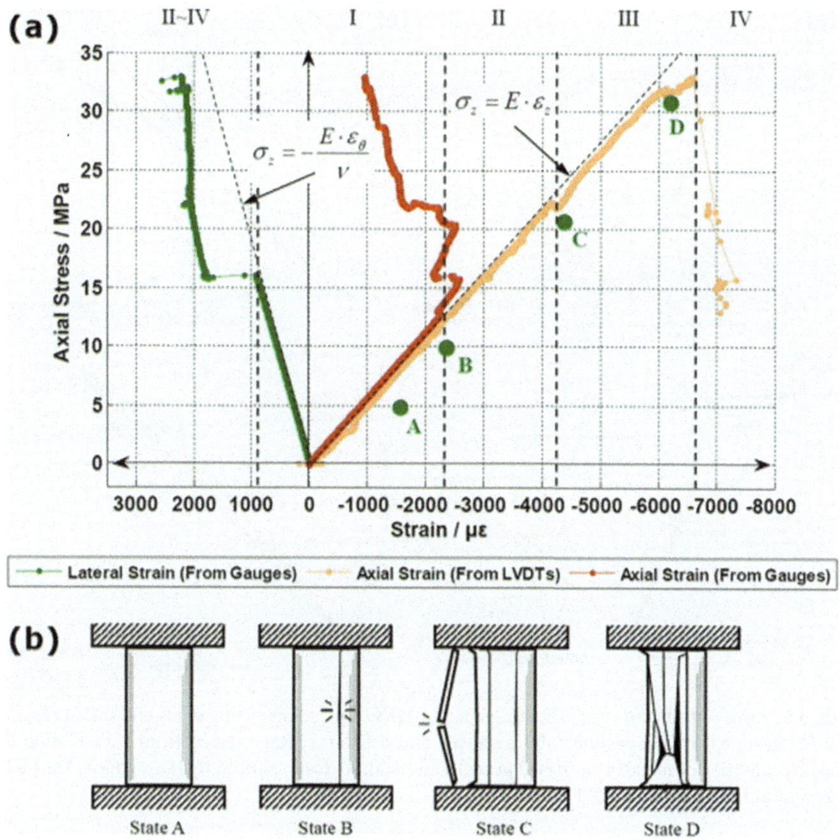

Fig. 3.4 Typical mechanical behaviors of a MKG paste under uniaxial compression: **a** stress–strain curves and **b** configurations of the different fracture states. Reproduced from [Correlating the elastic properties of metakaolin-based geopolymer with its composition] by [Dongming Yan] with permission from [Materials & Design]

cracks is likely to increase with an increase in the compression loadings. Identifying the fracture extents of MKG specimens under uniaxial loadings based on visible superficial cracks is challenging. However, analyzing the development of axial and lateral strains measured by gauges with an increase in compression loads can aid in determining the fracture behaviors of materials that exhibit obvious and direct transition points in the curves. For example, when the axial stress is increased to approximately 16 MPa, the lateral strain is abruptly changed from 900 to 1800 μ; for example, when the axial strain reaches roughly − 2200 μ, it transitions from a linear plot against the stress to a non-linear shape with additional loading; see Fig. 3.4a for more information. The observed sudden and significant shifts in strain are likely attributed to the propagation and penetration of macro-cracks in regions adjacent to the lateral gauges. This study provides evidence of damage resulting from crack initiation and propagation within the solid matrix which biases both the

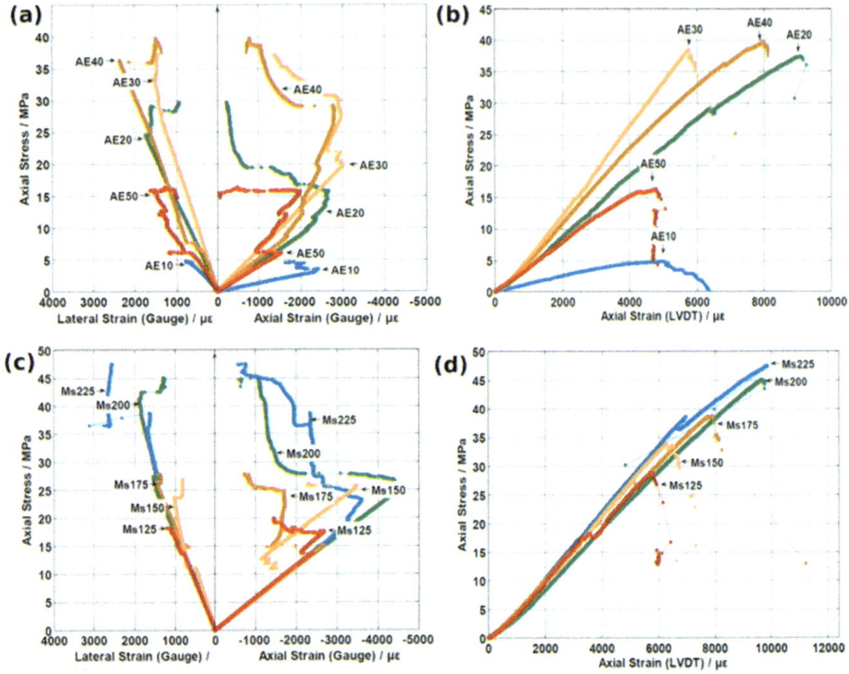

Fig. 3.5 Stress–strain curves of the AE-grouped MKG specimens by **a** gauges and **b** LVDTs, and the Ms-grouped MKG specimens by **c** gauges and **d** LVDTs. Reproduced from [Correlating the elastic properties of metakaolin-based geopolymer with its composition] by [Dongming Yan] with permission from [Materials & Design]

axial and lateral stress–strain curves. Although it was anticipated that the stresses corresponding to the axial and lateral strain shift locations would be consistent, our findings showed erratic data with no statistically significant differences. Despite the local distribution of cracks, the axial stress–strain curves recorded by LVDTs remained linear, as the undamaged matrix could sustain further loadings.

Stage III: Crack Propagation

In this stage, cracks penetrate the material matrix resulting in superficial spalling and moderate to advanced splitting fracture damage, as shown in diagram State C of Fig. 3.4b. The axial strain values obtained from the gauges decrease, indicating that the superficial portion of the material has been fractured and separated from the main matrix due to compression. Meanwhile, the minor increase in lateral strain with increasing compression loading suggests significant fracture damage that cannot be captured by local strain measurements using gauges, thereby limiting their representativeness for the overall mechanical behavior of the material. Additionally, due to the material's fracture behaviors, stress–strain curves obtained by LVDTs diverge from a linear trend and show abrupt decreases. Crack penetration causes instability in the mechanical behavior of the material, with several random shifts in both axial and lateral strains occurring prior to specimen breakdown. These shifts divide the stress–strain curves into discrete parts, as illustrated in Fig. 3.5.

Stage IV: Failure Completing

At this stage, the specimens undergo splitting cracks that connect through shear failures within the material leading to their breakdown into several parts, as illustrated in State D of Fig. 3.4b. Thereafter, the failure process enters the post-peak stage. At this point, the local axial and lateral gauges adhered with the superficial MKG materials spall out of the matrix, making the gauge data meaningless. As specimens fail completely, loads experience sudden release, as shown by the axial stress–strain curves measured by LVDTs (Figs. 3.4, 3.5b and d). This sudden stress drop after peak load for geopolymer-based materials has also been reported by other studies (Heah et al. 2012; He et al. 2013; Sarker et al. 2013).

3.2.2.3 Elastic Properties

The stress–strain curves measured at Stage I by both strain gauges and LVDTs revealed the linear elastic behavior of MKG pastes. However, due to small perturbations observed in the linear stress–strain curves recorded by LVDTs during initial loading periods, only gauge strains from Stage I were utilized to calculate Young's, bulk, shear moduli, and Poisson's ratio for both AE-grouped and Ms-grouped MKG pastes.

Effect of AE Dosage

Young's moduli of AE-grouped MKG pastes range from 1.1 to 6.1 GPa, which are consistently lower than those of conventional OPC pastes (10 to 40 GPa depending on water-cement ratio, curing conditions, and age). The computed elastic characteristics of these pastes are shown in Fig. 3.6. However, compressive strengths of AE-grouped MKG pastes fall within the range of 4–36 MPa, comparable to those of typical OPC pastes. The high strength and relatively low Young's modulus observed in MKG-type materials represents a notable mechanical feature. Plots of Young's and shear moduli against AE dosage display a 'Λ' shape curve, with maximum values for AE30 specimens (Fig. 3.6a and d). As AE dosage increases from 10 to 30%, MKG specimens' Young's and shear moduli rise from 1.3 GPa to 6.1 GPa and 0.55 GPa to 2.4 GPa, respectively, representing increases of around 490% and 450%. However, further increase in AE dosage leads to a decrease in Young's and shear moduli. In particular, the Young's and shear moduli of MKG specimens decline from 6.1 GPa

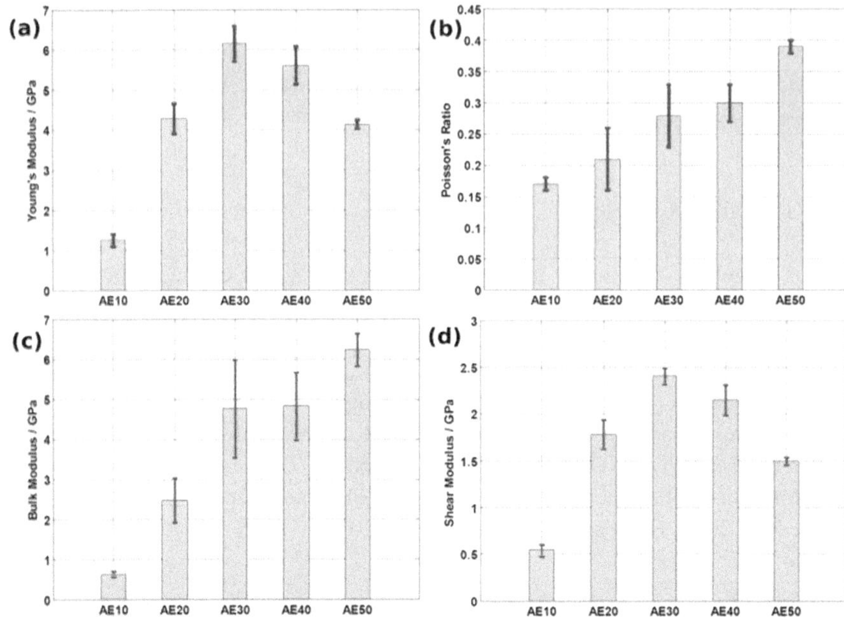

Fig. 3.6 Elastic properties of the AE-grouped MKG specimens: **a** Young's modulus, **b** Poisson's ratio, **c** bulk modulus and **d** shear modulus. Reproduced from [Correlating the elastic properties of metakaolin-based geopolymer with its composition] by [Dongming Yan] with permission from [Materials & Design]

to 4.2 GPa and 2.4 GPa to 1.5 GPa, respectively, at a rate of roughly 33% and 38% when AE dose is increased from 30 to 50%.

Figure 3.6b and c reveal simple and monotonous increases of Poisson's ratio and bulk modulus with increasing AE dosage for MKG specimens. Specifically, as AE dosage increases from 10 to 50%, Poisson's ratio and bulk modulus increase respectively from 0.17 to 0.38 (by about 230%) and 0.6 to 6.3 GPa (by about 980%). While the range of Poisson's ratio values measured for MKG pastes is wider than those reported in literature (e.g. 0.2–0.27 for a fly-ash-based geopolymer under different curing temperatures), they are comparable to those of OPC pastes at varying maturities or hydration levels (e.g., from 0.15 with advanced maturities to roughly 0.5 with very low maturities for an OPC paste simulated using finite element computation). In contrast, MKG specimens' bulk and shear moduli are significantly lower than those of OPC pastes (for example, an OPC paste with w/c = 0.5 has a Young's modulus of 22 GPa and a bulk modulus of 14 GPa). These results suggest that MKG pastes are a type of 'soft' material.

The SEM–EDS technique can be used to analyze the microscopic correlations between elastic properties and composition of MKG specimens. Figure 3.7 presents micro-scaled morphology of fractured surfaces of AE10, AE30, and AE50 samples. Discrete prism-shaped particles are widely dispersed and loosely constitute the

Fig. 3.7 SEM microstructures of **a** AE10, **b** AE30 and **c** AE50 MKG pastes. Reproduced from [Correlating the elastic properties of metakaolin-based geopolymer with its composition] by [Dongming Yan] with permission from [Materials & Design]

matrix in AE10 sample. Geopolymer gel is hardly detected as most of it forms only on particle surfaces with weak binding effect. These observations corroborate previous findings that both compressive strength and elastic modulus are low. He et al. (2012) also reported a MKG paste with Si/Al ratio of 1.75 possessing similar patches of porous, discrete, and prism-shaped particles along with a dense and continuous gel-like matrix. As AE content increases to 30%, particles become wrapped or embedded in geopolymer gel making it the dominant phase in the matrix. The densification leads to a tough and stiff matrix capable of sustaining high compression loading and elastic deformation, consistent with the findings of Steveson and Sagoe-Crentsil (2005). Micro-cracks around particles and in the continuous matrix are also observed in Fig. 3.7, possibly induced by compressive loading and shrinkage during geopolymerization process, as suggested by He et al. (2012).

From a chemical perspective, the alkali content of activator solution may limit the reaction for low AE dosage cases. During geopolymerization process, alkali cations incorporate into (N)–A–S–H gel structure to balance net negative charge generated by Al^{3+} replacing Si^{4+} in tetrahedral chain sites (Provis and Bernal 2014). Increasing AE dosage promotes extent of geopolymerization reaction and production of more geopolymer gels binding together to form the material matrix, particularly when AE dosage is relatively low. The two impacts of gel products on the matrix are bonding (which hardens loosely packed particles into a framework) and filling (which closes pore gaps and compacts porous solid). While the latter impact improves incompressibility, the former effect increases the matrix's strength and stiffness. These methods explain why MKG pastes' Young's modulus, Poisson's ratio, bulk modulus, shear modulus, and compressive strength increase when AE dose rises.

A mixture's maximum AE consumption for the geopolymerization process exists, after which adding more alkali wouldn't increase the reaction's extent. Instead, excess alkali and silicate accumulate in the matrix, potentially weakening bond strength and homogeneity of solid matrix. SEM observation of AE40 MKG specimen in Fig. 3.7c indicates looser microstructure than AE30 sample (Figs. 3.7b and 3.8a). SEM–EDS analysis reveals crystal- or particle-like solids with high Na/Si ratio and/or Na/Al ratio in AE40 sample (Fig. 3.8b). High sodium content typically weakens material

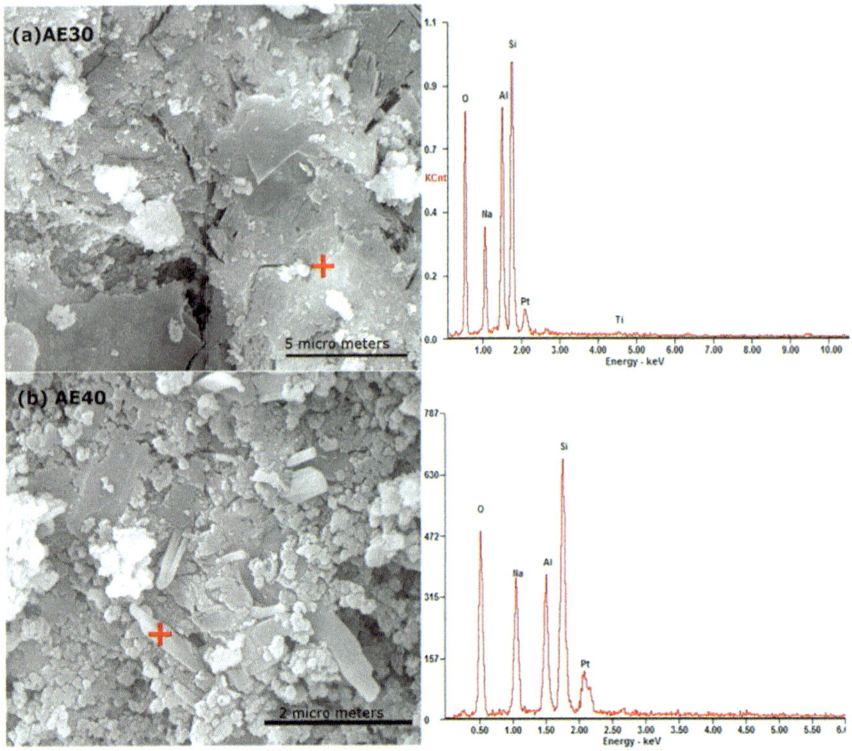

Fig. 3.8 SEM–EDS results of **a** AE30 and **b** AE40 MKG pastes. Reproduced from [Correlating the elastic properties of metakaolin-based geopolymer with its composition] by [Dongming Yan] with permission from [Materials & Design]

structure as low silica gels form into discrete particulate units with poor polymerization instead of generating continuous gel network, as reported in previous studies (Barbosa et al. 2000; Rowles et al. 2003; He et al. 2012; Provis and Bernal 2014). Our trials' results support this tendency, with AE40 pastes having lower Young's and shear moduli and compressive strengths than AE30 pastes. The incompressibility of the matrix is improved by the filling action of hydrated aluminate-silicate gels and inclusions. As a result, the Poisson's ratio and bulk modulus of MKG pastes grow steadily and monotonously when AE dose is increased.

Effect of Ms Level

Figure 3.9 shows the computed elastic characteristics of Ms-grouped MKG pastes. Strangely, the Ms level has little effect on the elastic properties of MKG specimens. Specifically, Young's modulus values range from 5.5 to 5.8 GPa, Poisson's ratio from 0.27 to 0.33, bulk modulus from 4.6 to 5.3 GPa, and shear modulus from 2.1 to

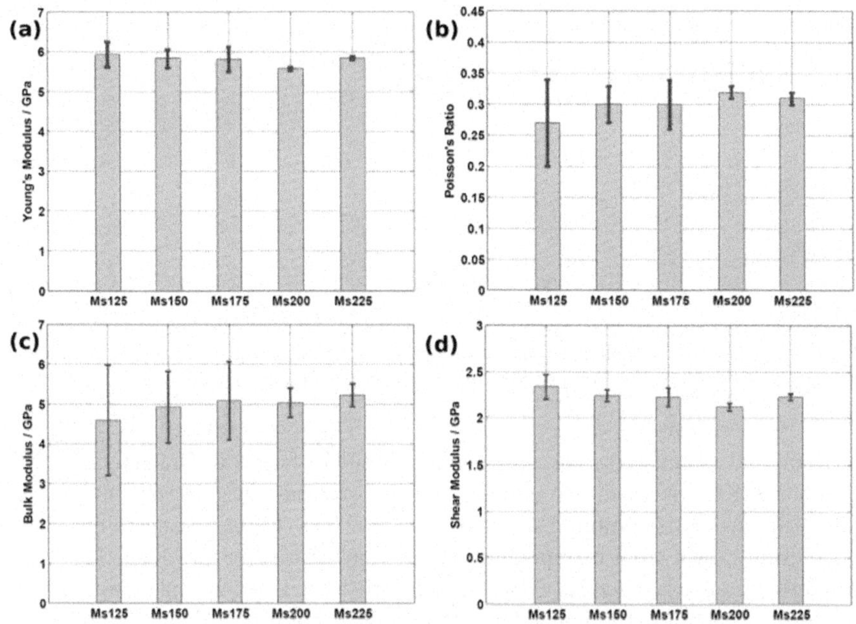

Fig. 3.9 Elastic parameters of the Ms-grouped MKG specimens: **a** Young's modulus, **b** Poisson's ratio, **c** bulk modulus and **d** shear modulus. Reproduced from [Correlating the elastic properties of metakaolin-based geopolymer with its composition] by [Dongming Yan] with permission from [Materials & Design]

2.3 GPa. The near-constant elastic property values of Ms-grouped MKG specimens depicted in Fig. 3.9 differ somewhat from compressive strength values illustrated in Fig. 3.3b. These results suggest a "soft but tough" mechanical feature of the MKG materials prepared in this study. To conduct a more reasonable evaluation, variance analysis method (Fisher 1954) is used to analyze statistical significance of elastic parameters versus Ms level. F-test statistic of each parameter can be calculated using Eq. (3.5).

$$F = \frac{MSD_{level}}{MSD_{error}} = \frac{SSD_{level}/(N_{level} - 1)}{SSD_{error}/(N_{sample} - 1)} \tag{3.5}$$

In Eq. (3.5), the terms MSD_{error} and SSD_{Error} reflect the mean square deviations and sum of square deviations within these levels, respectively, whereas MSD_{level} and SSD_{Level} define the mean square deviations and total of square deviations between different levels of Ms. N_{level} signifies the number of Ms levels, and N_{sample} refers to the total number of samples in the Ms group. Significance of a relationship can be assessed by comparing critical F value obtained from F-distribution with analysis results. In this study, all critical F values are acquired from F-distribution with degrees of freedom equal to ($N_{level} - 1$, $N_{sample} - 1$) and significance level equal to 0.05.

Table 3.3 The results of analysis of variance for the Ms-grouped MKG specimens

Property	N_{level}	N_{sample}	MSD_{level}	MSD_{error}	Calculated F	Critical F	Significance
Young's modulus	5	15	0.0186	0.0502	0.3710	3.4780	Insignificance
Poisson's ratio	5	15	0.0003	0.0017	0.2009	3.4780	Insignificance
Bulk modulus	5	15	0.0575	0.7836	0.0733	3.4780	Insignificance
Shear modulus	5	15	0.0062	0.0069	0.9067	3.4780	Insignificance

Table 3.3 summarizes variance analysis results of elastic parameters for Ms-grouped MKG specimens. In conclusion, Ms level has insignificant influence on all elastic properties within tested range (1.25–2.25). These findings align with a conclusion drawn from extrapolation analysis in Duxson et al. (2005, 2007) that Young's modulus is essentially constant when Si/Al $\geq$ 1.65 (equivalent to Ms $\geq$ 1.0 in this study) within experimental uncertainty. However, when Ms level exceeds test range, influences on elastic parameters may become significant (Duxson et al. 2005; Lizcano et al. 2012), necessitating further extensive experimental and theoretical investigations on relevant issues in the future.

Although Ms level may have negligible influence on elastic properties of MKG pastes within experimental range, it is not authentic to state the same correlation between Ms level and microstructure of geopolymer materials. Based on SEM observations, our results show observable microstructure changes when the Ms level varies (see Fig. 3.10 for sample SEM images of specimens with Ms125, Ms175, and Ms225). Since chosen SEM images do not appear to be acceptable in illustrating changes in microstructure caused by differences in Ms level, detailed microstructure information (such as specific surface area, mean pore radius, and nanoscaled elastic characteristics) for all Ms-grouped samples is not supplied. As seen in Fig. 3.10, all samples exhibit relatively tight compacted microstructure with some obvious cracks present in material matrix due to preparation of SEM samples from fractured specimens and heterogeneous texture shrinkage during geopolymerization. Given relatively dense microstructure, all Ms-grouped specimens display constant elastic parameters (Fig. 3.9). But a closer look at the SEM images in Fig. 3.10 and the EDS results of the Ms225 paste reveals the presence of a poly-aluminate-silicate phase (or macro-aluminate-silicate "crystal" phase) in the geopolymer gels, with the size and quantity of this phase rising with the Ms level. Interactions between macro-crystals and geopolymer gels appear strong enough to sustain large compressive loading with no visible cracks along interfaces. Consequently, poly-aluminate-silicate solids acting as aggregates may enhance material strength under compression testing, as demonstrated in Fig. 3.3b.

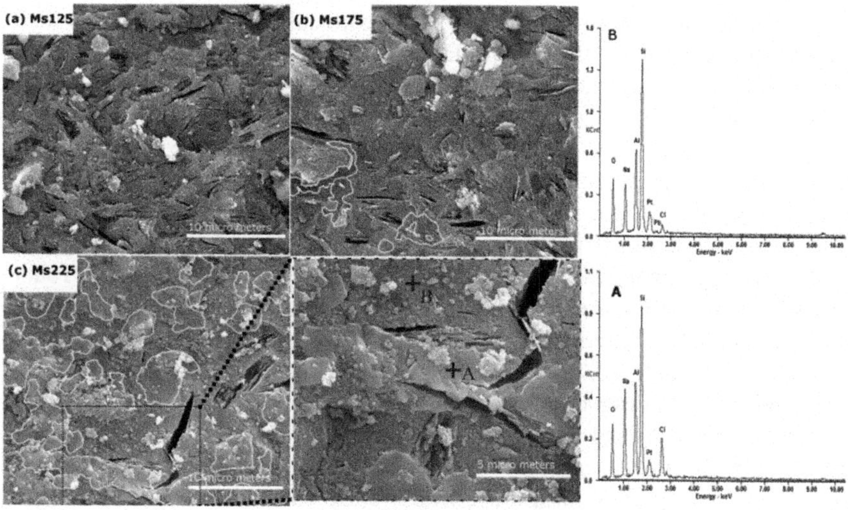

Fig. 3.10 SEM–EDS results of the selected Ms-grouped MKG pastes: **a** Ms125, **b** Ms175 and **c** Ms225. Reproduced from [Correlating the elastic properties of metakaolin-based geopolymer with its composition] by [Dongming Yan] with permission from [Materials & Design]

3.2.2.4 Quantitative Assessment

Qualitative evaluations of compositional factors that affect mechanical properties of geopolymer-type materials provide insight into mechanisms of strength development during composition-associated geopolymerization. However, such evaluations are inadequate for achieving excellent designs and engineering uses of materials. Therefore, a quantitative analysis of the relationships between the composition of MKG material and its elastic characteristics is required. This contribution examines and discusses data of Young's modulus and Poisson's ratio obtained from AE-grouped MKG pastes in light of the statistically insignificant influences of Ms level on elastic properties of MKG specimens and the fact that bulk and shear moduli can be derived from Young's modulus and Poisson's ratio.

As mentioned earlier, AE dosage has a dual effect on Young's modulus of MKG pastes. This relationship can be represented by a quadratic double-logarithmic function, expressed in Eq. (3.6)

$$\mathrm{E} = \mathrm{E}^* K_{AE\%} = \mathrm{E}^* \exp\left(-\kappa \ln^2(\chi)\right), \quad \chi = \frac{AE\%}{AE\%^*}, \quad \kappa > 0 \tag{3.6}$$

,

The quadratic double-logarithmic function expresses Young's modulus as the product of specific Young's modulus (E*) and influential coefficient of AE dosage (KAE%). The relationship between the logarithm of KAE% and the logarithm of the nondimensionalized component, which represents the relative ratio of AE dose to specific AE dosage (AE%), is quadratic. Young's modulus of MKG material is similar to its particular value (E = E*) when AE% equals AE% ($\chi = 1$). When AE% deviates from AE%, i.e., χ approaches zero or infinity, Young's modulus of MKG material trends towards zero. Positive shape factor κ serves as reference index describing sensitivity of Young's modulus to variation of AE content. Large κ indicates rapid decrease of Young's modulus when AE% deviates from AE%. Small κ implies mild tolerance of Young's modulus to deviation of AE% from AE%*.

Figure 3.11 illustrates measured Young's moduli compared to AE dosages and fitting curve. Equation (3.6) effectively models dual effect of AE dosage on Young's moduli of MKG pastes, with majority of experimental results falling within prediction bound of estimation function at 10% uncertainty. For AE-grouped MKG pastes, the estimated values of E, AE%*, and are 5.89 GPa, 30.52%, and 1.29, respectively. This formulation's correlation coefficient is 0.95.

Selected experimental results from literature are used for comparison. Young's moduli of MKG pastes with Ms = 1.0–2.0, $Na_2O/Al_2O_3 = 1$ (AE% ≈ 25%), and $H_2O/Na_2O \approx 11$ ($H_2O\% = 32$–35% with average value of 33%) were discovered by Duxson et al. (2005, 2007) to be around 4.8–5.9 GPa. Lizcano et al. (2012) discovered Young's modulus of Na-based MKG pastes with $SiO_2/Al_2O_3 = 1.5$–2 (Ms = 0.9–1.9), $Na_2O/Al_2O_3 = 1$ (AE% ≈ 26%), $H_2O/Na_2O = 11$ ($H_2O\% = 32$–36% with average value of 34%) to be around 3.4–6.1 GPa. Experimental results

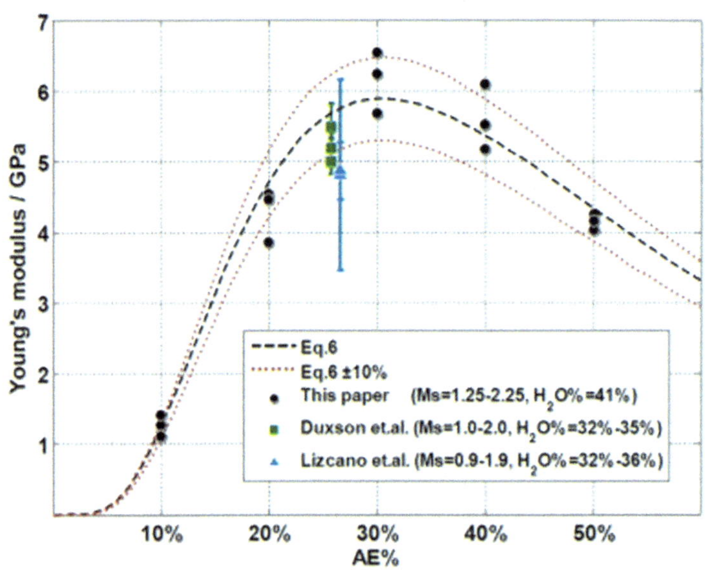

Fig. 3.11 Quantitative correlations between the Young's modulus and the AE dosage of the MKG pastes by the estimation formulation of Eq. (3.6). For comparison purpose, the experimental data in (Duxson et al. 2005, 2007; Lizcano et al. 2012) are adopted. Reproduced from [Correlating the elastic properties of metakaolin-based geopolymer with its composition] by [Dongming Yan] with permission from [Materials & Design]

in previous studies (Duxson et al. 2005, 2007; Lizcano et al. 2012) demonstrate acceptable consistency with estimations by Eq. (3.6), despite different experimental schemes (e.g. raw materials, curing conditions, test constraints) utilized. Curing temperature increase tends to enlarge pore size and volume of geopolymer-type porous material, altering mechanical properties of the material (Rovnaník 2010). For example, in Lizcano et al.'s study (2012), some MKG specimens were cured at 80 °C, potentially weakening connections of formed geopolymers. It is reasonable that Young's moduli of MKG pastes in this study fall below estimations. According to Mo et al. (2014), temperature has a significant impact on the mechanical properties of MKG materials. Additionally, Lizcano's experiment's short drying time (2–3 days) might have decreased MKG pastes' Young's moduli.

Experimental tests indicate that Poisson's ratio values of MKG pastes progressively increase as AE dosage increases from 10 to 50%. However, most solids are volumetrically compressible, indicating bulk modulus of material cannot be infinite. As a result, the theoretical limit of the Poisson's ratio for a real elastic solid is 0.5. This implies that linear or exponential simple rising functions are not advised for depicting the link between the Young's modulus and the AE dose found in this study. To capture connection, a sigmoid function given by Eq. (3.7) is suggested:

$$\nu = (\nu_0 - 0.5)\exp\left(-\lambda\,\mathrm{AE}\%^{\,2}\right) + 0.5, \quad \lambda > 0 \tag{3.7}$$

This function defines Poisson's ratio (v) between two limits: initial value ν_0 at $\mathrm{AE}\% = 0$, and theoretical limit ($\nu = 0.5$) as AE% approaches infinity. A sigmoid smooth curve between these bounds represents formulation. Poisson's ratio growth rate in relation to AE dosage is controlled by shape factor λ.

Figure 3.12 presents experimental data of Poisson's ratios of MKG pastes with different AE dosages and correlation curves fitted by Eq. (3.7). Proposed sigmoid function effectively fits the experimental data with a correlation coefficient of 0.87. Results indicate $v_0 = 0.16$ and $\lambda = 4.17$, indicating that the clean sodium silicate solution's Poisson's ratio is 0.5 and that of pure MK solid is 0.16. Significant changes of Young's modulus and Poisson's ratio with AE dosage warrant further attention regarding practical uses of geopolymers, thus requiring rigorous investigation in the future.

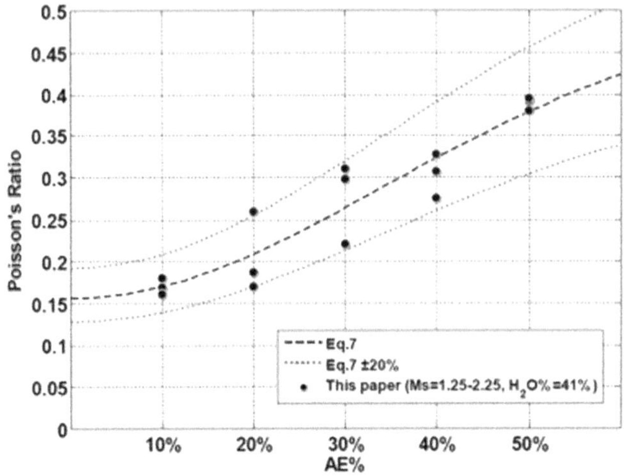

Fig. 3.12 Quantitative correlations between the Poisson's ratio and the AE dosage of the MKG pastes by the estimation formulation of Eq. (3.7). Reproduced from [Correlating the elastic properties of metakaolin-based geopolymer with its composition] by [Dongming Yan] with permission from [Materials & Design]

3.3 Creep Behavior of MKG

3.3.1 Experiment

3.3.1.1 Materials

A geopolymer test sample was produced from metakaolin powder and alkali activating solution. Commercially available MetaMax (Basf Co.) was utilized as the metakaolin powder. The chemical composition of MetaMax was determined via X-ray fluorescence analysis (Shimadzu-XRF-1800) and presented in Table 3.4. Molar ratio between silicate and aluminate content in MetaMax was approximately 1.08, close to its theoretical composition (Si/Al = 1).

A lab-synthesized alkali solution was utilized for chemical activation of MK powder, comprising commercial sodium silicate solution and sodium hydroxide pellets. Hengli Chem. Co., Ltd supplied the sodium silicate solution, consisting of 26.0% silicate and 8.2% sodium hydroxide (equivalent oxide mass). Analytical

Table 3.4 Chemical composition of metakaolin powder

Component	Al_2O_3	SiO_2	CaO	TiO_2	MgO	Na_2O	K_2O	LOI
Mass content[%]	41.64	53.29	2.78	1.04	0.43	0.27	0.21	0.34

Reproduced from [Binder-scale creep behavior of metakaolin-based geopolymer] by [Shikun Chen] with permission from [Cement and Concrete Research]

Table 3.5 Mix proportions

	MK [g]	Sodium silicate [g]	NaOH [g]	H_2O [g]	Ms	Si/Al	Na/Al	H_2O wt%
Mix 1	515.59	0	176.33	483.10	0	1.0	1.0	44.5%
Mix 2	435.11	439.85	102.70	219.82	1.02	1.6	1.0	44.5%
Mix 3	415.51	543.39	85.09	158.72	1.32	1.8	1.0	44.5%
Mix 4	319.42	512.42	55.23	81.52	1.62	1.9	1.0	44.5%
Mix 5	381.15	724.96	53.96	48.58	1.92	2.1	1.0	44.5%

reagent grade sodium hydroxide pellets (NaOH, Sinopharm Chemical Reagent Co., Ltd) were used. In order to create solutions, sodium hydroxide was dissolved in sodium silicate solution at the proper concentration, resulting in a range of molar ratios of SiO_2/Na_2O (modulus of silicate, Ms) from 1.02 to 1.92. Binder with lowest Si/Al ratio (Si/Al = 1.0) was synthesized using only sodium hydroxide and water ($SiO_2/Na_2O = 0$). In order to cool and stabilize chemical species and reactivity, these solutions were created, contained, and kept at room temperature (20, 5 °C) for 24 h.

3.3.1.2 Synthesis of Geopolymer Binder

By combining MK powder, the appropriate activating solution, and more water into a slurry combination at desired ratios, geopolymer binder was created. Geopolymer binders with mix proportions had water content of 44.5%, a Si/Al ratio that ranged from 1.0 to 2.1, and a constant Na/Al ratio of 1. Final mix proportion of samples is outlined in Table 3.5.

Synthesis was performed using an automated mixer (ASM-DA600). Powder was gradually added into alkali solution within 1 min while stirring, and mixing continued for 4 min at a stirring speed of 300 rpm. Slurry was de-aired using a 600 W ultrasonic defoamer (Cheersonic CS5000D) and all macroscopic air bubbles were eliminated by cooling the mixture over ice for about 3 min. The final liquid was put into 16 mm-diameter cylindrical plastic molds and covered with a plastic lid. Specimens were placed in a curing chamber at 60 °C and 90% relative humidity for 24 h, then placed in the curing chamber at room temperature for continuous curing over a period of 14 days.

3.3.1.3 Nanoindentation Experiments

Nanoindentation samples were obtained by cutting cured geopolymer specimens. The samples had diameters of 16 mm, were 8 mm tall, and were embedded in epoxy resin (EpoThin™ 2 Buehler). They were polished using silicon carbide abrasive paper

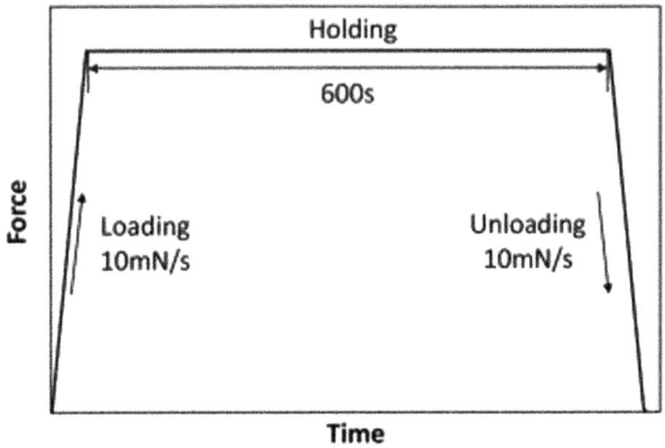

Fig. 3.13 Loading procedure of nanoindentation. Reproduced from [Binder-scale creep behavior of metakaolin-based geopolymer] by [Shikun Chen] with permission from [Cement and Concrete Research]

(NKC waterproof), with mesh numbers ranging from 800# to 2000#. Automated polisher (EcoMet™300 Pro Buehler) was used for polishing. Pre-polished surface was finely ground with diamond suspension containing 0.1 μm particles for 10 min.

Using an Agilent Nano Indenter G200 equipment with a Berkovich tip and according to the steps outlined by El-Safty et al. (2012), nanoindentation tests were carried out (Fig. 3.13). The three stages of the indentation were quick loading (10 mN/s), force holding (600 s), and rapid unloading (10 mN/s). Holding force was set to 5, 50, 200, and 400 mN, except for mix 1 where indentation depth under 400 mN exceeded equipment's testing ability. Only 5, 50, and 200 mN were tested for mix 1. Thermal drift of the system was recorded and ensured to be below 0.1 nm/s before each experiment. Three or more indentations were performed at different locations for each sample.

3.3.1.4 SEM and EDS

A field emission environmental scanning electron microscopy (SEM, FEI Quanta FEG650) and energy dispersive spectrometer (EDS, FEI EDXA 560) were used to describe the microscale morphology of MKG. For the investigation of energy dispersion, a point-type probe with a working distance of 10.5–12.1 mm was utilized. Each measurement point's spectrum was gathered for 30 s.

3.3.1.5 Mercury Intrusion Porosimetry

Mercury intrusion porosimetry (MIP) with a Micrometrics AutoPore IV 9510 equipment was used to analyze the pore structure of the samples. Cutting and crushing cylindrical specimens into large (0.9–1.2 mm) particles, they were then dried in an oven for 24 h at 60 °C. MIP measurements were made between 0.5 and 60,000 psia of pressure. The duration and pressure required for evacuation were 50 mHg and

5 min, respectively. The mercury was filled to a pressure of 0.5 psia and took 10 s to re-adjust. Mercury's material constants (contact angle of 130°, surface tension of 485 dynes/cm, and density of 13.5335 g/mL) were used to translate pressure to pore size.

3.3.2 *Analysis*

3.3.2.1 Hardness and Young's Modulus

Figure 3.14 depicts typical nanoindentation response consisting of three stages corresponding to three loading segments (Fig. 3.13). Young's modulus (E) and hardness (H) were obtained from the following equations, respectively:

$$E = \left(1 - v^2\right)M \tag{3.8}$$

$$H = \frac{P_H}{A_C} \tag{3.9}$$

where P_H is the holding force (50 mN), M and A_c are the indentation modulus and contact area, respectively, and v is the Poisson's ratio of the probed material, respectively. These were calculated using Oliver-Pharr method for analyzing unloading response (Pharr and Oliver 2004). Detailed calculations of M and A_c are presented below. Our previous report (Yan et al. 2016) indicated that various Si/Al ratios typically caused Poisson's ratio of geopolymer to range from 0.27 to 0.33. This variation is not significant; therefore, average Poisson's ratio (0.3) was used in further calculations.

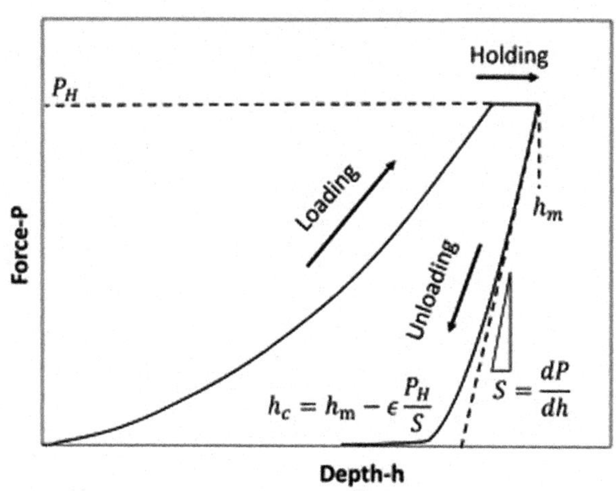

Fig. 3.14 Typical force–depth response of nanoindentation. Reproduced from [Binder-scale creep behavior of metakaolin-based geopolymer] by [Shikun Chen] with permission from [Cement and Concrete Research]

Figure 3.15 displays indentation load versus displacement responses, and Fig. 3.16 shows typical post-indentation sites. Using the Oliver-Pharr method (Pharr and Oliver 2004), the indentation modulus M can be determined using the following equation:

$$M = \frac{1}{\frac{1}{E_r} - \frac{1}{M_i}} \tag{3.10}$$

where E_r is the effective modulus of tested material and M_i is the indentation modulus of indenter tip.

The indentation modulus M_i was determined using Young's modulus (1141 GPa) and Poisson's ratio (0.07) of diamond (Fischer-Cripps 2006) and Eq. (3.8).

The effective modulus, denoted as E_r, can be determined using Eq. 3.11:

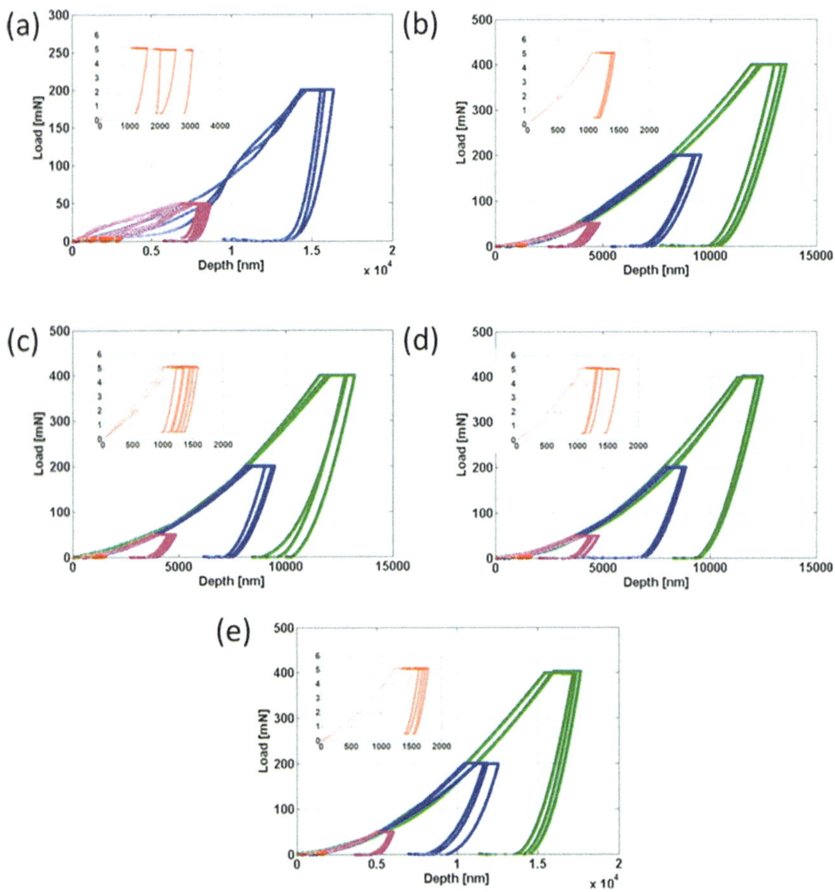

Fig. 3.15 Typical load-depth response of **a** Mix 1, **b** Mix 2, **c** Mix 3, **d** Mix 4, and **e** Mix 5. Reproduced from [Binder-scale creep behavior of metakaolin-based geopolymer] by [Shikun Chen] with permission from [Cement and Concrete Research]

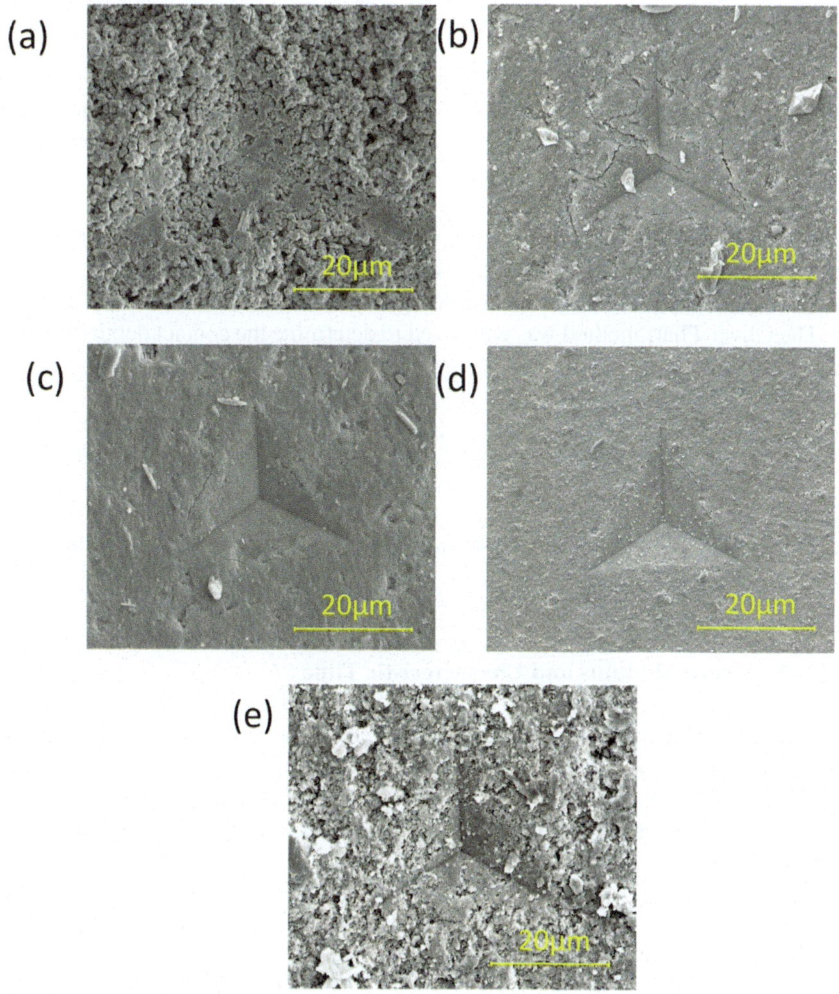

Fig. 3.16 Post-indentation (50 mN) sites of **a** Mix 1, **b** Mix 2, **c** Mix 3, **d** Mix 4, and **e** Mix 5. Reproduced from [Binder-scale creep behavior of metakaolin-based geopolymer] by [Shikun Chen] with permission from [Cement and Concrete Research]

$$E_r = \frac{\sqrt{\pi}}{2\beta}\frac{S}{\sqrt{A_c}} \tag{3.11}$$

Here, S represents the elastic unloading stiffness, A_c denotes the contact area, and β is a correction coefficient that considers the shape of the tip used. For the specific case of the Berkovich tip utilized in this study, Oliver and Pharr (2004) suggested a value of $\beta = 1.034$.

The calculation of the elastic unloading stiffness S can be performed utilizing Eq. (3.12):

$$S = \frac{dP}{dh}\bigg|_{h=h_m} \tag{3.12}$$

Here, h_m corresponds to the indentation depth at the initial unloading stage. In this investigation, the value of S was approximated by fitting the linear function $P = c_1 * h + c_2$ to the first 20% of the unloading curve, whereby the slope of the fitted equation corresponds to $S = c_1$.

Moreover, the contact area A_c was determined based on the contact depth h_c, and for the Berkovich tip employed in this study, the relationship between A_c and h_c is represented as $A_c = 24.479 * {h_c}^2$ according to Jin and Ebenstein (2017).

The Oliver-Pharr method was employed to determine the contact depth h_c, which can be expressed as:

$$h_c = h_m - \epsilon \frac{P_H}{S} \tag{3.13}$$

Here, h_m denotes the maximum indentation depth during the initial unloading, P_H represents the holding force during the holding stage, and S corresponds to the elastic unloading stiffness.

3.3.2.2 Creep Modulus and Characteristic Time

In the case of indenting a viscoelastic material, characterization of the creep behavior of the probed material can be accomplished by utilizing the contact creep compliance L(t), as described by Zhang et al. (2014). For step-loaded indentation using a Berkovich tip, the relationship between L(t) and the indentation depth rate can be expressed as:

$$L(t) = \frac{2r_c(t)h(t)}{P_H} \tag{3.14}$$

Here, $r_c(t)$ represents the radius of the projected area of the contact interface. Assuming that the contact radius remains constant during the holding phase, as suggested by Vandamme (2008), the change in contact creep compliance $\Delta L(t) = L(t) - L(0)$ can be linked to the change in indentation depth with time $\Delta h(t) = h(t) - h(0)$, as shown below:

$$\Delta L(t) = \frac{2r_{cu}(t)\Delta h(t)}{P_H} \tag{3.15}$$

Here, r_{cu} denotes the contact radius at the end of the holding stage.

In general, the creep behavior of cement can exhibit two types of behaviors: power-law and logarithmic-law, as reported by Tamtsia and Beaudoin (2000), Königsberger et al. (2016), and Irfan-ul-Hassan et al. (2017) for the former, and Vandamme and Ulm

(2013) and Zhang et al. (2014) for the latter. As a result, various data interpretation approaches have been proposed to comprehend the creep behavior probed by nanoindentation experiments, as described by Jones and Grasley (2011) and Vandamme and Ulm (2013).

In this study, both the power-law and logarithmic-law approaches were employed to fit the experimental outcomes; however, only the logarithmic function proposed by Vandamme and Ulm (2013) demonstrated satisfactory results. Therefore, the logarithmic-law approach was utilized, and the fitting function is given by:

$$\Delta L(t) = \frac{\ln(\frac{t}{\tau} + 1)}{C} \quad (3.16)$$

Here, C represents the contact creep modulus, while τ denotes a characteristic time. A large value of C corresponds to low creep compliance, indicating less creep strain. Similarly, a large characteristic time τ suggests a slow creep rate, as suggested by Vandamme (2008) and Zhang et al. (2014).

By combining Eqs. 3.2.4 and 3.2.5, a logarithmic relationship can be established between the change in indentation depth $\Delta h(t)$ and the holding time:

$$\Delta h(t) = \alpha \ln(\frac{t}{\tau} + 1) \quad (3.17)$$

$$\alpha = \frac{P_H}{2r_{cu}C}$$

In this study, a modified equation suggested by Vandamme (2008) and Vandamme and Ulm (2013) was employed to enhance the fitting of the experimental data:

$$\Delta h(t) = \alpha \ln\left(\frac{t}{\tau} + 1\right) + \gamma t + \delta \quad (3.18)$$

The first term in Eq. (3.18) describes the logarithmic creep behavior of the analyzed material, and its creep properties can be characterized by the parameters α and τ, as explained earlier. Conversely, the other terms and corresponding parameters (γ and δ) are solely dependent on the testing equipment and fitting errors, according to Vandamme and Ulm (2013), Zhang et al. (2014), and Lee et al. (2018). By fitting Eq. (3.18) to the obtained $\Delta h(t)$ curves (refer to Fig. 3.17), the contact creep modulus and characteristic time of the geopolymer can be determined.

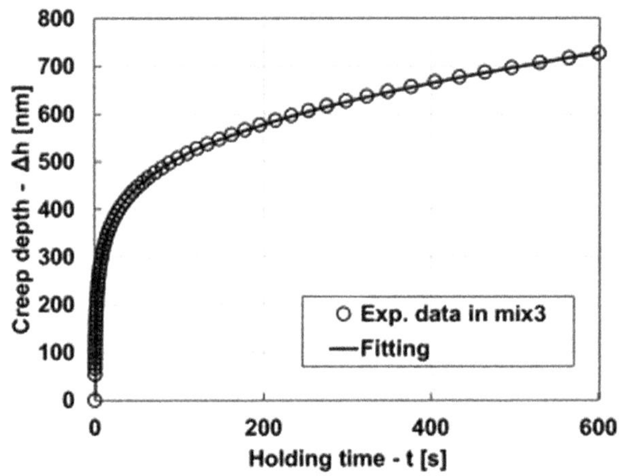

Fig. 3.17 Typical creep depth versus holding time response. Reproduced from [Binder-scale creep behavior of metakaolin-based geopolymer] by [Shikun Chen] with permission from [Cement and Concrete Research]

3.3.3 Results

3.3.3.1 Young's Modulus and Hardness

In Fig. 3.18, the values of Young's modulus (E) and hardness (H) extracted from the nanoindentation experiment are presented against the contact depth h_c. It is observed that the inhomogeneity and polished-surface effect lead to a considerable scattering of the extracted properties when using a 5 mN indentation load. However, as the indentation load increases to 50 mN and beyond, the variation in results decreases. This suggests that the impact of inhomogeneity and polished-surface effect reduces gradually with an increase in probed volumes. For Mix 1, an indentation load of 200 mN was used, while for other compositions, it was 400 mN. The extracted properties stabilize as the indentation load increases, and thus representative values were taken for each geopolymer composition.

The hardness (0.1–0.14 GPa) and Young's modulus (2.57–3.16 GPa) of Mixes 2, 3, and 4 are comparable, but Mix 1 has the lowest hardness (0.033–0.037 GPa) and Young's modulus (1.72–1.97 GPa). Mix 5 exhibits intermediate Young's modulus (2.05–2.32 GPa) and hardness (0.062–0.064 GPa). Compared to earlier investigations on geopolymers based on metakaolin (Němeček et al. 2011; Pelisser et al. 2013; Subaer et al. 2016), the Young's modulus and hardness values measured in this study are relatively low. For instance, Němeček et al. (2011) reported a range of 17–18 GPa for Young's modulus, and Pelisser et al. (2013) reported 8.21–11.50 GPa. The hardness ranged from 0.17 to 0.42 GPa (Pelisser et al. 2013) and 0.05–0.6 GPa (Subaer et al. 2016). The relatively lower measured Young's modulus and indentation hardness values in this study are primarily attributed to the higher water content (44.5%) in comparison to previous research [approximately 22–38% in Subaer et al. (2016), 33–35% in Pelisser et al. (2013), and 38–39% in Němeček et al. (2011)]. The results of Pelisser et al. (2013), who found that an excessively low silicate in the

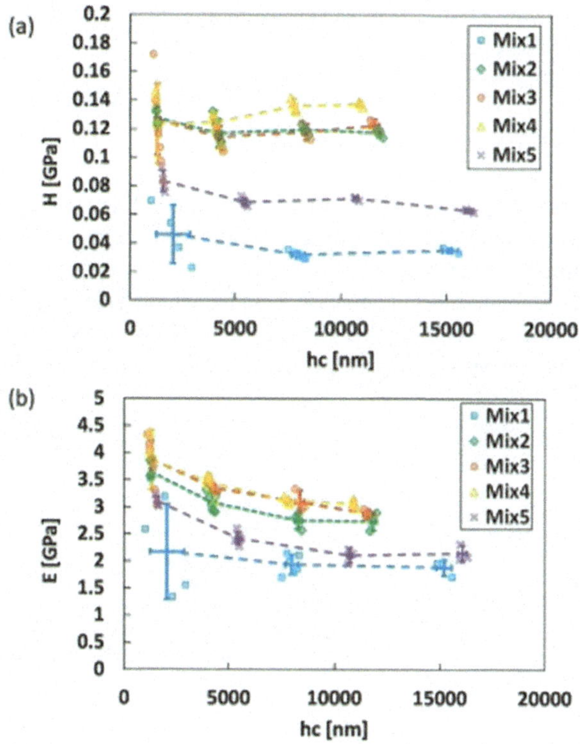

Fig. 3.18 Results of **a** hardness, **b** Young's modulus. Reproduced from [Binder-scale creep behavior of metakaolin-based geopolymer] by [Shikun Chen] with permission from [Cement and Concrete Research]

activator reduces the Young's modulus and hardness of the matrix, are consistent with the trends in the chemical composition of the activator examined in our experiment, despite the fact that the high water content decreased the absolute value of E and H. Contrary to our earlier findings from macroscale compressive testing, where the Young's modulus stayed nearly constant as the Si/Al ratio changed from 1.9 to 2.4, the lower Young's modulus brought on by the high Si/Al ratio (Yan et al. 2016). This deviation could be due to the curing age of the sample; the relatively shorter curing age (14 days) in this study might have limited the reaction extent of the sample with the highest Si/Al ratio (Mix 5). Previous studies (Singh et al. 2005; Sagoe-Crentsil et al. 2007) have shown that the low pH value in high silicate-contained activators might slow down the reaction of raw particles, and thus may require more curing time to achieve a close-to-full reaction. However, the SEM photos in the next sections show that for various Si/Al ratios, the curing age is adequate to generate a homogenous developed gel with few unreacted raw components. This phenomenon also reveals that the reaction rate of the geopolymer significantly changes for an Si/

Al ratio ranging from 2.0 to 2.2; however, extending the curing time could alleviate this delayed reaction effect.

3.3.3.2 Creep Properties from Nanoindentation

Figure 3.19 displays the indentation curves during the holding stage, while Fig. 3.20 showcases the extracted creep properties, which were obtained by following the procedure outlined in Sect. 3.3.

The contact-creep modulus of Mixes 2, 3, and 4, which have an Si/Al ratio ranging from 1.6 to 2.0, are very comparable (25–32 GPa). However, Mixes 1 and 5, which have lower and higher Si/Al ratios, respectively, exhibit a reduction of approximately

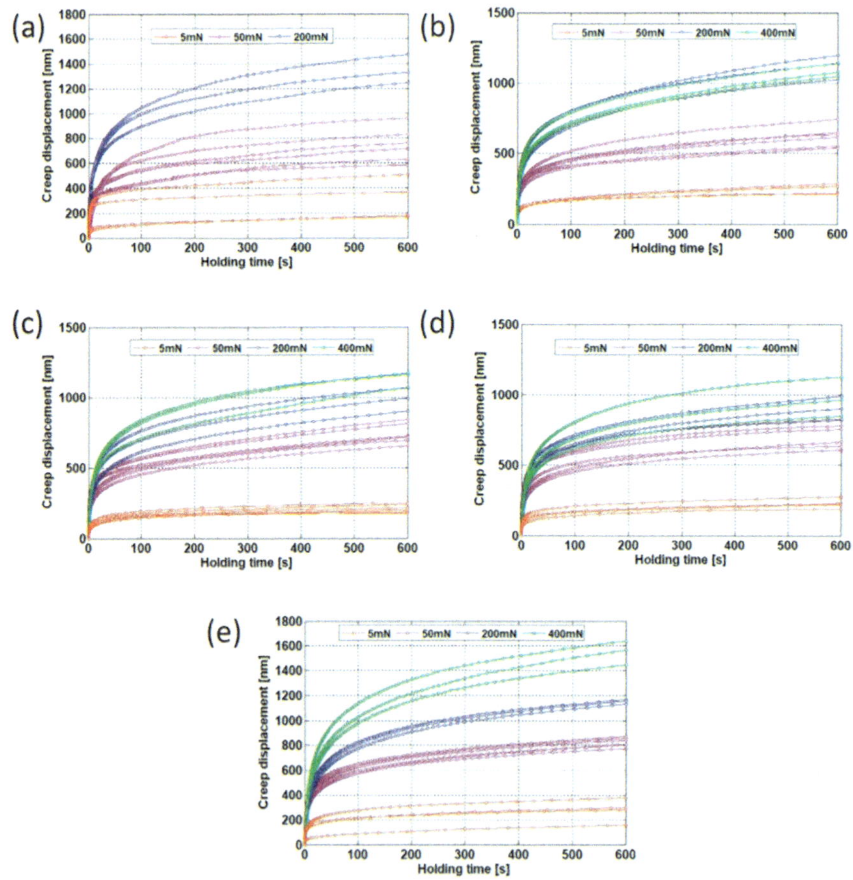

Fig. 3.19 Creep displacement versus holding time of **a** Mix 1, **b** Mix 2, **c** Mix 3, **d** Mix 4 and **e** Mix 5. Reproduced from [Binder-scale creep behavior of metakaolin-based geopolymer] by [Shikun Chen] with permission from [Cement and Concrete Research]

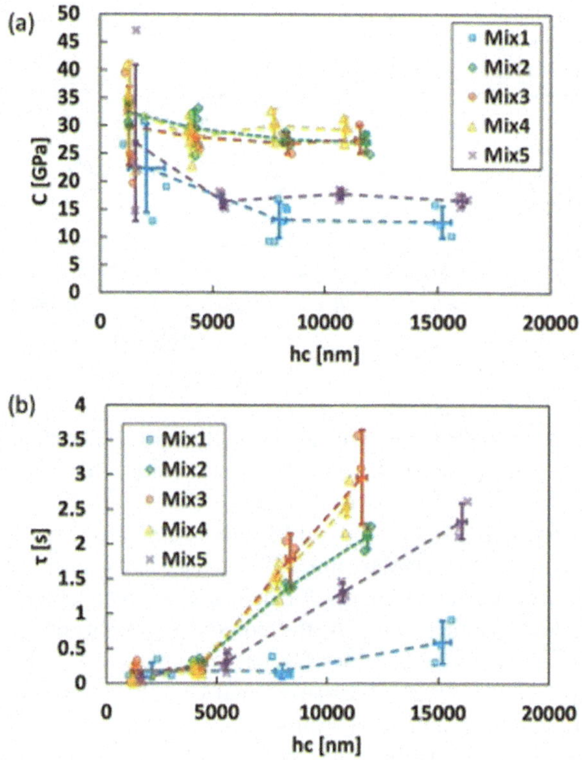

Fig. 3.20 Results of **a** creep modulus, **b** characteristic time. Reproduced from [Binder-scale creep behavior of metakaolin-based geopolymer] by [Shikun Chen] with permission from [Cement and Concrete Research]

50% in the creep modulus compared to the aforementioned mixes (10–18 GPa). The variation in creep modulus with respect to the Si/Al ratio follows a similar trend as that observed for Young's modulus and hardness. The creep behavior in the geopolymer appears to be influenced by the same structural factors that dictate E and H; this will be discussed in later sections.

Furthermore, when compared to the results (40–105 GPa) obtained by Lee et al. (2018) using a similar indentation method for fly-ash-based geopolymer samples with and without silicate fumes, the creep modulus measured in our samples is relatively lower. As previously mentioned, this deviation could be partially attributed to the high water content in our samples. Another contributing factor may be the difference in holding force, with our research utilizing forces between 5 and 400 mN, whereas Lee et al. used 2 mN. According to Němeček (2009), a higher holding force usually results in larger creep deformation and a low creep modulus. However, our experiments indicate that higher holding forces (above 50 mN) are required to obtain a stable representative creep property of the binder by involving a sufficient volume of tens of the characteristic pore size in length.

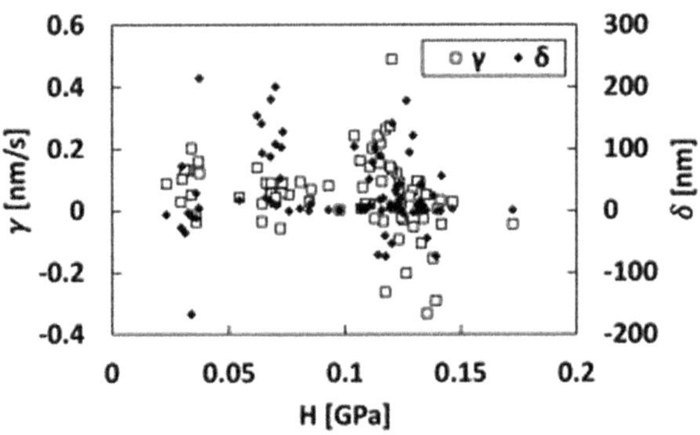

Fig. 3.21 Fitting parameter γ and δ versus the extracted hardness. Reproduced from [Binder-scale creep behavior of metakaolin-based geopolymer] by [Shikun Chen] with permission from [Cement and Concrete Research]

On the other hand, as the indentation load rises, the extracted characteristic timings do not converge. The characteristic times of several binders likewise grew as indentation load increased. The characteristic timings of the various binders varied significantly for high indentation loads (200 and 400 mN), with binders with lower hardness often having a shorter characteristic time. This is reasonable since weaker materials usually require less activation energy for deformation; as a result, external stress accelerates the deformation process. As the required activation energy of the material is low, the characteristic time becomes shorter.

Similar to previous research (Vandamme and Ulm 2013), the parameters γ and δ utilized in fitting the creep curves exhibit little correlation with composition and mechanical properties; this is illustrated in Fig. 3.21.

In this study, test results were compared with data on geopolymer concretes reported in the literature (Hardjito et al. 2004; Wallah 2010; Islam 2015; Castel et al. 2016). The creep modulus of concrete materials was extracted from uniaxial creep experiments, whereby the basic creep compliance of a concrete material could be evaluated using Eq. (3.19):

$$L(t) = \frac{\epsilon_{tot}(t) - \epsilon_s(t)}{\sigma} \tag{3.19}$$

Here, $\varepsilon_{tot}(t)$ and $\varepsilon_s(t)$ denote the total strain measured under the sustained stress σ and the shrinkage strain measured during the test period without external loading, respectively.

The creep function, which is the difference between the elastic response and the basic creep compliance, is given by Eq. (3.20):

$$\Delta L(t) = L(t) - \frac{1}{E_0} = \frac{\epsilon_{tot}(t) - \epsilon_s(t)}{\sigma} - \frac{\epsilon_e}{\sigma} = \frac{\Delta\epsilon(t)}{\sigma} \tag{3.20}$$

Here, $\Delta\varepsilon(t) = \varepsilon_{tot}(t) - \varepsilon_s(t) - \varepsilon_e$ represents the strain increase purely due to creep and is usually provided in the literature.

Similar to the indentation creep, the creep strain increase during a uniaxial creep test could also be fitted to a logarithmic kinetics equation (Eq. 3.21):

$$\Delta\epsilon(t) = \alpha \ln(\frac{t}{\tau} + 1) \tag{3.21}$$

The creep modulus of the concrete material is given by Eq. (3.22):

$$C_{con} = \frac{\sigma}{\alpha} \tag{3.22}$$

where τ is the characteristic time.

According to earlier studies (Vandamme and Ulm 2013; Zhang et al. 2014), the creep modulus of the binder material may be calculated using a Mori–Tanaka technique by subtracting the creep modulus of the concrete material. In this homogenization, the aggregate creep is disregarded, but the matrix creep is deemed to be deviatoric. Additionally, it is assumed that the adhesion between the binder and the aggregate is perfect. This yields the estimation equation shown in Eq. (3.23):

$$C_{bin} = \frac{2(1 - f_{agg})}{2 + 3f_{agg}} C_{con} \tag{3.23}$$

where f_{agg} denotes the volume fraction of the aggregate.

Although different raw materials, binder compositions, and test details were utilized in the studies compared, the results provide a basis for comparing indentation creep tests and bulk creep tests in geopolymer-based materials. The results of the retreatment are presented in Table 3.6. The findings show that the contact creep modulus calculated by nanoindentation in this work (10–32 GPa) is lower than the bulk uniaxial creep test-derived creep modulus (20–60 GPa) in earlier publications; however, their magnitudes remain within the same order. Conversely, the characteristic times obtained from the nanoindentation test (0.1–3.5 s) are significantly shorter than those determined from bulk creep tests (0.06–1.06 d). This short characteristic time is also observed in earlier studies investigating the contact creep behavior of OPCs (Vandamme and Ulm 2013) and fly ash-based geopolymers (Lee et al. 2018) using nanoindentation.

It is dubious to directly compare the extracted binder scale creep properties with those from traditional bulk scale creep tests without taking into account these inherent differences, as the probed volume is much smaller and the effective stress level is much higher in the indentation creep test. However, the indentation creep test can quickly provide details on the composition-dependency of the creep behaviors. This knowledge might support rational bulk scale design through calibrations based on bulk scale outcomes. This, however, is outside the purview of this article and will be investigated in further research.

Table 3.6 Creep properties extracted from uniaxial creep test data reported in the literature

Data source	Sample	σ (MPa)	C_{con} (GPa)	τ (d)	f_{agg}	C_{bin} (GPa)
Hardjito et al. (2004)	A-2	22.00	562.37	0.06	0.77	60.02
Wallah (2010)	1CR	27.00	508.47	0.13	0.77	54.27
	2CR	23.00	348.06	0.12	0.77	37.15
	3CR	19.00	227.68	0.52	0.77	24.30
	4CR	16.00	276.00	0.10	0.77	29.46
Islam (2015)	GP4000	11.03	171.11	0.54	0.74	21.01
	GP8000	20.68	455.00	0.40	0.74	55.88
	GP35	14.07	316.57	0.50	0.74	38.88
	GP45	12.41	219.54	0.53	0.74	26.96
	GP55	11.03	168.17	0.30	0.74	20.65
	GP65	10.34	169.60	0.21	0.74	20.83
Castel et al.(2016)	3D40	10.00	29.29	0.13	0.79	2.88
	7D80	20.00	434.88	1.06	0.79	42.71

3.3.3.3 Composition of the Gel Phase from EDS

Davidovits (1991) highlighted that the gel phase of the geopolymer consists of three distinct aluminosilicate monomers, where alkali ions balance the negative charge of the aluminate center. Various Si/Al ratios would result in different fractions of monomers and connecting coordinates in the aluminosilicate network (White et al. 2012). Energy dispersive spectrometry (EDS) was used to determine the actual elemental composition of hardened geopolymer binder samples in order to determine if the variation in the Si/Al ratio in the gel phase of the geopolymer was caused by varying starting Si/Al ratios in the mixture design. The geopolymer matrix, as shown in Fig. 3.22, could include a number of phases, including the gel phase and unreacted MK particles. The spectra was collected and fitted to determine the relative atomic fraction of different elements at locations where the gel phase made up a significant component of the sample. The Na/Al ratio and Si/Al ratio in each phase were calculated using Eqs. (3.24) and (3.25), respectively:

$$\frac{\text{Na}}{\text{Al}} = \frac{\text{At\%(Na)}}{\text{At\%(Al)}} \tag{3.24}$$

$$\frac{\text{Si}}{\text{Al}} = \frac{\text{At\%(Si)}}{\text{At\%(Al)}} \tag{3.25}$$

Here, At% (*) represents the relative atomic (molar) fraction of a given element.

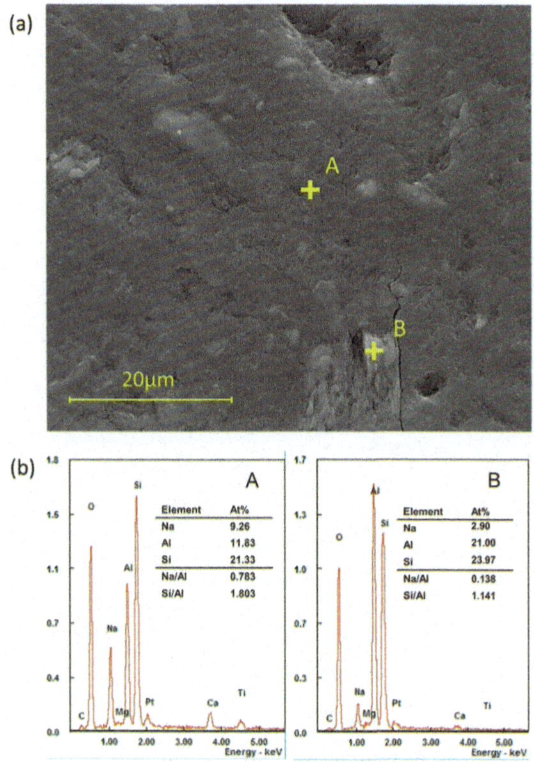

Fig. 3.22 **a** SEM and **b** EDS characterization of specimens with Mix 3. Reproduced from [Binder-scale creep behavior of metakaolin-based geopolymer] by [Shikun Chen] with permission from [Cement and Concrete Research]

The chemical composition of the gel phase was statistically determined using data from 11 to 16 locations. The outcomes reveal that the average Na/Al ratio of the gel phase ranges from 0.70 to 0.98 (Fig. 3.23a). These measured Na/Al ratios are relatively lower than the target value of 1.0. This deviation may be attributed to washing during polishing. Some sodium ions are still unbound in the structure of the N–A–S–H gel because the reaction rate of the MK powder could not reach 100% and some calcium impurity was present in the MK powder. Alkali ions will dissolve during water washing since they are highly leachable (Aly et al. 2012), which will skew the readings.

The silicate and aluminate levels in the hardened geopolymer are more persistent and challenging to dissolve when compared to leachable alkali ions. Thus, the measured Si/Al ratio is closer to the target value of the designed mixture, as demonstrated in Fig. 3.23b. This result provides a useful baseline for further analysis.

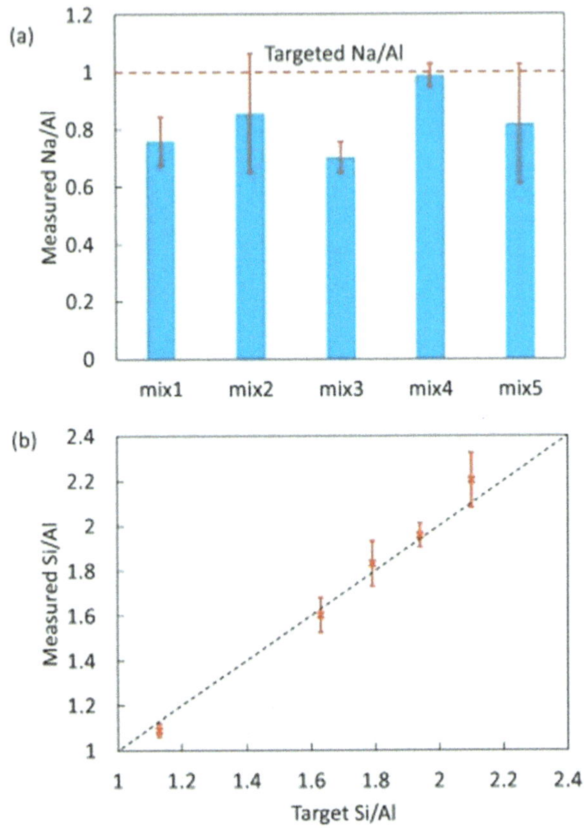

Fig. 3.23 The **a** Na/Al ratio and **b** Si/Al ratio measured by EDS analysis. Reproduced from [Binder-scale creep behavior of metakaolin-based geopolymer] by [Shikun Chen] with permission from [Cement and Concrete Research]

3.3.3.4 Gel Morphology from SEM and XRD

Figure 3.24 illustrates the microstructure of geopolymer binders with varying Si/Al ratios.

The gel phase in Mix 1 exhibits a granular character after being activated with pure sodium hydroxide solution and a Si/Al ratio close to 1.09. Smaller gel particles cover and bind several cubic crystals. From the EDS results (Fig. 3.25), the elemental composition of the crystal phase consists mainly of O, Na, Al, and Si, with a Si/Al ratio close to 1.1. The XRD outcomes reveal that Mix 1 presents significant crystal peaks belonging to the zeolite-A phase. This observation was also reported in previous studies (Ozer and Soyer-Uzun 2015; Subaer et al. 2016; Wan et al. 2017), where the composition's Si/Al ratio ranged from 1.0 to 1.25. The surrounding amorphous gel phase's elemental composition did not deviate significantly from that

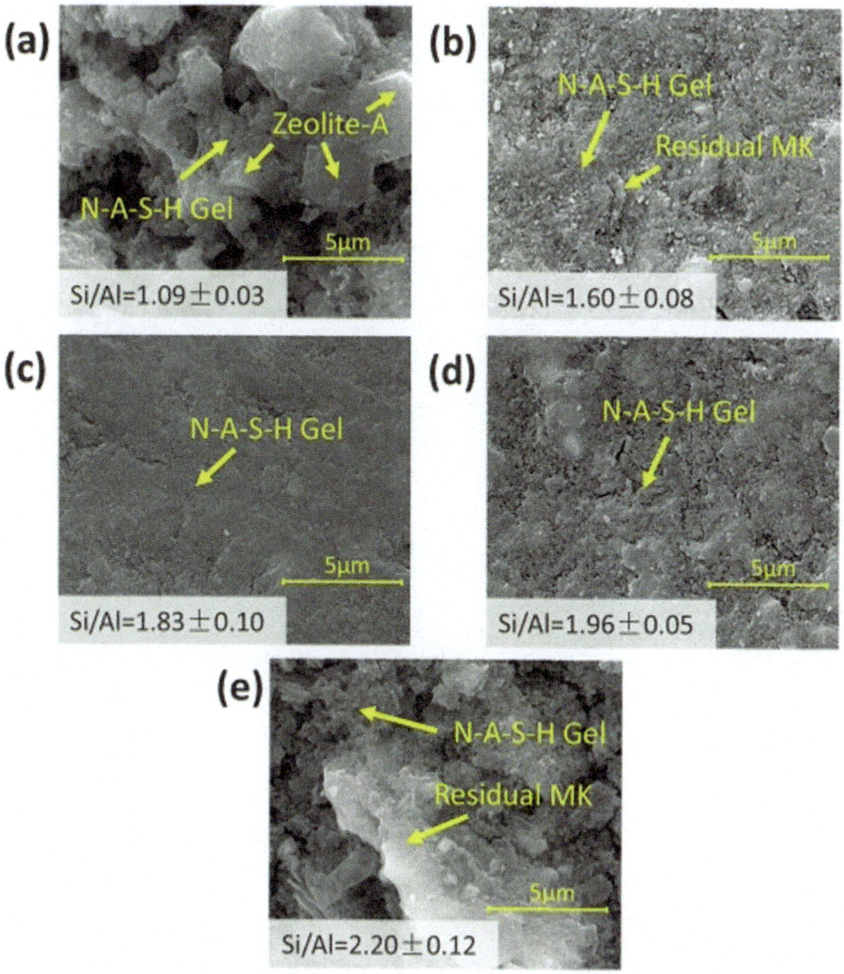

Fig. 3.24 Comparison of microstructure of **a** Mix 1, **b** Mix 2, **c** Mix 3, **d** Mix 4 and **e** Mix 5. Reproduced from [Binder-scale creep behavior of metakaolin-based geopolymer] by [Shikun Chen] with permission from [Cement and Concrete Research]

of the crystal phase. The porous gel structure is the main reason why the amorphous phase has a somewhat lower Na/Al ratio than the solid phase does. Unbound Na ions dissolve out due to the greater contact area with the liquids in the pores, which further biases the observed composition. The crystal phase in Mix 1 enhances the skeletal matrix's density, while the surrounding gel phase has a lot of big holes.

The other geopolymer binders (Mixes 2, 3, 4, and 5), activated using the sodium silicate solution with Ms > 1.0, do not contain a similar crystal phase. As shown in Fig. 3.26, their major XRD pattern comprises wide humps spread from 15° to 40° and centered at 27°–28°. There are only peaks for quartz and anatase, which

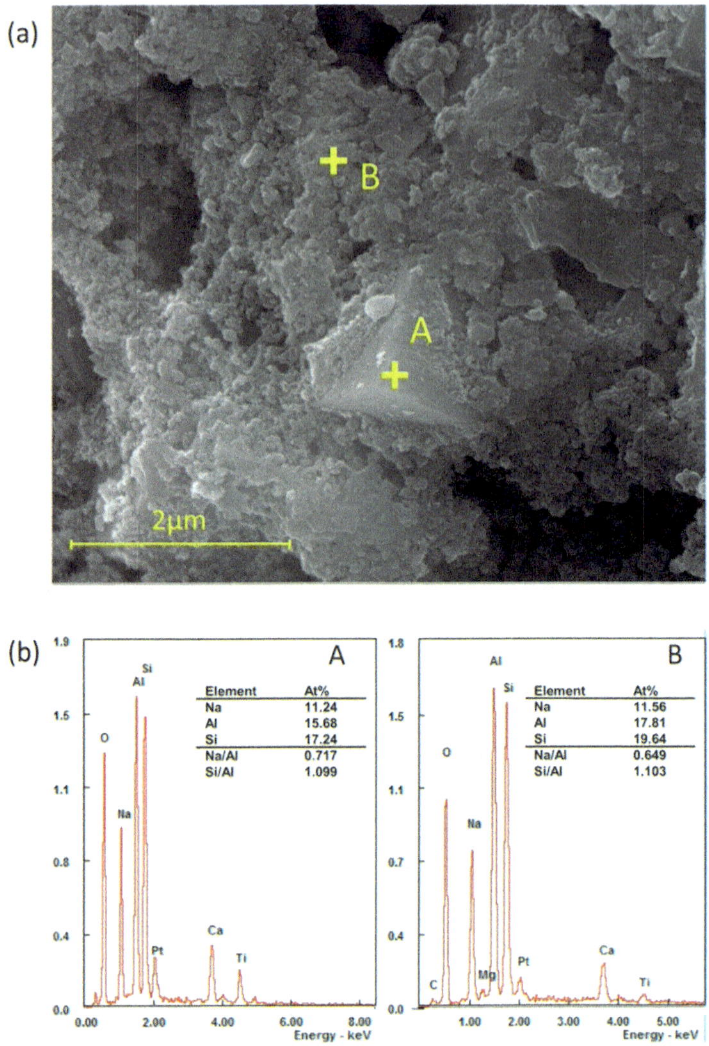

Fig. 3.25 **a** SEM and **b** EDS characterization of the crystal phase in mix 1 sample. Reproduced from [Binder-scale creep behavior of metakaolin-based geopolymer] by [Shikun Chen] with permission from [Cement and Concrete Research]

are predominantly brought on by crystal SiO_2 and TiO_2 impurities in the source material. With the exception of a minor drop in the hump's center, this hump pattern is featureless for various mixes. It shows that amorphous gel makes up the majority of the matrix of geopolymers with Si/Al ratios greater than 1.6. Previous findings from MK-based geopolymers with Si/Al ratios ranging from 1.5 to 4.0 also support this phase characterization (Wan et al. 2017). The SEM images shown in Fig. 3.24b–d illustrate that the primary phases in these mixtures are homogeneous matured gels

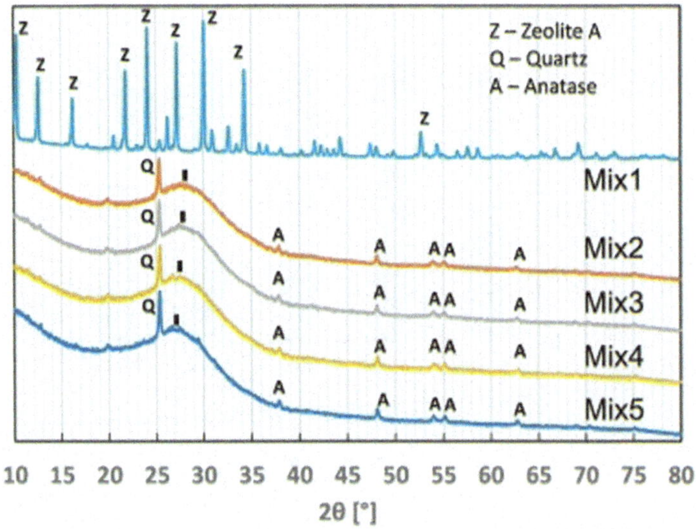

Fig. 3.26 XRD patterns. Reproduced from [Binder-scale creep behavior of metakaolin-based geopolymer] by [Shikun Chen] with permission from [Cement and Concrete Research]

along with numerous micro or nano pores. He et al. (2016) showed that variations in the degrees of geopolymerization rather than changes in the Si/Al ratio were the main effects of the change in the Si/Al ratio on the phase's structural characteristics. Figure 3.24e shows that Mix 5 has bigger holes in the surrounding gel phase and has substantial residual metakaolin particles in the matrix. Nonetheless, the major gel phase remains amorphous and contains no crystals, as observed in Mix 1.

3.3.3.5 Pore Structure from MIP

MIP measurements provide a clearer distinction of pore structures among different mixtures. The outcomes indicate that the porosity of the five mixtures barely changes (approximately 3% relative standard deviation) (Fig. 3.27a). Averaged porosity derived from the MIP test ranges from 40.74 to 44.27%, with a mean value of 43%. No particular trend is observed concerning the Si/Al ratio. Moreover, both bulk density (including MIP pores) and skeletal density (excluding MIP pores) of samples are independent of their compositions (Fig. 3.27b). Despite the fact that certain extremely tiny pore volumes could not be identified using MIP methods, it is safe to believe that the sample's porosity is consistent throughout a range of Si/Al ratios. This slight variation in porosity can be explained to the intended mixture's consistent water content. A constant water content would provide a constant porosity because pores develop when water is taken out of the matrix.

Conversely, the pore distribution among different mixtures displays distinct features. Figure 3.28 reveals that the characteristic pore size in Mix 1 is about

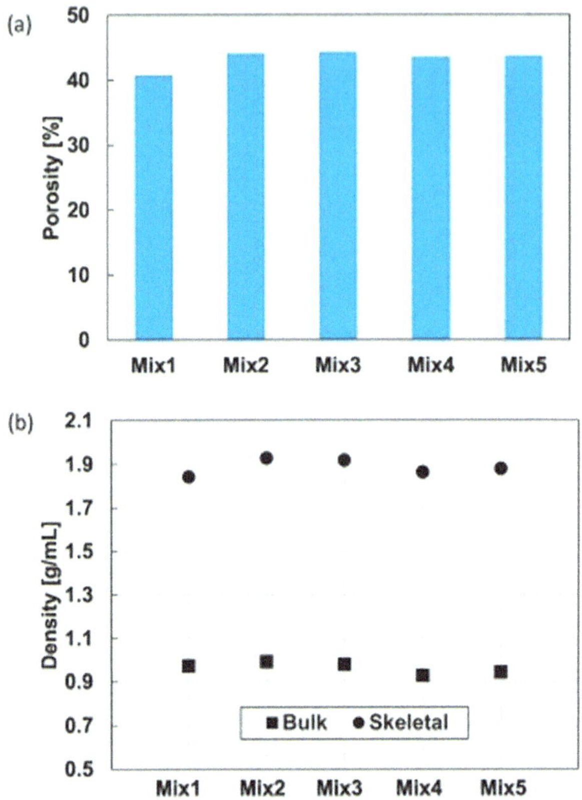

Fig. 3.27 **a** Porosity and **b** density of each mixture. Reproduced from [Binder-scale creep behavior of metakaolin-based geopolymer] by [Shikun Chen] with permission from [Cement and Concrete Research]

1300 nm, with major pore distributions in the micro (several micrometers) and sub-micro (hundreds of nanometers) scales. In contrast, Mixes 2, 3, and 4 have main pore sizes that range from few to tens of nanometers, with typical pore diameters that are far below 50 nm. Although Mix 5 has a characteristic pore size that is a little bit bigger (around 100 nm), the primary pores are still dispersed at the nanoscale.

3.3.4 Discussion

3.3.4.1 Composition-Dependent Pore Structure

The gel phase morphology of geopolymer binders with varying Si/Al ratios can be divided into three classes based on previous microstructure characterizations, as

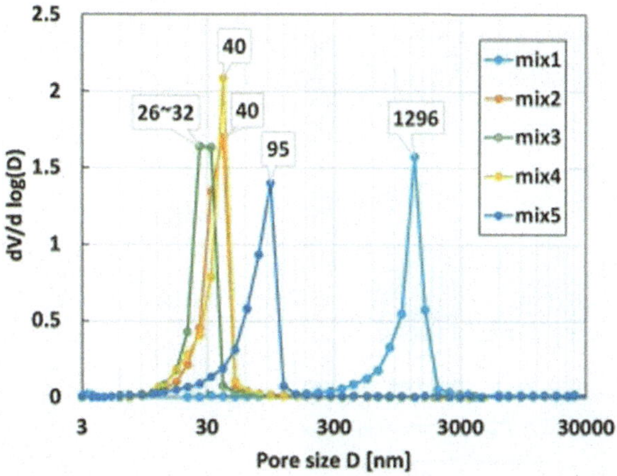

Fig. 3.28 The pore size distribution of each mixture. Reproduced from [Binder-scale creep behavior of metakaolin-based geopolymer] by [Shikun Chen] with permission from [Cement and Concrete Research]

listed in Table 3.7. The most prominent feature of these three classes is the pore size distribution. Figure 3.29 provides a conceptual explanation of the impact of the Si/Al ratio on the pore structure.

The Ms of the activator decreases to zero as the Si/Al ratio gets closer to 1.0. As a result, the alkaline solution does not give any external silicate. All the necessary silicate and aluminate species to create a gel dissolve from the metakaolin particle in the absence of soluble silicate in solution. Additionally, granular crystal nuclei with sizes ranging from 3 to 5 μm are present within the gel due to the low Si/Al ratio in gel composition, leaving behind pores between them with sizes of similar order (several micrometers) (Fig. 3.29a). The formation of large pores due to the creation of zeolitic nuclei has been reported in a previous publication (Wan et al. 2017).

As the Si/Al ratio increases, the number of silicate species from the external activator also increases. The gel precipitates widely and forms a more homogeneous phase, resulting in N–A–S–H gel with numerous nanopores and some distributed residual metakaolin particles forming the primary phase in the binder. The micro pores between particles are largely eliminated (Fig. 3.29b).

Table 3.7 Classification of the gel phase morphologies

Morphology	Si/Al	Structural order	Pore size (nm)
Class I	1.0–1.25	Crystal + amorphous	180–3500
Class II	1.5–2.0	Amorphous	8–90
Class III	> 2.1	amorphous	10–180

Reproduced from [Binder-scale creep behavior of metakaolin-based geopolymer] by [Shikun Chen] with permission from [Cement and Concrete Research]

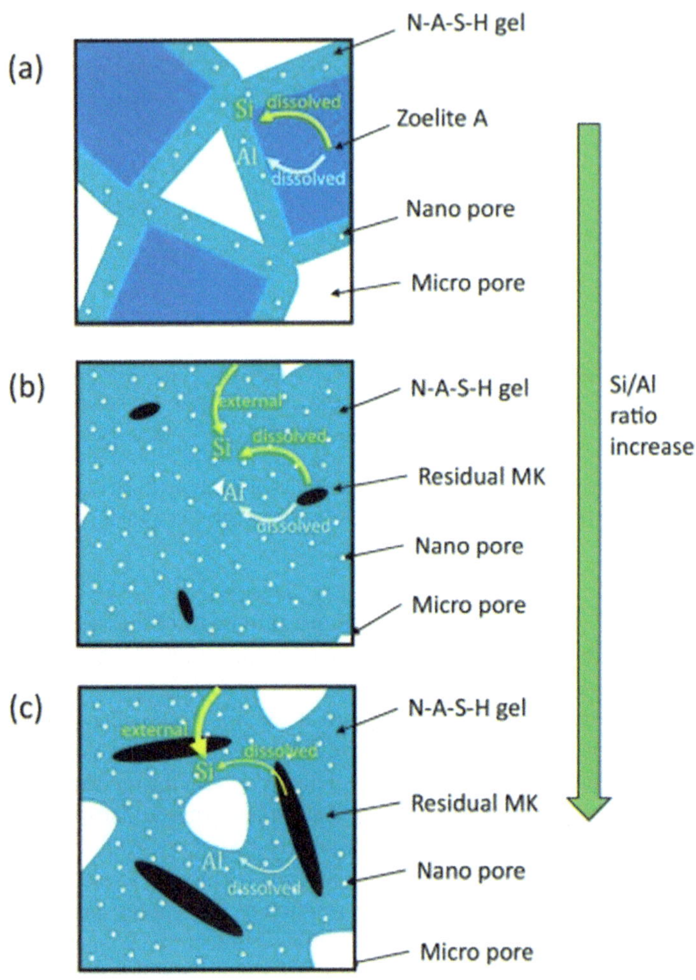

Fig. 3.29 Conceptual explanation of influence of Si/Al ratio on the pore structures. Reproduced from [Binder-scale creep behavior of metakaolin-based geopolymer] by [Shikun Chen] with permission from [Cement and Concrete Research]

When the Si/Al ratio further increases, the silicate modulus of the alkali solution increases, and the alkalinity of the solution decreases with the formation of a large number of silicate species (Provis et al. 2005), which significantly affects the dissolution process of metakaolin particles. When the alkalinity dropped, the amount of raw material dissolution would as well (Cho et al. 2017). The binder would contain some of the big metakaolin particles. The paucity of aluminate species necessary to create the gel is caused by the drop in dissolved metakaolin. The void between residual particles cannot be fully filled by the gel, leading to the formation of some micro pores (Fig. 3.29c).

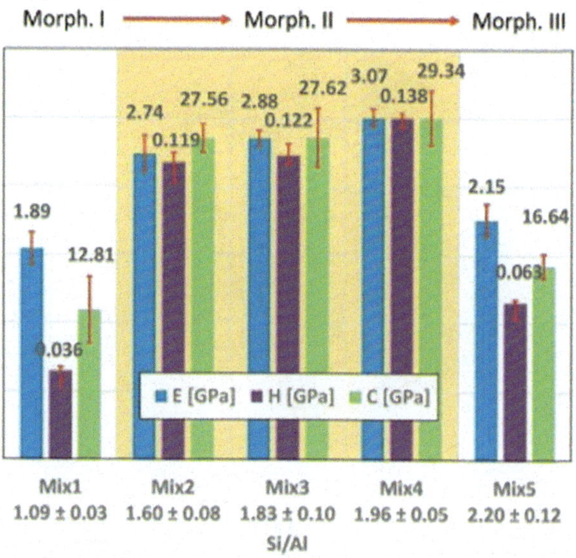

Fig. 3.30 Influence of the Si/Al ratio. Reproduced from [Binder-scale creep behavior of metakaolin-based geopolymer] by [Shikun Chen] with permission from [Cement and Concrete Research]

3.3.4.2 Correlation Between Pore Structure and Mechanical Behavior

Figure 3.30 illustrates that the Si/Al ratio divisions, based on the mechanical properties obtained from the highest indentation load, correspond with the gel phase morphology classifications. The MIP results reveal that the primary difference in morphology among different mixes is their pore size distribution. The porosity (ranging from 40.74–44.27%) is almost identical for all mixes. As a result, a pore size effect has been observed from the nanoscale mechanical properties of MKG extracted via nanoindentation.

A "soft" porous solid with a "hard" pore surface layer might help to partially explain this pore size effect. An easy analytical model is presented to objectively assess this pore size impact. Two-phase materials (the matrix and the surface layer around the pores) were taken into account while evaluating the effective mechanical characteristics.

A regular foam with an orthogonal pore network and surrounding surface layers was used to simulate the mechanical behavior of a porous solid with randomly distributed pores and surface layers (Fig. 3.31a). For ease of calculation, the thickness of the surface layer t was assumed to be constant (as long as it does not exceed the threshold of $d + 2tl$) and the pore section was assumed to be a square with size d. The typical foam's unit cell is seen in Fig. 3.31c. The connection between cell size (l), pore size (d), and porosity (p) is as follows:

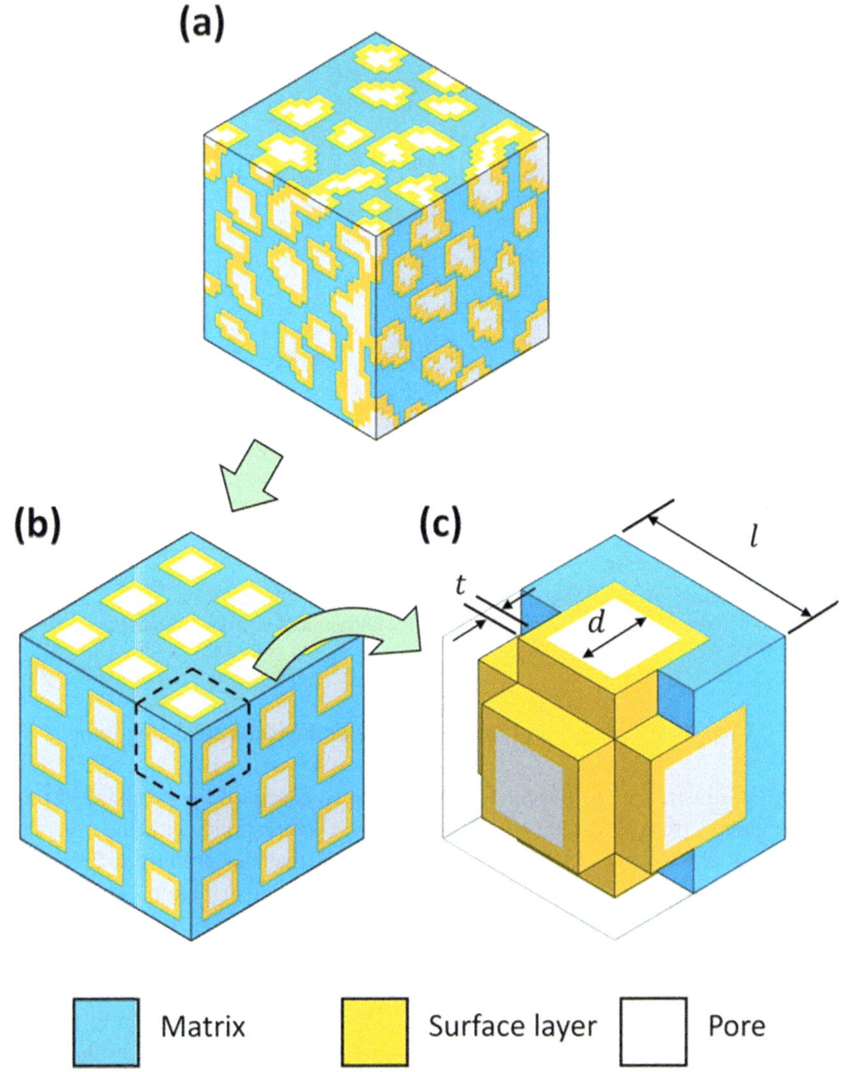

Fig. 3.31 Schematic of modeling **a** random porous solid with pore size effect with, **b** regular foam with orthogonal square pore network and surrounding surface layers **c** unit cell of the regular foam model. For clarity, the near half of the matrix is removed to show the inner structures. Reproduced from [Binder-scale creep behavior of metakaolin-based geopolymer] by [Shikun Chen] with permission from [Cement and Concrete Research]

$$\phi_p = 3\left(\frac{d}{l}\right)^2 - 2\left(\frac{d}{l}\right)^3 \tag{3.26}$$

The ratio d/l = 0.456 if $\phi_p = 0.43$. The volume fraction of the solid matrix could also be calculated using the following:

$$\phi_m = \begin{cases} 3\left(1 - \frac{d}{l} - 2\frac{t}{l}\right)^2\left(\frac{d}{l} + 2\frac{t}{l}\right) + \left(1 - \frac{d}{l} - 2\frac{t}{l}\right)^3, & d + 2t < l \\ 0, & d + 2t \geq l \end{cases} \tag{3.27}$$

The volume fraction of the surface layer is:

$$\phi_s = 1 - \phi_p - \phi_m \tag{3.28}$$

When the thickness of surface layer t exceeds the limit d + 2t < l, the foam would only be composed as a harder phase and pores, thus $\phi_s = 1 - \phi_p$.

The effective mechanical properties could be estimated using the simple mixture law (Haecker et al. 2005). The Voigt method (Fig. 3.32a), which considers components bearing the same strain, provides an upper bound on the effective properties:

$$E_{eff} = E_m\phi_m + E_s\phi_s \tag{3.29}$$

where, E_{eff}, E_m, and E_s are the Young's modulus of the regular foam, the matrix, and the surface layer, respectively. The contribution from the pore is neglected.

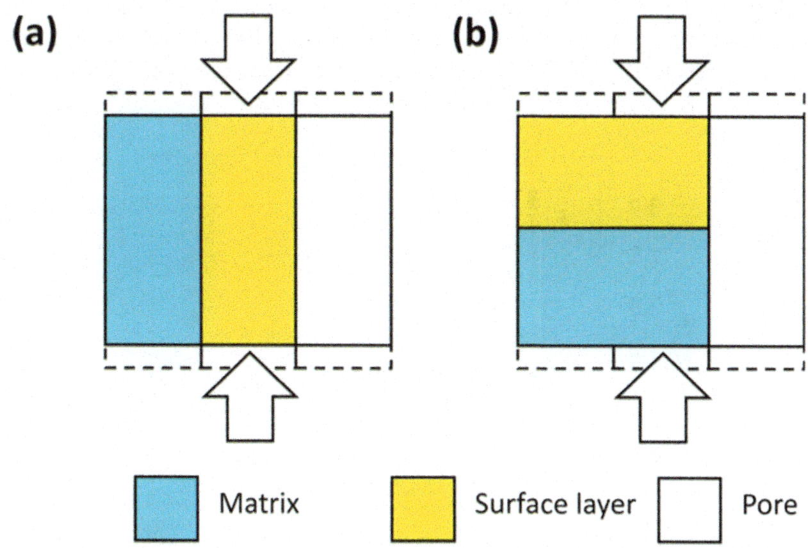

Fig. 3.32 Schematics of mixture law based on **a** Voigt, and **b** mixed Voigt-Reuss method. Reproduced from [Binder-scale creep behavior of metakaolin-based geopolymer] by [Shikun Chen] with permission from [Elsevier]

Table 3.8 Estimated properties of the inner matrix and surface layer

	Thickness [nm]	Young's modulus [GPa]
Matrix	–	3.0–3.5
Surface layer	15–20	5.0–5.8

Reproduced from [Binder-scale creep behavior of metakaolin-based geopolymer] by [Shikun Chen] with permission from [Cement and Concrete Research]

Simultaneously, a mixed Voigt-Reuss method (Fig. 3.32b), which considers solid components (matrix and surface layer) bearing the same stress while the void (pores) sustains no stress and experiences the same deformation, then provides a lower bound on the effective properties:

$$E_{eff} = \left(1 - \phi_p\right)\left(\left(\frac{1}{\phi_s + \phi_m}\right)\left(\frac{\phi_s}{E_s} + \frac{\phi_m}{E_m}\right)\right)^{-1} \tag{3.30}$$

The upper and lower bounds of the effective properties obtained using Eqs. (3.29) and (3.30) with experimental results can roughly estimate the properties of the matrix and surface layer.

The estimated effective Young's modulus is presented in Table 3.8, which is consistent with the corresponding test results (± 1 std.) shown in Fig. 3.33. For comparison, estimates without considering the surface layer are also displayed in Fig. 3.33, and there is no dependence on pore size. This mechanism is analogous to the surface-elasticity-induced size-effects observed in stiffness of nanowires (Cuenot et al. 2004; Chen et al. 2006; Silva et al. 2006) and nanoporous materials (Mathur

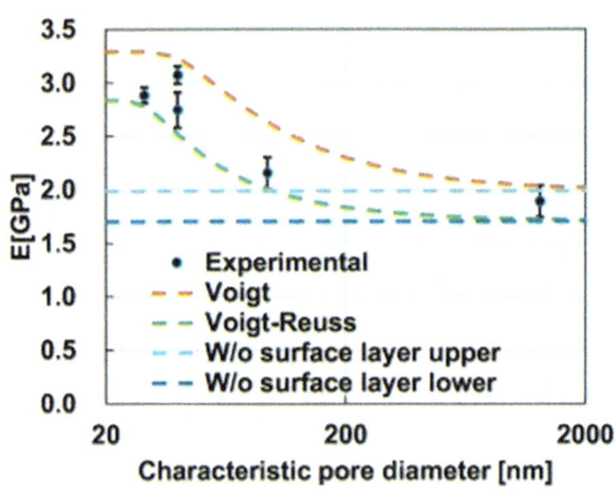

Fig. 3.33 Comparison between the model evaluation and the experimental results of Young's modulus. Reproduced from [Binder-scale creep behavior of metakaolin-based geopolymer] by [Shikun Chen] with permission from [Cement and Concrete Research]

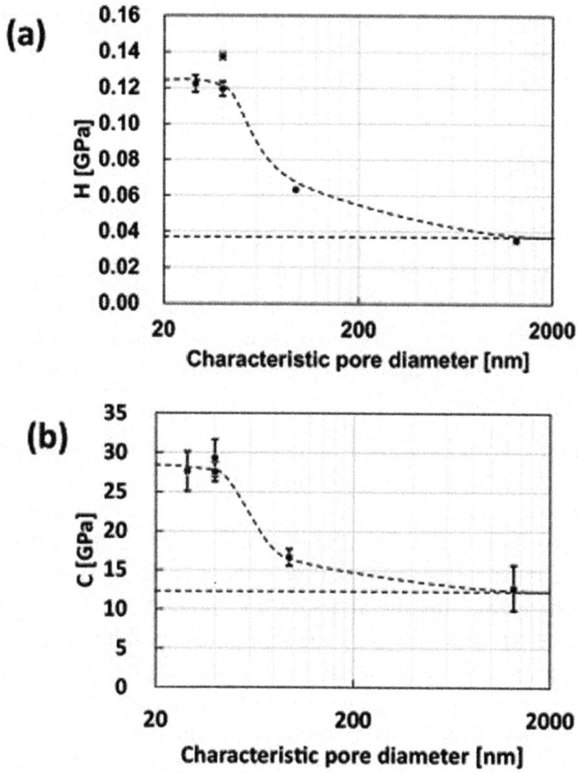

Fig. 3.34 Variation trend of **a** hardness and **b** creep modulus versus the characteristic pore diameter. Reproduced from [Binder-scale creep behavior of metakaolin-based geopolymer] by [Shikun Chen] with permission from [Cement and Concrete Research]

and Erlebacher 2007; Feng et al. 2009). The tendency may also be seen in Fig. 3.34, which plots the change of hardness and creep modulus against the typical pore size. Because of this, the surface layer structure existing close to the walls of the pores in MKG may have an impact on behavior other than elastic, such as yielding and viscous behavior.

How these surface effects arise from the material's structure remains an open question due to the lack of techniques to probe the material properties at the very surface. However, there is proof that the density of the surface of sintered ceramics differs from that of the interior cores (German 2014), which might have an impact on the surface. In this work, it is still not established if the surface layer of ambient-derived geopolymer exists at a scale of tens of nanometers. The alteration in mechanical properties of MKG with various Si/Al ratios might possibly be explained by additional factors, such as intrinsic molecular structural alterations (Lolli et al. 2018) or nanoscale interactions between gel networks and confined water (Sadat et al. 2016).

Thoroughly elucidating these fundamental mechanisms requires further studies at the molecular scale.

3.4 Conclusions

In this study, AE-grouped and Ms-grouped MKG pastes were synthesized, and their axial and lateral stress-strain curves were measured while their microstructures were characterized. The objective was to determine the elastic properties of the MKG pastes, including the Young's, bulk, and shear moduli and the Poisson's ratio, and correlate the data with the composition and microstructure. The obtained results lead to the following conclusions:

(1) While the Ms level exhibits a monotonous boosting impact on compressive strength, the AE dose has a dual effect on compressive strength with the ideal dosage at 40%.
(2) A four-stage failure process can be used to describe the fracture behavior of MKG pastes: (a) elastic deformation without cracks; (b) slight fracture deformation with visible superficial cracks due to crack initiation and propagation; (c) moderate fracture damage with superficial spalling of the material due to crack penetration and percolation; and (d) fracture failure.
(3) Λ-like curve with highest values at AE30 is also seen in the plots of the Young and shear moduli of the MGK pastes to the AE dose. An equation with a quadratic double-logarithmic function can describe this relationship.
(4) The Poisson's ratio and bulk modulus of the MKG pastes exhibit simple and monotonous increases with increasing AE dosage. This relationship can be characterized by a sigmoid function.
(5) Variance analysis shows that the Young's, bulk, and shear moduli as well as the Poisson ratio of MKG are all unchanged and lack statistical significance.
(6) The MKG paste has a low elastic modulus and high compressive strength, making it a soft but robust substance. The right composition design is necessary for a well-densified and polymerized network of geopolymer, such as by regulating the ratios of Si/Na and/or Si/Al.
(7) Our findings show that the creep modulus of geopolymer determined from nanoindentation measurements is comparable to values measured with conventional macro-scale tests but requires substantially less time.
(8) The Si/Al ratio significantly influences the microstructure and creep behavior in MKG, as well as the Young's modulus and hardness. According to a thorough examination, it is significantly responsible for changing the binder's microstructure, particularly the primary size of the matrix's pores. Despite practically constant porosity, the observed Young's modulus, hardness, and creep modulus rise when the average pore size is reduced by optimizing the Si/Al ratio. Furthermore, the presence of a zeolitic phase was seen when the Si/Al ratio was close to 1.0, and the presence of significant residual metakaolin particles was observed

when the Si/Al ratio was close to 2.2. The binder's pore size increased as a result of these microstructural changes, which also resulted in decreased Young's modulus, hardness, and creep modulus.
(9) The MKG exhibited optimum mechanical properties and holes that were largely spread throughout tens of nanometers when the Si/Al ratio was between 1.6 and 2.0. It could therefore be inferred that the pore size changes with varying Si/Al ratio may be related to the varying amount of silicate species and alkalinity of the alkaline activating solution, while the change of mechanical properties due to pore size might be related to a surface layer around the pores based on micromechanical models and analysis. Further research is required to properly analyze this process and rule out findings from alternative hypotheses because the presence of such a structure is not yet proved.

References

Aly Z, Vance ER, Perera DS (2012) Aqueous dissolution of sodium aluminosilicate geopolymers derived from metakaolin. J Nucl Mater 424:164–170. https://doi.org/10.1016/j.jnucmat.2012.02.027

Andini S, Cioffi R, Colangelo F et al (2008) Coal fly ash as raw material for the manufacture of geopolymer-based products. Waste Manag 28:416–423. https://doi.org/10.1016/j.wasman.2007.02.001

Barbosa VFF, MacKenzie KJD, Thaumaturgo C (2000) Synthesis and characterisation of materials based on inorganic polymers of alumina and silica: sodium polysialate polymers. Int J Inorg Mater 2:309–317. https://doi.org/10.1016/S1466-6049(00)00041-6

Barpi F, Valente S (2003) Creep and fracture in concrete: a fractional order rate approach. Eng Fract Mech 70:611–623. https://doi.org/10.1016/S0013-7944(02)00041-3

Bažant ZP (2001) Prediction of concrete creep and shrinkage: past, present and future. Nucl Eng Des 203:27–38. https://doi.org/10.1016/S0029-5493(00)00299-5

Bazant ZP, Panula L (1980) Creep and shrinkage characterization for analysing prestressed concrete structures. PCI J 25:86–122

Benboudjema F, Meftah F, Torrenti JM (2005) Interaction between drying, shrinkage, creep and cracking phenomena in concrete. Eng Struct 27:239–250. https://doi.org/10.1016/j.engstruct.2004.09.012

Castel A, Foster SJ, Ng T et al (2016) Creep and drying shrinkage of a blended slag and low calcium fly ash geopolymer concrete. Mater Struct Constr 49:1619–1628. https://doi.org/10.1617/s11527-015-0599-1

Chen CQ, Shi Y, Zhang YS et al (2006) Size dependence of Young's modulus in ZnO nanowires. Phys Rev Lett 96.https://doi.org/10.1103/PhysRevLett.96.075505

Chen JH, Huang JS, Chang YW (2011) Use of reservoir sludge as a partial replacement of metakaolin in the production of geopolymers. Cem Concr Compos 33:602–610. https://doi.org/10.1016/j.cemconcomp.2011.03.002

Cheng H, Lin KL, Cui R et al (2015a) The effects of SiO_2/Na_2O molar ratio on the characteristics of alkali-activated waste catalyst-metakaolin based geopolymers. Constr Build Mater 95:710–720. https://doi.org/10.1016/j.conbuildmat.2015.07.028

Cheng H, Lin KL, Cui R et al (2015b) Effect of solid-to-liquid ratios on the properties of waste catalyst-metakaolin based geopolymers. Constr Build Mater 88:74–83. https://doi.org/10.1016/j.conbuildmat.2015.01.005

Cho YK, Yoo SW, Jung SH et al (2017) Effect of Na_2O content, SiO_2/Na_2O molar ratio, and curing conditions on the compressive strength of FA-based geopolymer. Constr Build Mater 145:253–260. https://doi.org/10.1016/j.conbuildmat.2017.04.004
Cuenot S, Frétigny C, Demoustier-Champagne S, Nysten B (2004) Surface tension effect on the mechanical properties of nanomaterials measured by atomic force microscopy. Phys Rev B—Condens Matter Mater Phys 69.https://doi.org/10.1103/PhysRevB.69.165410
Davidovits J (1991) Geopolymers. J Therm Anal Calorim 37:1633–1656
Duxson P, Provis JL, Lukey GC et al (2005) Understanding the relationship between geopolymer composition, microstructure and mechanical properties. Colloids Surfaces A Physicochem Eng Asp 269:47–58. https://doi.org/10.1016/j.colsurfa.2005.06.060
Duxson P, Mallicoat SW, Lukey GC et al (2007) The effect of alkali and Si/Al ratio on the development of mechanical properties of metakaolin-based geopolymers. Colloids Surfaces A Physicochem Eng Asp 292:8–20. https://doi.org/10.1016/j.colsurfa.2006.05.044
El-Safty S, Silikas N, Akhtar R, Watts DC (2012) Nanoindentation creep versus bulk compressive creep of dental resin-composites. Dent Mater 28:1171–1182. https://doi.org/10.1016/j.dental.2012.08.012
Feng XQ, Xia R, Li X, Li B (2009) Surface effects on the elastic modulus of nanoporous materials. Appl Phys Lett 94.https://doi.org/10.1063/1.3067999
Fischer-Cripps AC (2006) Critical review of analysis and interpretation of nanoindentation test data. Surf Coatings Technol 200:4153–4165. https://doi.org/10.1016/j.surfcoat.2005.03.018
Fisher R (1954) Discussion of the analysis of variance with various binomial transformations. Biometrics 10:140. https://doi.org/10.2307/3001668
Gao K, Lin KL, Wang D et al (2014) Effects SiO_2/Na_2O molar ratio on mechanical properties and the microstructure of nano-SiO_2 metakaolin-based geopolymers. Constr Build Mater 53:503–510
German R (2014) Sintering: from empirical observations to scientific principles
Haecker CJ, Garboczi EJ, Bullard JW et al (2005) Modeling the linear elastic properties of Portland cement paste. Cem Concr Res 35:1948–1960. https://doi.org/10.1016/j.cemconres.2005.05.001
Hardjito D, Wallah SE, Sumajouw DMJ, Rangan BV (2004) On the development of fly ash-based geopolymer concrete. ACI Mater J 101:467–472. https://doi.org/10.14359/13485
He J, Zhang J, Yu Y, Zhang G (2012) The strength and microstructure of two geopolymers derived from metakaolin and red mud-fly ash admixture: a comparative study. Constr Build Mater 30:80–91. https://doi.org/10.1016/j.conbuildmat.2011.12.011
He J, Jie Y, Zhang J et al (2013) Synthesis and characterization of red mud and rice husk ash-based geopolymer composites. Cem Concr Compos 37:108–118. https://doi.org/10.1016/j.cemconcomp.2012.11.010
He P, Wang M, Fu S et al (2016) Effects of Si/Al ratio on the structure and properties of metakaolin based geopolymer. Ceram Int 42:14416–14422. https://doi.org/10.1016/j.ceramint.2016.06.033
Heah CY, Kamarudin H, Mustafa Al Bakri AM et al (2012) Study on solids-to-liquid and alkaline activator ratios on kaolin-based geopolymers. Constr Build Mater 35:912–922. https://doi.org/10.1016/j.conbuildmat.2012.04.102
Hung TC, Huang JS, Wang YW, Fan YC (2013) Microstructure and properties of metakaolin-based inorganic polymer foams. J Mater Sci 48:7446–7455. https://doi.org/10.1007/s10853-013-7559-3
Irfan-ul-Hassan M, Königsberger M, Reihsner R et al (2017) How water-aggregate interactions affect concrete creep: multiscale analysis. J Nanomechanics Micromechanics 7:04017019. https://doi.org/10.1061/(asce)nm.2153-5477.0000135
Islam MR (2015) Creep and shrinkage behavior of fly ash based geopolymer concrete
Islam A, Alengaram UJ, Jumaat MZ, Bashar II (2014) The development of compressive strength of ground granulated blast furnace slag-palm oil fuel ash-fly ash based geopolymer mortar. Mater Des 56:833–841. https://doi.org/10.1016/j.matdes.2013.11.080
Jin C, Ebenstein DM (2017) Nanoindentation of compliant materials using Berkovich tips and flat tips. J Mater Res 32:435–450. https://doi.org/10.1557/jmr.2016.483

Jones CA, Grasley ZC (2011) Short-term creep of cement paste during nanoindentation. Cem Concr Compos 33:12–18. https://doi.org/10.1016/j.cemconcomp.2010.09.016

Königsberger M, Irfan-ul-Hassan M, Pichler B, Hellmich C (2016) Downscaling based identification of nonaging power-law creep of cement hydrates. J Eng Mech 142:04016106. https://doi.org/10.1061/(asce)em.1943-7889.0001169

Lee NP (2007) Creep and shrinkage of inorganic polymer concrete. Judgeford, New Zealand

Lee H, Vimonsatit V, Chindaprasirt P et al (2018) Creep properties of cement and alkali activated fly ash materials using nanoindentation technique. Constr Build Mater 168:547–555. https://doi.org/10.1016/j.conbuildmat.2018.02.166

Lizcano M, Kim HS, Basu S, Radovic M (2012) Mechanical properties of sodium and potassium activated metakaolin-based geopolymers. J Mater Sci 47:2607–2616. https://doi.org/10.1007/s10853-011-6085-4

Lolli F, Manzano H, Provis JL et al (2018) Atomistic simulations of geopolymer models: the impact of disorder on structure and mechanics. ACS Appl Mater Interfaces 10:22809–22820. https://doi.org/10.1021/acsami.8b03873

Lyu SJ, Hsiao YH, Wang TT et al (2013) Microstructure of geopolymer accounting for associated mechanical characteristics under various stress states. Cem Concr Res 54:199–207. https://doi.org/10.1016/j.cemconres.2013.09.007

Mathur A, Erlebacher J (2007) Size dependence of effective Young's modulus of nanoporous gold. Appl Phys Lett 90.https://doi.org/10.1063/1.2436718

Measurement of hardness and elastic modulus by instrumented indentation (2004) Advances in understanding and refinements to methodology. J Mat Res 19(1):3-20.https://doi.org/10.1557/jmr.2004.19.1.3

Metha PK, Monterio PJM (2006) Concrete, microstructure, properties and materials. McGraw-Hill, London

Mo BH, Zhu H, Cui XM et al (2014) Effect of curing temperature on geopolymerization of metakaolin-based geopolymers. Appl Clay Sci 99:144–148. https://doi.org/10.1016/j.clay.2014.06.024

Němeček J (2009) Creep effects in nanoindentation of hydrated phases of cement pastes. Mater Charact 60:1028–1034. https://doi.org/10.1016/j.matchar.2009.04.008

Němeček J, Šmilauer V, Kopecký L (2011) Nanoindentation characteristics of alkali-activated aluminosilicate materials. Cem Concr Compos 33:163–170. https://doi.org/10.1016/j.cemconcomp.2010.10.005

Ozer I, Soyer-Uzun S (2015) Relations between the structural characteristics and compressive strength in metakaolin based geopolymers with different molar Si/Al ratios. Ceram Int 41:10192–10198. https://doi.org/10.1016/j.ceramint.2015.04.125

Pelisser F, Guerrino EL, Menger M et al (2013) Micromechanical characterization of metakaolin-based geopolymers. Constr Build Mater 49:547–553. https://doi.org/10.1016/j.conbuildmat.2013.08.081

Pharr GM, Oliver WC (2004) Measurement of hardness and elastic modulus by instrumented indentation: advances in understanding and refinements to methodology. J Mater Res 19:3–20. https://doi.org/10.1557/jmr.2004.19.1.3

Posi P, Teerachanwit C, Tanutong C et al (2013) Lightweight geopolymer concrete containing aggregate from recycle lightweight block. Mater Des 52:580–586. https://doi.org/10.1016/j.matdes.2013.06.001

Provis JL, Bernal SA (2014) Geopolymers and related alkali-activated materials. Annu Rev Mater Res 44:299–327

Provis JL, Duxson P, Lukey GC et al (2005) Modeling speciation in highly concentrated alkaline silicate solutions. Ind Eng Chem Res 44:8899–8908. https://doi.org/10.1021/ie050700i

Rashad AM (2013) Alkali-activated metakaolin: a short guide for civil engineer—an overview. Constr Build Mater 41:751–765. https://doi.org/10.1016/j.conbuildmat.2012.12.030

Rovnaník P (2010) Effect of curing temperature on the development of hard structure of metakaolin-based geopolymer. Constr Build Mater 24:1176–1183. https://doi.org/10.1016/j.conbuildmat.2009.12.023

Rowles M, O'Connor B, Matthew R et al (2003) Chemical optimisation of the compressive strength of aluminosilicate geopolymers synthesised by sodium silicate activation of metakaolinite. J Mater Chem 13:1161–1165. https://doi.org/10.1039/b212629j

Sadat MR, Bringuier S, Asaduzzaman A et al (2016) A molecular dynamics study of the role of molecular water on the structure and mechanics of amorphous geopolymer binders. J Chem Phys 145:134706. https://doi.org/10.1063/1.4964301

Sagoe-Crentsil K, Weng L, Sagoe-Crentsil K (2007) Dissolution processes, hydrolysis and condensation reactions during geopolymer synthesis: part II. High Si/Al ratio systems. J Mater Sci 42:2997–3006. https://doi.org/10.1007/s10853-006-0820-2

Sagoe-Crentsil K, Brown T, Taylor A (2013) Drying shrinkage and creep performance of geopolymer concrete. J Sustain Cem Mater 2:35–42. https://doi.org/10.1080/21650373.2013.764963

Sarker PK, Haque R, Ramgolam KV (2013) Fracture behaviour of heat cured fly ash based geopolymer concrete. Mater Des 44:580–586. https://doi.org/10.1016/j.matdes.2012.08.005

Sicard V, Francois R, Ringot E, Pons G (1992) Influence of creep and shrinkage on cracking in high strength concrete. Cem Concr Res 22:159–168. https://doi.org/10.1016/0008-8846(92)90146-M

Silva ECCM, Tong L, Yip S, Van Vliet KJ (2006) Size effects on the stiffness of silica nanowires. Small 2:239–243. https://doi.org/10.1002/smll.200500311

Singh PS, Trigg M, Burgar I, Bastow T (2005) Geopolymer formation processes at room temperature studied by ^{29}Si and ^{27}Al MAS-NMR. Mater Sci Eng A 396:392–402

Steveson M, Sagoe-Crentsil K (2005) Relationships between composition, structure and strength of inorganic polymers. J Mater Sci 40:2023–2036. https://doi.org/10.1007/s10853-005-1226-2

Subaer S, Haris A, Nurhayati N et al (2016) The influence of Si:Al and Na:Al on the physical and microstructure characters of geopolymers based on metakaolin. Mater Sci Forum 841:170–177.https://doi.org/10.4028/www.scientific.net/MSF.841.170

Tamtsia BT, Beaudoin JJ (2000) Basic creep of hardened cement paste. A re-examination of the role of water. Cem Concr Res 30:1465–1475. https://doi.org/10.1016/S0008-8846(00)00279-9

Vandamme M (2008) The nanogranular origin of concrete creep : a nanoindentation investigation of microstructure and fundamental properties of calcium-silicate-hydrates

Vandamme M, Ulm FJ (2013) Nanoindentation investigation of creep properties of calcium silicate hydrates. Cem Concr Res 52:38–52. https://doi.org/10.1016/j.cemconres.2013.05.006

Wallah SE (2010) Creep behaviour of fly ash-based geopolymer concrete. Civ Eng Dimens 12.https://doi.org/10.9744/ced.12.2.73-78

Wallah SE, Rangan BV (2006) Low-calcium fly ash-based geopolymer concrete: long-term properties

Wan Q, Rao F, Song S et al (2017) Geopolymerization reaction, microstructure and simulation of metakaolin-based geopolymers at extended Si/Al ratios. Cem Concr Compos 79:45–52. https://doi.org/10.1016/j.cemconcomp.2017.01.014

Wang H, Li H, Yan F (2005) Synthesis and mechanical properties of metakaolinite-based geopolymer. Colloids Surfaces A Physicochem Eng Asp 268:1–6. https://doi.org/10.1016/j.colsurfa.2005.01.016

White CE, Provis JL, Proffen T, van Deventer JSJ (2012) Molecular mechanisms responsible for the structural changes occurring during geopolymerization: multiscale simulation. AIChE J 58:2241–2253. https://doi.org/10.1002/aic.12743

Wongpa J, Kiattikomol K, Jaturapitakkul C, Chindaprasirt P (2010) Compressive strength, modulus of elasticity, and water permeability of inorganic polymer concrete. Mater Des 31:4748–4754. https://doi.org/10.1016/j.matdes.2010.05.012

Yan D, Chen S, Zeng Q et al (2016) Correlating the elastic properties of metakaolin-based geopolymer with its composition. Mater Des 95:306–318. https://doi.org/10.1016/j.matdes.2016.01.107

Zhang Y, Sun W, Li Z (2010) Composition design and microstructural characterization of calcined kaolin-based geopolymer cement. Appl Clay Sci 47:271–275. https://doi.org/10.1016/j.clay.2009.11.002

Zhang Q, Le Roy R, Vandamme M, Zuber B (2014) Long-term creep properties of cementitious materials: comparing microindentation testing with macroscopic uniaxial compressive testing. Cem Concr Res 58:89–98. https://doi.org/10.1016/j.cemconres.2014.01.004

Chapter 4
Drying Shrinkage of MKG

Abstract The drying shrinkage of geopolymers represents a significant constraint on their potential application as construction materials. This chapter investigates the drying shrinkage of MKG with different initial water/solid ratios and pore structures. Experiments using mini-bar shrinkage techniques reveal a two-stage behavior in the relationship between drying shrinkage and water loss of MKG. Although the initial water/solid ratio affects the crucial water loss and length of the shrinkage curves' stopping phase, it has little impact on the overall trend. Through microstructure analysis and physical estimate, the underlying dependence of the shrinkage on the pore structure of the binder is clarified. During different stages of water loss, capillary pressure, surface energy changes, and gel densification are the dominant factors that contribute to the drying shrinkage of MKG. The findings indicate that in addition to porosity control, finer tuning of the pore size distribution is necessary to regulate the drying shrinkage of MKG.

4.1 Introduction

The environmental impact of the ordinary Portland cement (OPC) industry necessitates the development of more sustainable construction materials. Geopolymer, a newly developed binder system, has garnered attention from scientists and engineers for its eco-friendly benefits, including low CO_2 emissions and energy consumption (Collins and Turner 2013; Habert and Ouellet-Plamondon 2016). Various pozzolanic aluminosilicate sources, including metakaolin, fly ash, slag, and bottom ash, are alkali-activated to create geopolymers (Antunes Boca Santa et al. 2017; Panda and Tan 2018; Li et al. 2019; Yousefi Oderji et al. 2019). High early strength (Davidovits 1994), fire resistance (Duxson et al. 2007), and corrosion durability (Thokchom et al. 2009) are just a few of the numerous advantages geopolymers have over OPC. These desirable properties are primarily attributed to the unique gel structure of the binder, which comprises of 3D aluminosilicate networks with covalent bonds, balanced alkali ions (Na, K, etc.), and nanopores (Davidovits 1991; White et al. 2011). This gel structure does, however, have several drawbacks, including the possibility for

D. Yan et al., *Metakaolin-Based Geopolymers*,
https://doi.org/10.1007/978-981-97-0652-5_4

breaking and significant shrinkage after drying. The practical use of geopolymers is considerably hampered by these shortcomings. As a result, these challenges have been the subject of substantial investigation.

In earlier studies, Hardjito et al. (2004) concluded that fly ash-based geopolymer concrete exhibits extremely low drying shrinkage. However, subsequent research has shown that the drying shrinkage is dependent on the curing conditions. Wallah and Rangan (2006) found that ambiently cured fly ash-based geopolymer had significantly higher drying shrinkage than OPC. Nevertheless, heat treatment was found to significantly reduce the drying shrinkage (Khan et al. 2019). Castel et al. (2016) also found that the compositional characteristics of geopolymers, such as the NaOH concentration, liquid/solid ratio, sodium silicate-to-NaOH ratio, silicate content of the solution, and calcium content, are strongly connected to the shrinkage of geopolymers. The utilization of fibers and fillers could also alleviate the risk of cracking due to drying shrinkage (Punurai et al. 2018; Xiang et al. 2019; Si et al. 2020). Furthermore, Kuenzel et al. (2012) identified a critical residual water content for metakaolin-based geopolymer (MKG), below which the shrinkage significantly increased during drying. Additionally, they noted that this essential water content, known as "structural water content" in earlier investigations, changes depending on composition. Recent studies have shown that drying shrinkage of geopolymers is significantly influenced by the pore structure (Ma and Ye 2015; Mobili et al. 2016; Yang et al. 2017). According to theories put out by Mastali et al. (2018) and Mosale Vijayakumar (2014), strong tensile stresses within the tiny pores of geopolymers may be the root of these substantial drying shrinkages. Geopolymers' porosity and drying shrinkage behaviors can be significantly altered by varying their initial water contents, as demonstrated in previous studies (Boca Santa et al. 2018; Novais et al. 2018). Moreover, the use of other materials, such as calcium bentonite or shrinkage reducing admixtures, can also modify the pore characteristics and shrinkage behavior of geopolymers (Ling et al. 2019; Huang 2020). Although these studies have disclosed some crucial features of drying shrinkage of geopolymers, the quantitative relationship between drying shrinkage and microstructure remains undetermined. An understanding of this relationship is critical for the long-term durability design of geopolymers (Amran et al. 2021). In this chapter, we conducted both experimental and modeling investigations to reveal the underlying mechanism of drying shrinkage for MKG. Physical modeling, microstructural characterisation, and a mini-bar shrinkage test were used to quantitatively examine the drying shrinkage-water loss relations of MKGs with various starting water/solid ratios (WSRs). This study discusses the correlation between the unique microstructure of MKG and its drying shrinkage.

4.2 Experiments

4.2.1 Materials and Synthesis

To generate MKGs with different porosities and pore structures, the WSR was modified. The geopolymer binder was created using commercial metakaolin (Metamax, BASF Co., Germany) and a laboratory-prepared activator. X-ray fluorescence analysis (XRF-1800, Shimadzu, Japan) was used to ascertain the chemical make-up of the metakaolin powder, as given in Table 4.1. Scanning electron microscopy (Quanta FEG650, FEI, USA) was employed to characterize particle morphology, as displayed in Fig. 4.1. The particle size distribution was evaluated using a laser particle size analyzer (LS-230, Beckman Coulter, USA), as depicted in Fig. 4.2.

The activating solution used in this study comprised of a commercial water-glass (WG) solution obtained from Hengli Chemical Co., Ltd. (China), and sodium hydroxide pellets procured from Sinopharm Chemical Reagent Co., Ltd. (China). The WG comprised 26.0% (in weight) SiO_2 and 8.2% (in weight) Na_2O, while the sodium hydroxide pellet contained analytical reagent level chemical reagents with 98% purity. In order to produce a molar SiO_2/Na_2O ratio of 1.55, the sodium

Table 4.1 Chemical composition of the metakaolin powder (LOI: loss on ignition)

Component	Content (%)	Component	Content (%)
Al_2O_3	39.68	Fe_2O_3	0.42
SiO_2	57.26	Na_2O	0.28
TiO_2	1.78	K_2O	0.22
CaO	0.03	LOI*	0.33

Reproduced from [Relation between drying shrinkage behavior and the microstructure of metakaolin-based geopolymer] by [Shikun Chen] with permission from [Journal of Zhejiang University-SCIENCE A]

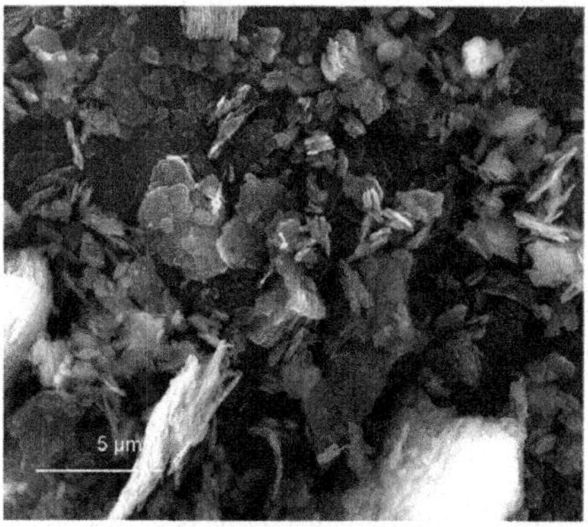

Fig. 4.1 Micromorphology of metakaolin powder as received. Reproduced from [Relation between drying shrinkage behavior and the microstructure of metakaolin-based geopolymer] by [Shikun Chen] with permission from [Journal of Zhejiang University-SCIENCE A]

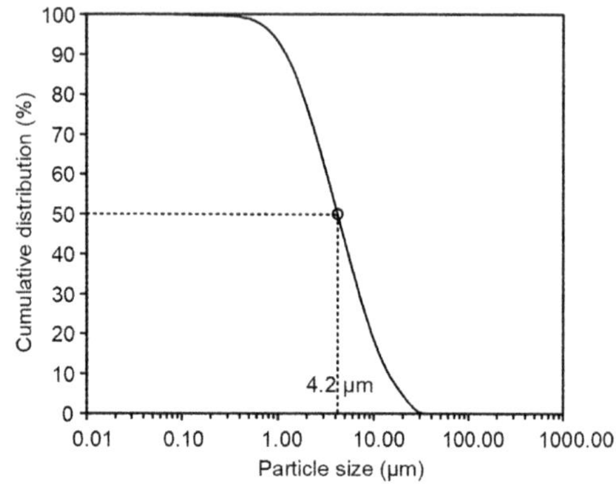

Fig. 4.2 Particle size distribution of metakaolin powder as received. Reproduced from [Relation between drying shrinkage behavior and the microstructure of metakaolin-based geopolymer] by [Shikun Chen] with permission from [Journal of Zhejiang University-SCIENCE A]

Table 4.2 Compositions of the synthesized geopolymer binders

Mixture	Metakaolin powder (g)	WG solution (g)	Additional water (g)	Si/Al (mol/mol)	Na/Al (mol/mol)	WSR
WSR65	100	155.26	8.66	2	1	0.65
WSR70	100	155.26	17.11	2	1	0.70
WSR75	100	155.26	25.05	2	1	0.75
WSR80	100	155.26	33.24	2	1	0.80

Reproduced from [Relation between drying shrinkage behavior and the microstructure of metakaolin-based geopolymer] by [Shikun Chen] with permission from [Journal of Zhejiang University-SCIENCE A]

hydroxide pellet was dissolved in the WG solution at a ratio of 0.12:1 to create the activating solution. The binder paste was created by combining the solution with metakaolin powder and more water. Based on the proportions given in Table 4.2, the WSR of various pastes varied. Once the paste was thoroughly mixed, it was cast into steel molds and sealed in plastic bags before curing in a chamber at a temperature of (20 ± 1) °C with humidity above 95% until testing.

4.2.2 *Mini-Bar Shrinkage Test*

Following 7 days of curing, the specimens were extracted from the molds and wiped with absorbent paper until achieving a surface dry state before testing. Linear shrinkage during the drying process was measured on a mini-bar specimen measuring 25 mm × 25 mm × 280 mm, as depicted in Fig. 4.3. This size is compliant with test standards JC/T 603 (NDRC, 2004) and ASTM C490 (ASTM, 2017) to minimize drying nonuniformity (with a small cross section size) and reduce relative errors (with a long measurement length).

Fig. 4.3 Mini-bar specimens. Reproduced from [Relation between drying shrinkage behavior and the microstructure of metakaolin-based geopolymer] by [Shikun Chen] with permission from [Journal of Zhejiang University-SCIENCE A]

Initially, the weight and length of specimens were measured with an electrical balance and Vernier caliper, respectively. Subsequently, the samples were sealed in plastic bags with silica gel desiccant and stored in a drying oven at (20 ± 1) °C to imitate low humidity drying conditions. The use of silica gel desiccant establishes an environment with 21–24% relative humidity (RH) upon sealing and 45–55% RH upon opening the bags. Nastic et al. (2019) employed the silica gel desiccant method to assess concrete shrinkage in low RH environments. Weight and length changes were carefully recorded after removing the samples from the bags. After measurement, the desiccant was replaced, and the specimens were resealed in plastic bags and returned to the same oven for continued curing, as depicted in Fig. 4.4.

On the 4th and 11th days, the drying temperature was increased to 60 °C and 80 °C, respectively, when weight and length changes slowed down. These increased temperatures accelerated the drying process and mimicked harsher drying conditions that would occur at high temperatures (Brue et al. 2017; Nastic et al. 2019). Due to unstable damage to the MKG, all tests stoped on the 17th day.

Linear shrinkage (ε_L) was computed based on the length change of specimens using Eq. (4.1), while the water loss (w) was evaluated from the weight change utilizing Eq. (4.2). Each group was tested with three parallel specimens.

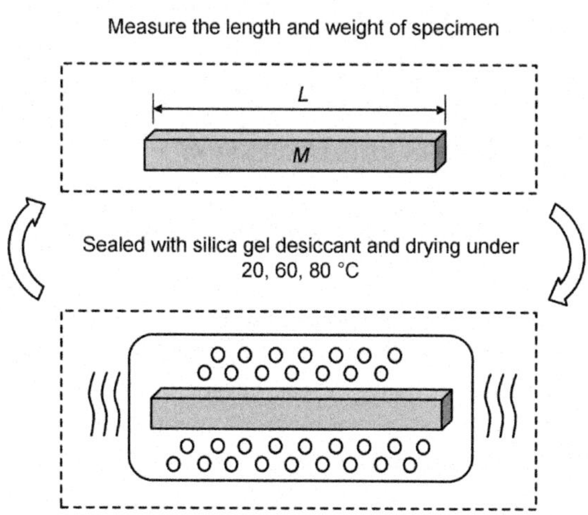

Fig. 4.4 Illustration of measurement and drying process. Reproduced from [Relation between drying shrinkage behavior and the microstructure of metakaolin-based geopolymer] by [Shikun Chen] with permission from [Journal of Zhejiang University-SCIENCE A]

$$\varepsilon_L = \frac{L - L_0}{L_0} \tag{4.1}$$

$$w = \frac{M - M_0}{V_0} \tag{4.2}$$

In the drying process, the measured length and weight of the specimen are denoted by L and M, respectively, while the initial length and weight of the sample are represented by L_0 and M_0, respectively. V_0 represents the bulk volume of the sample.

4.2.3 Scanning Electron Microscopy (SEM) Characterization

Morphological analysis of dried MKGs at various length scales was conducted using a field emission environmental scanning electron microscope (SEM) (Quanta FEG650, FEI, USA) operating at an accelerating voltage of 20 kV. Following the shrinkage test, samples were extracted from the tested mini-bars and manually split with a sharp knife. To retain crucial details, SEM investigation of naturally formed fracture surfaces was executed instead of polishing.

4.2.4 Mercury Intrusion Porosimetry (MIP) Characterization

Using a Micrometrics AutoPore IV 9510 equipment with a pressure range of 3.45 103–4.14 108 Pa, the mercury intrusion porosimetry (MIP) technique was used to examine the pore structure of the binder. Pores larger than 3 nm can be included in the porosity measurement using this equipment. MIP is recognized as a convenient and effective technique that provides relatively reliable information about pore structures in cementitious materials (Gallé 2001; Ma 2014), albeit its application remains a topic of debate (Diamond 2000).

4.3 Results and Discussion

4.3.1 Water Loss and Drying Shrinkage

As the solute in the pore solution of hardened MKGs is not evaporable and carbonation is restricted by plastic bag sealing, the primary factor contributing to specimen weight change is water evaporation from the contained pore solution. Consequently, water loss was regarded as equivalent to the weight change of the specimens. Figure 4.5 depicts changes in water loss and drying shrinkage relative to drying time.

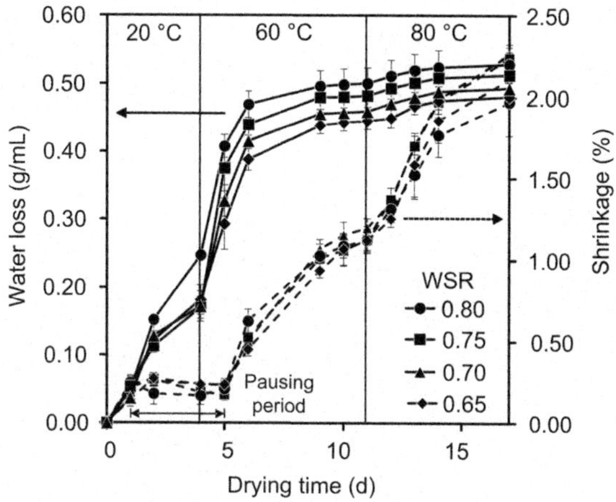

Fig. 4.5 Water loss (solid lines) and drying shrinkage (dashed lines) variation versus drying time. Reproduced from [Relation between drying shrinkage behavior and the microstructure of metakaolin-based geopolymer] by [Shikun Chen] with permission from [Journal of Zhejiang University-SCIENCE A]

The drying shrinkage and water loss exhibited varying trends throughout the three consecutive drying stages.

Step 1 involved drying at 20 °C. Water loss gradually increased from 0 to 0.17–0.25 g/mL (depending on WSR) during this phase. The average loss rate was roughly 0.043–0.063 g/(mL d). Drying shrinkage increased to 0.18–0.21% within the first day or two before entering a pausing period with minimal fluctuation for the rest of the stage.

In Step 2, specimens were dried at 60 °C. With the increase in drying temperature, water loss rates rose significantly. The average loss rate increased to approximately 0.105–0.135 g/(mL d) following the temperature elevation and then gradually declined to 0.0032–0.0063 g/(mL d) over the next seven days of drying. The total water loss during this phase ranged in 0.25–0.31 g/mL. Despite this, drying shrinkages remained in a pause phase during the first day of this step. Thereafter, they began to rise again at rates of 0.127–0.141% per day. Shrinkage increased to 1.12–1.20% by the end of this step.

Step 3 entailed drying at 80 °C. Although the temperature was elevated to 80 °C, there was no significant water loss during this drying phase (less than 0.05 g/mL). Nonetheless, drying shrinkages increased significantly (with average rates of 0.13–0.19% daily) during this step. The drying shrinkages reached around 1.97–2.26% at the end of the test.

The test data was rearranged to depict the changes in drying shrinkage relative to water loss, as displayed in Fig. 4.6. An critical water loss threshold (approximately 0.29–0.41 g/mL, contingent on the WSR) was detected.

The shrinkage response can be divided into two stages as follows according to our experimental results:

(1) Stage I: This stage begins with water loss below the critical water loss value, during which drying shrinkage increases and enters a pausing period (typically

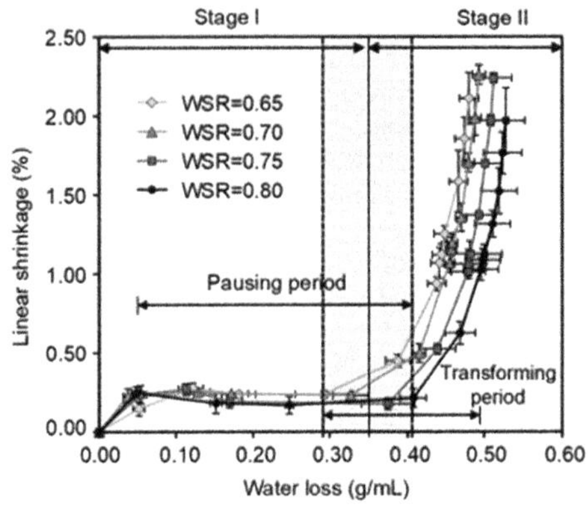

Fig. 4.6 Drying shrinkage versus water loss relation. Reproduced from [Relation between drying shrinkage behavior and the microstructure of metakaolin-based geopolymer] by [Shikun Chen] with permission from [Journal of Zhejiang University-SCIENCE A]

ranging from 0.05 to 0.40 g/mL). The critical value and the duration of the pausing period are roughly proportional to the initial WSR.

(2) Stage II: This stage is characterized by water loss above the critical water loss value. When the water loss surpasses the threshold amount, the drying shrinkages exhibit a significant rise after a changing period (about 0.29–0.50 g/mL). Interestingly, the shrinkage curves for different mixtures exhibit similar trends once water loss surpasses the critical value. Moreover, the drying shrinkages of all mixtures demonstrate a linear relationship with water loss beyond the critical value.

The results of our study demonstrate a consistent trend in the drying shrinkage of MKG with the findings reported by Kuenzel et al. (2012). Specifically, they also observed two different stages of drying shrinkage in MKG before and after reaching a critical state. However, we note that the values of shrinkage measured prior to the critical state in our experiment (0.17–0.27%) were significantly higher than those reported in the earlier study (approximately 0.04–0.11%) by Kuenzel et al. (2012), which may be attributed to different reference times during the drying shrinkage test (i.e., the 7th day in our experiment and the 56th day in their previous experiments). The higher reported values resulted from measurements of early-stage shrinkage made possible by our earlier reference time. We found an interesting relationship between the critical water loss and the length of the stopping time, which suggests that MKG with higher WSR may have more stability in terms of drying shrinkage under variable humidity circumstances. The underlying mechanisms of these behaviors will be further elucidated in subsequent sections.

4.3.2 Pore Systems in MKG

Consistent with prior research (Kriven et al. 2003; Kong et al. 2007; Maitland et al. 2011), the majority of pores in geopolymers were found to be on the microscale

Fig. 4.7 Multi-scale pore systems in MKG: **a** level I, meso pores; **b** level II, micro pores; **c** level III, nanopores. Reproduced from [Relation between drying shrinkage behavior and the microstructure of metakaolin-based geopolymer] by [Shikun Chen] with permission from [Journal of Zhejiang University-SCIENCE A]

(< 5 μm), which was confirmed by our SEM and MIP data. Our SEM analysis revealed that all samples exhibited similar pore systems comprising pores at various scales. Figure 4.7 illustrates a representative set of results obtained from the WSR70 sample.

Figure 4.7 illustrates a three-leveled hierarchical pore system observed in the binder, which can be described as follows:

Level I—meso pores (void size > 5 μm): The binder at the sub-millimeter scale appeared to be homogeneous but contained air bubbles and meso cracks, which are often introduced during mixing or due to fractures. Their presence was found to be random with a low volume fraction (< 1.6%) that was considered to have little impact on binder shrinkage.

Level II—micro pores (void size between 5 nm and 5 μm): The matrix at the micrometer to sub-micrometer scale exhibited a highly porous nature due to the clustered aluminosilicate gel. According to MIP data in Fig. 4.8a, these granular clusters developed a sizeable volume percentage of interstitial micro pores and micro fractures, leading to high interconnected porosity (33.29–43.34% in our trials). Both Duxson et al. (2005) and Maitland et al. (2011) found similar findings.

Level III—nanopores (void size < 5 nm): High-resolution SEM images in Fig. 4.7c revealed that the gel clusters consist of nano-sized globules producing interconnected nanopores. Prior studies using TEM (Kriven et al. 2003) and molecular dynamics simulations (Sadat et al. 2016) have reported similar findings. These networked cavities within gel clusters may contain chemically bound water in the form of silanol or aluminol groups (Duxson et al. 2005; Zheng et al. 2010). However, these pore sizes typically fall outside the detection range of MIP and are therefore not commonly acknowledged in other studies, potentially leading to underestimation of their contributions to total porosity.

Upon closer examination of the level II micro pores in MKGs with varying WSRs (Fig. 4.9), we observed a slight shift in characteristic pore size and porosity as the WSR increased, which was confirmed by MIP results. As depicted in Fig. 4.10, the characteristic pore size increased from 11.05 to 21.09 nm with an increase in

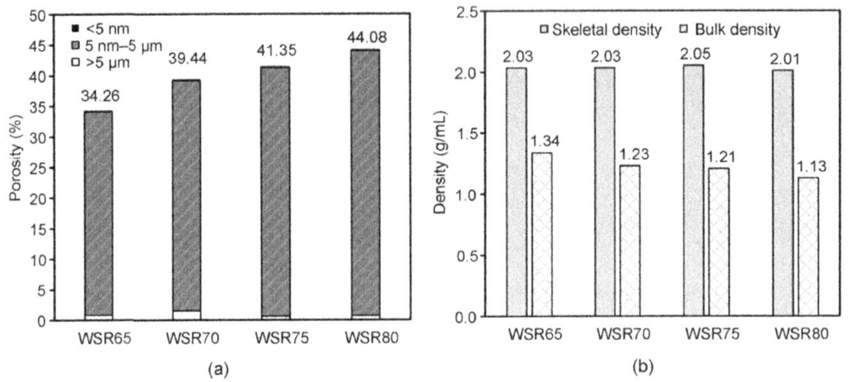

Fig. 4.8 Total porosity (**a**) and density (**b**) measured by MIP. Reproduced from [Relation between drying shrinkage behavior and the microstructure of metakaolin-based geopolymer] by [Shikun Chen] with permission from [Journal of Zhejiang University-SCIENCE A]

WSR from 0.65 to 0.80. Correspondingly, the total porosity measured by MIP also increased from 34.26 to 44.08% (Fig. 4.8a). This increase in porosity was primarily attributed to level II micro pores (5 nm to 5 μm), resulting in a significant decrease in bulk densities (Fig. 4.8b). Despite fluctuations in WSR, skeletal densities computed by deducting the micro and meso porosities determined by MIP were mostly consistent (between 2.01 and 2.05 g/mL), demonstrating that the nano-sized pore structure was independent of initial WSR.

The observed nano-sized pores were found to be closely associated with the polycondensation of aluminosilicate networks on a molecular scale, as reported in previous studies (Sadat et al. 2016). This observation is in agreement with findings reported by Kuenzel et al. (2012), who suggested that "structural water" (water contained within nanopores) is an intrinsic property primarily determined by the chemical compositions of aluminosilicate networks.

4.3.3 Modeling Drying Shrinkage

We have identified three primary factors driving the observed shrinkage behavior, which are detailed below. We also provide a quantitative expression of their relationship with volumetric shrinkage strain.

1. Capillary pressure-induced shrinkage

The first driving force contributing to shrinkage is capillary pressure resulting from the loss of water from micro pores. As water within MKG progressively evaporates during drying, air infiltrates the pore space within the sample. The interface between the air and pore water forms a curved liquid–gas meniscus, which generates internal

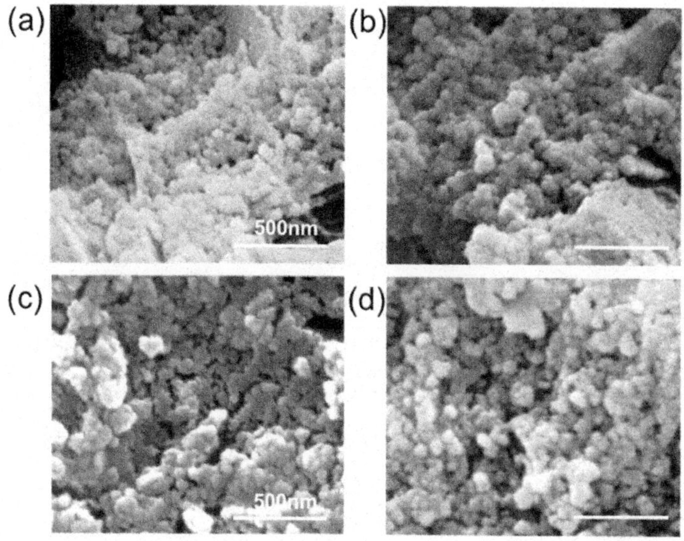

Fig. 4.9 Micro-morphology of WSR65 (**a**), WSR70 (**b**), WSR75 (**c**), and WSR80 (**d**) specimens. Reproduced from [Relation between drying shrinkage behavior and the microstructure of metakaolin-based geopolymer] by [Shikun Chen] with permission from [Journal of Zhejiang University-SCIENCE A]

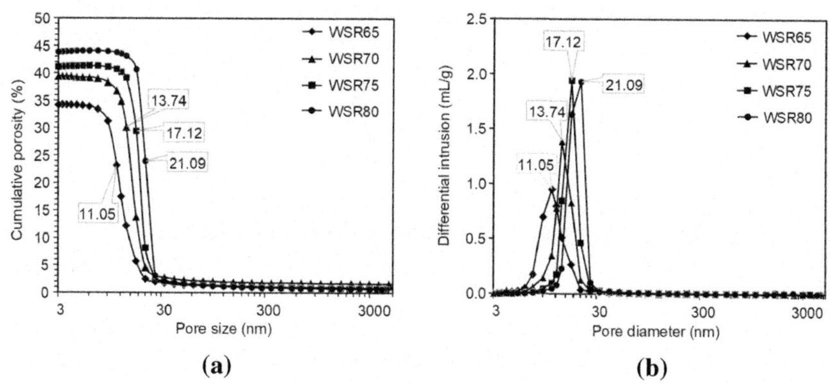

Fig. 4.10 Cumulative (**a**) and differential (**b**) pore size distribution measured by MIP. Reproduced from [Relation between drying shrinkage behavior and the microstructure of metakaolin-based geopolymer] by [Shikun Chen] with permission from [Journal of Zhejiang University-SCIENCE A]

pressure between the two phases due to surface tension along the interface (de Gennes et al. 2004). This internal pressure, also known as Laplace or capillary pressure, is a primary factor leading to MKG shrinkage.

As seen in Fig. 4.11b, the random level II micro pore systems seen in geopolymers (Fig. 4.11a) may be roughly represented as a network of cylindrical pores. Due to

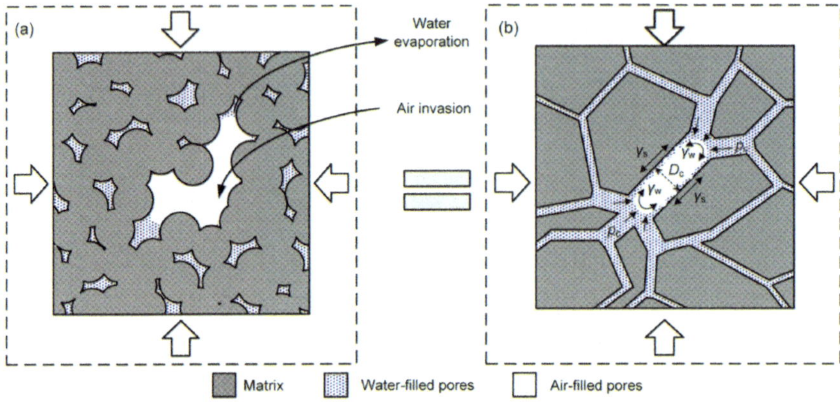

Fig. 4.11 Simplified model for the shrinkage mechanism during loss of micro pore water: **a** random pores system; **b** equivalent cylindrical pore system. Reproduced from [Relation between drying shrinkage behavior and the microstructure of metakaolin-based geopolymer] by [Shikun Chen] with permission from [Journal of Zhejiang University-SCIENCE A]

its low volume fraction (1.6%) and little effect on capillary pressure (0.1 MPa), the level I meso porous system (> 5 m) is ignored. As water is gradually substituted by air flowing from larger to smaller pores, the pore space is partitioned into air- and water-filled components.

The capillary pressure caused by the formation of menisci inside water-filled pores can be calculated using the Young–Laplace equation (1832; Wittmann 1973):

$$P_c = P_a - P_l = -\gamma\left(\frac{1}{R_1} + \frac{1}{R_2}\right) = -\frac{2\gamma\cos\theta}{R_c}, \tag{4.3}$$

where P_c is the Laplace pressure or capillary pressure, P_a is the gas pressure, P_l is the liquid pressure, γ is the surface tension of the liquid, R_1 and R_2 are the curves radii of two perpendicular directions, R_c is the Kelvin radius, and ψ is the contact angle.

The calculation of Kelvin radius is conventionally executed through utilization of the Kelvin equation, as demonstrated by Pinson et al. (2015):

$$R_c = -\frac{2\gamma_w a^3}{kT\ln(h)}, \tag{4.4}$$

where γ_w is the surface tension of water, a^3 is the characteristic volume of a water molecule in the liquid state, k is the Boltzmann constant, T is the absolute temperature, and h is the RH.

In this chapter, the Kelvin radius was calculated by employing data obtained from the pore structure within the MKG binder. As shown in Fig. 4.12, our research first focused on the cumulative pore volume distribution (CPVD), which was defined by

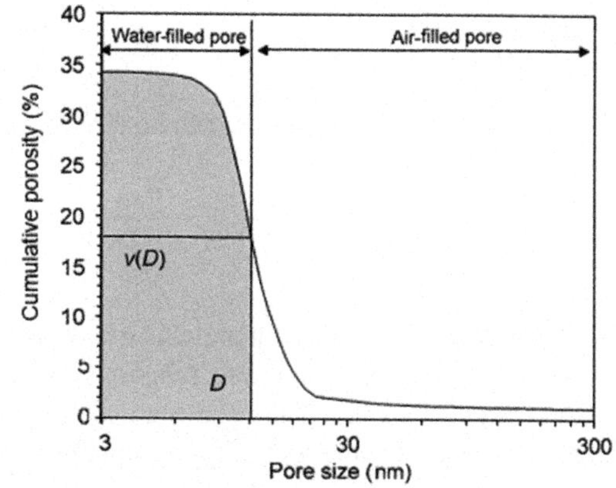

Fig. 4.12 Illustration of cumulative volume distribution (CPVD) function. Reproduced from [Relation between drying shrinkage behavior and the microstructure of metakaolin-based geopolymer] by [Shikun Chen] with permission from [Journal of Zhejiang University-SCIENCE A]

the relationship between cumulative porosity and pore diameters,

$$v(D) = \frac{V_D}{V_0}, \tag{4.5}$$

where D is a given pore diameter, V_D is the cumulative volume of pores with a diameter larger than D.

The process of pore water evaporation initiates from larger pores and subsequently proceeds towards smaller ones. Accordingly, the loss of water can be associated with the volume of air-filled pores via application of Eq. (4.6) whereby "air-filled" denotes a state lacking bulk water whilst encompassing a thin layer of adsorbed water on the surface of the pore wall,

$$w = \rho_w v(D_c), \tag{4.6}$$

where ρ_w is the density of water, and D_c is the corresponding minimum micro pore size that is currently filled with air.

Inversely, we could also calculate the currently air-filled pore size, D, from Eq. (4.6):

$$D_c = v^{-1}\left(\frac{w}{\rho_w}\right), \tag{4.7}$$

where $v^{-1}()$ is the inverse function of the CPVD $v()$, which can be linearly interpolated from the CPVD curves.

On the assumption that the capillary menisci exhibit a contact angle of zero degrees in relation to the pore wall as a consequence of the existence of an adsorbed water layer, the Kelvin radius can be defined as half of the minimum diameter of air-filled pores:

$$R_c = \frac{D_c}{2}. \tag{4.8}$$

Then, the capillary pressure can be directly calculated from the water loss as

$$p_c = -\frac{4\gamma_w}{D_c} = -\frac{4\gamma_w}{v^{-1}\left(\frac{w}{\rho_w}\right)}. \tag{4.9}$$

As indicated by prior theoretical investigations (MacKenzie 1950; Thokchom et al. 2010), capillary pressure functions as an effective stress within a partially saturated porous solid. The resulting deformation arising from this effective stress can be computed using the following equation:

$$\varepsilon_{V,c} = S_c p_c \left(\frac{1}{K_b} - \frac{1}{K_s}\right), \tag{4.10}$$

where K_b is the bulk modulus of a porous solid, K_s is the bulk modulus of the solid component, and $\varepsilon_{V,c}$ is the volumetric strain caused by the capillary pressure.

S_c is the saturation (the water-filled pore volume fraction), which can be calculated as

$$S_c = \frac{\phi_c - \frac{w_c}{\rho_w}}{\phi_c}, \tag{4.11}$$

where w_c is the water loss of capillary pore, and ϕ_c is the capillary porosity, which is the sum of the level I and level II porosities.

Because K_s is usually much greater than K_b for a medium with a large porosity, Eq. (4.10) can be simplified as:

$$\varepsilon_{V,c} = \frac{S_c p_c}{K_b}. \tag{4.12}$$

2. Surface energy change-induced shrinkage

Alongside the deformations resulting from capillary pressure, the absence of water within micro pores affects the surface energy of gel clusters. This alteration is tantamount to an elevation in surface tension that acts upon the micro pore walls, hence causing drying shrinkage (Ye and Radlińska 2016). The variation in surface energy ($\Delta\Pi$) can be evaluated by considering the alterations in surface area of air-filled pores alongside the surface tension exhibited by adsorbed water layers as expounded in existing literature (Bangham and Razouk 1937; Hansen 1987; Pinson et al. 2015):

$$\Delta\Pi = \kappa_s \gamma_s - \kappa_{s,0}\gamma_{s,0}, \tag{4.13}$$

where κ_s is the specific pore surface area (sum of the pore surface area divided by the sample volume) that is currently filled by air, $\kappa_{s,0}$ are the corresponding values of the saturated state and therefore are considered to be zero (Pinson et al. 2015), γ_s is the surface tension of the water layer at the partially covered state, and $\gamma_{s,0}$ is the surface tension at the fully covered state.

The cumulative pore surface area distribution (CPSD), s(D), may be used to directly calculate the particular pore surface area, κ_s:

$$\kappa_{s,c} = \frac{s(D)}{V_0}, \tag{4.14}$$

The value of s(D) is determined by calculating the differential relation between the volume and side surface area of a cylindrical pore based on the CPVD v(D).

$$\frac{\mathrm{d}s(D)}{\mathrm{d}D} = \frac{4}{D}\frac{\mathrm{d}v(D)}{\mathrm{d}D}. \tag{4.15}$$

The surface tension of the adsorbed water layer, γ_s, can be calculated by the Gibbs equation (Feldman and Sereda 1964; Pinson et al. 2015):

$$\gamma_s = \gamma_{s,0} - \frac{kT}{a^2}\int_{h_0}^{h} \theta \frac{\mathrm{d}h}{h}, \tag{4.16}$$

where h_0 is the relative humidity of the fully covered condition (100% RH), θ is the surface coverage, and an is the typical length of a water molecule.

The Kelvin equation and the current empty pore diameter may be used to inversely determine the current relative humidity:

$$\ln(h) = -\frac{4\gamma_w a^3}{kTD_c} \tag{4.17}$$

Pinson et al. (2015) have presented evidence to support the claim that surface tension in a fully covered state is equivalent to the surface tension between air and bulk water; namely, $\gamma_{s,0} = \gamma_w$. Furthermore, the coverage of surfaces at specific relative humidity (RH) values can be computed via utilization of the Langmuir equation as comprehensively outlined by Langmuir (1918):

$$\theta = \frac{\alpha h}{1 + \alpha h}, \tag{4.18}$$

where α is a constant associated with the energy of adsorption, which is taken to be 65 in accordance with the commonly used number in the Brunauer–Emmett–Teller (BET) method. At a RH of 11%, this value equates to the existence of a single monolayer of water adsorbing on the pore surface (Hagymassy et al. 1969).

The Bangham equation, as proposed by Bangham and Razouk (1937) and subsequently elaborated by Pinson et al. (2015), provides a comprehensive account of the deformation caused by shrinkage that arises from alterations in the surface energy of a solid:

$$\varepsilon_{V,s} = -\frac{\Delta\Pi}{K_b(1 - 2\mu)}, \tag{4.19}$$

where $\varepsilon_{V,s}$ is the volumetric strain due to the change in surface energy, and μ is Poisson's ratio of the medium.

3. Gel densification-induced shrinkage

During the process of micro pore water drainage, the water that was deeply trapped within level III nanopores eventually moves towards the surface and undergoes evaporation. This fraction of nanopore water has been commonly designated as "structural water" in prior research (van Jaarsveld et al. 2002; Rahier et al. 2007; Kuenzel et al. 2012). The quantity of lost nanopore water (w_s) can be computed using the following equation:

$$w_s = w - \phi_c \rho_w, \tag{4.20}$$

The mechanism of deformation within a gel network can be represented as the compaction of inter-connected molecular spheres, as illustrated in Fig. 4.13. In our study, liquid spheres (depicted as light gray spheres in Fig. 4.13b) denote hydroxyls and confined water contained within nanopores, which were removed from the networks. Conversely, solid spheres (dark gray spheres in Fig. 4.13b) represent aluminosilicate backbones that undergo collapse during densification. Due to differences in neighborhood connections, not all voids left by the removal of liquid spheres are closed (as demonstrated in Fig. 4.13b). The decrease in volume of the gel network was considered to be proportional to the eliminated liquid volume. Consequently, volumetric strain stemming from loss of nanopore water can be expressed as follows:

$$\varepsilon_{V,g} = \frac{\Delta V_g}{V_0} = \frac{f \Delta V_w}{V_0} = f(w_s)\frac{w_s}{\rho_w}, \tag{4.21}$$

where ΔV_g is the volume change of the gel owing to the loss of nanopore water, $f()$ is the fraction function, and ΔV_w is the volume of the lost nanopore water.

The form and value of $f(w_s)$ could not be mathematically predicted because to the dearth of nanoscale data. This allowed for the creation of an empirical form by fitting the experimental data:

$$f(w_s) = 1 - \frac{w^*}{w_s}\operatorname{atan}\left(\frac{w_s}{w^*}\right), \tag{4.22}$$

where parameter w^* is related to the span of the transforming period and its influence on function shape is demonstrated in Fig. 4.14.

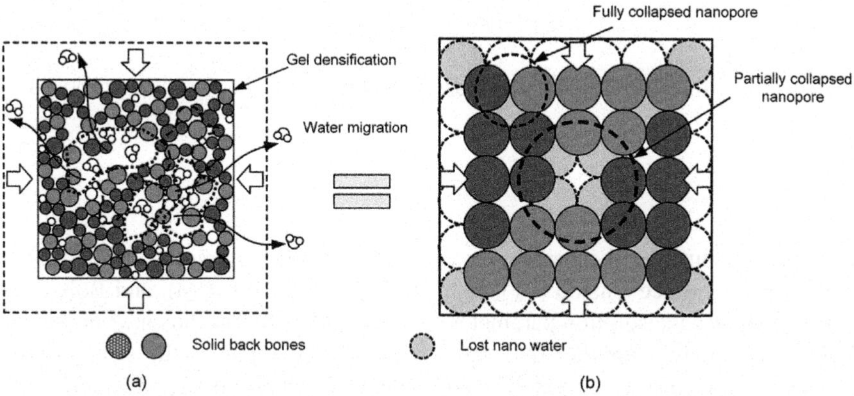

Fig. 4.13 Simplified model for the shrinkage mechanism during loss of nanopore water: **a** molecular network system; **b** equivalent sphere system. Reproduced from [Relation between drying shrinkage behavior and the microstructure of metakaolin-based geopolymer] by [Shikun Chen] with permission from [Journal of Zhejiang University-SCIENCE A]

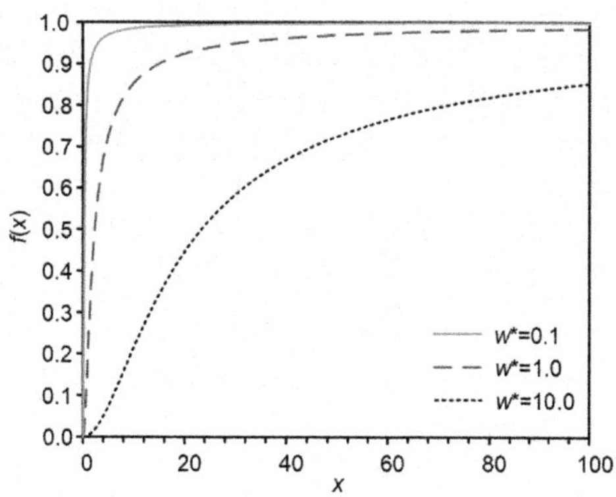

Fig. 4.14 One parameter fraction function with varied parameters. Reproduced from [Relation between drying shrinkage behavior and the microstructure of metakaolin-based geopolymer] by [Shikun Chen] with permission from [Journal of Zhejiang University-SCIENCE A]

Upon substituting Eq. (4.22) into Eq. (4.21), the shrinkage due to the nanopore water loss is:

$$\varepsilon_{V,g} = \frac{w_s}{\rho_w} - \frac{w^*}{\rho_w}\mathrm{atan}\left(\frac{w_s}{w^*}\right). \tag{4.23}$$

4.3.4 Comparisons and Discussion

Through integration of the shrinkages attributable to all three driving forces, including (1) capillary pressure, (2) surface energy alterations, and (3) gel densification, the complete volumetric shrinkage was determined as:

$$\varepsilon_V = \varepsilon_{V,c} + \varepsilon_{V,g} + \varepsilon_{V,s}. \quad (4.24)$$

Moreover, the linear shrinkage was evaluated using the formula proposed by Hobbs (1978):

$$\varepsilon_L = \frac{\varepsilon_V}{3}. \quad (4.25)$$

Table 4.3 outlines the parameters employed within our model, including the density (ρ_w), surface tension (γ_w), characteristic size of water (a), and Boltzmann constant (k). The adsorption parameter (α) was obtained from Pinson et al. (2015). The initial relative humidity (h_0) was assumed to be 100% as the samples were sealed and cured in a high-humidity environment following casting. A temperature of 20 °C (293.15 K) was utilized for estimation purposes to simplify calculations. While higher temperatures have been found to accelerate the drying process and intensify water loss during experiments, the impact of temperature is negligible if we solely consider the time-invariant relationship between drying shrinkage and water loss at a specific level of water loss. Calculated variations in surface tension and relative humidity corresponding to a given level of water loss display minimal fluctuations when temperature is increased from 20 to 80 °C. Such discrepancies only lead to a relative alteration in shrinkage of less than 0.04%, thus supporting the selection of simplified temperature. Poisson's ratio and bulk modulus used in estimation were determined through previous experimentation (Chen 2015). The parameter for gel densification was obtained using least square fitting and all coefficient determinations of the fitting curves exceeded 0.95.

Table 4.3 Parameters used to estimate the shrinkage

Parameter	Value			
	WSR65	WSR70	WSR75	WSR80
Bulk modulus, K_b (GPa)	3.5	2.8	2.2	2.0
Gel densification parameter, w^* (g/mL)	0.110	0.055	0.050	0.045
Coefficient of determination, R^2	0.965	0.977	0.967	0.986
Water density, ρ_w (g/mL)		1.0		
Water surface tension, γ_w (mN/m)		72.8		
Characteristic size of molecular water, a (nm)		0.278		
Boltzmann constant, k (J/K)		1.38×10^{-23}		
Adsorption parameter, α		65		
Initial RH, h_0 (%)		100		
Absolute temperature, T (K)		293.15		
Poisson's ratio, μ		0.3		

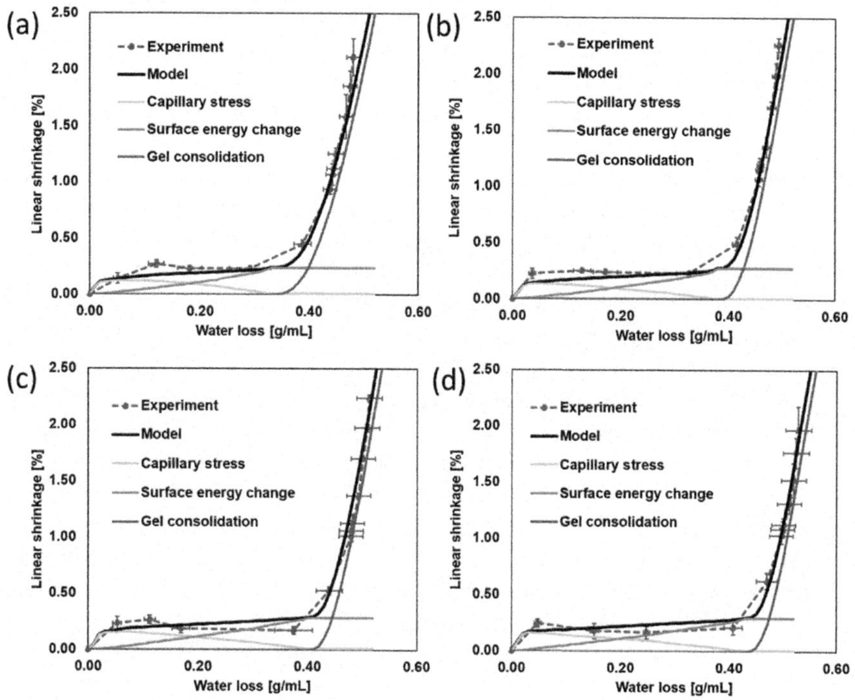

Fig. 4.15 Model estimation versus experiment results: **a** WSR65; **b** WSR70; **c** WSR75; **d** WSR80. Reproduced from [Relation between drying shrinkage behavior and the microstructure of metakaolin-based geopolymer] by [Shikun Chen] with permission from [Journal of Zhejiang University-SCIENCE A]

Figure 4.15 presents a comparison between the outcomes of our modeling and experimental results.

The overall trend of the predicted drying shrinkage versus water loss aligned with experimental data. Several key observations were made as follows:

(1) Capillary stress was the primary factor contributing to shrinkage during the early part of Stage I, with maximum impact occurring at a water loss of approximately 0.03–0.06 g/mL. As water within Level II micro pores gradually evaporated and pores became filled with air, the capillary stress effect began to decay. This pattern is consistent with previous predictions on C-S-H gel drying shrinkage (Pinson et al. 2015). However, this decay has yet to be experimentally confirmed in geopolymer materials and warrants further investigation.

(2) Toward the end of Stage I, shrinkage brought on by surface energy shift started to take control. Both the surface energy change-dominated and capillary stress-dominated regimes saw identical shrinkage values as a result of the pore size distribution of the tested materials and the test circumstances (h = 45%–55%, T = 20–60 °C). When the primary mechanism switched from the former to the

latter, there was an apparent halting period in the shrinkage-water loss curves. The length of the halting period correlated positively with the initial WSR and was proportional to micro porosity.

(3) Porosity and pore size distribution jointly controlled drying shrinkage during Stage I. For instance, because characteristic pore size in WSR65 was smaller than that of other mixtures (11.05 nm vs. 13.74–21.09 nm), capillary pressure was more intense for this mixture. However, lower porosity and larger bulk modulus of this sample mitigated high capillary pressure, resulting in similar shrinkage compared to other three mixtures. These results indicate that controlling drying shrinkage damage to MKG requires simultaneous optimization of both porosity and pore size distribution.

(4) During Stage II, drying shrinkage was mostly caused by gel densification. In Stage II, water loss started, and gel densification-induced shrinking accelerated. Gel densification-induced volumetric shrinkage was equal to the volume of lost nanopore water after a brief transition period (within a water loss of around w*). The resulting gel densification-induced shrinkage was 7–10 times that of capillary and surface shrinkage at a water loss of 0.5 g/mL. To avoid drying shrinkage damage to MKG, nanopore water loss must be prevented, and water loss should not go over a threshold level (about 0.29–0.41 g/mL for mixes with various WSRs, as shown in this chapter).

4.4 Conclusions

Based on the combined experimental and modeling study, the following conclusions were drawn:

(1) The two-stage drying shrinkage behavior of MKG binder is attributed to its three-leveled pore system. The effect of initial WSR on pore structure indirectly influences the shrinkage behavior of MKG.

(2) In Stage I, the major driving force switches from capillary stress to surface energy change, and shrinkage is regulated by the loss of micropore water. During this transition, drying shrinkage is regulated by both porosity and pore size distributions.

(3) In Stage II, loss of nanopore water and gel densification become the dominant factors, leading to significant volume reduction (up to 7–10 times the shrinkage during Stage I). To avoid this in applications, water loss must remain below the critical value, which also depends on micro porosity and initial WSR. Overall, the drying shrinkage problem of geopolymers is complex and multiscale in nature, and further research is needed to fully comprehend the basic mechanisms, particularly those on the nanoscale.

In addition to the aforementioned conclusions, it should be noted that the drying shrinkage problem of geopolymers is a complex and multi-scale phenomenon. Further elucidation is required for many shrinkage mechanisms, especially those on the nanoscale. Therefore, additional studies are necessary to fully comprehend these basic mechanisms.

References

Amran M, Debbarma S, Ozbakkaloglu T (2021) Fly ash-based eco-friendly geopolymer concrete: a critical review of the long-term durability properties. Constr Build Mater 270:121857. https://doi.org/10.1016/j.conbuildmat.2020.121857

Antunes Boca Santa RA, Soares C, Riella HG (2017) Geopolymers obtained from bottom ash as source of aluminosilicate cured at room temperature. Constr Build Mater 157:459–466. https://doi.org/10.1016/j.conbuildmat.2017.09.111

Bangham DH, Razouk RI (1937) The wetting of charcoal and the nature of the adsorbed phase formed from saturated vapours. Trans Faraday Soc 33:1463–1472. https://doi.org/10.1039/tf9373301463

Boca Santa RAA, Kessler JC, Soares C, Riella HG (2018) Microstructural evaluation of initial dissolution of aluminosilicate particles and formation of geopolymer material. Particuology 41:101–111. https://doi.org/10.1016/j.partic.2017.12.007

Brue FNG, Davy CA, Burlion N et al (2017) Five year drying of high performance concretes: effect of temperature and cement-type on shrinkage. Cem Concr Res 99:70–85. https://doi.org/10.1016/j.cemconres.2017.04.017

Castel et al. (2016) Creep and drying shrinkage of a blended slag and low calcium fly ash geopolymer Concrete Abstract Materials and Structures 49(5) 1619–1628. https://doi.org/10.1617/s11527-015-0599-1

Chen S (2015) Research on the mechanical properties and influencing factors of metakaolin base polymer foundation. Zhejiang University

Collins FG, Turner LK (2013) Carbon dioxide equivalent (CO_2-e) emissions: a comparison between geopolymer and OPC cement concrete. Constr Build Mater 43:125–130. https://doi.org/10.1016/J.CONBUILDMAT.2013.01.023

Davidovits J (1991) Geopolymers. J Therm Anal Calorim 37:1633–1656

Davidovits J (1994) Properties of geopolymer cements. In: First international conference on alkaline cements and concretes. Kiev State Technical University, Scientific Research Institute on Binders and Materials, Ukraine, pp 131–149

de Gennes P-G, Brochard-Wyart F, Quéré D (2004) Capillarity and wetting phenomena. Capillarity and wetting phenomena. Springer, New York, NY, pp 1–31

Diamond S (2000) Mercury porosimetry. An inappropriate method for the measurement of pore size distributions in cement-based materials. Cem Concr Res 30:1517–1525. https://doi.org/10.1016/S0008-8846(00)00370-7

Duxson P, Provis JL, Lukey GC et al (2005) Understanding the relationship between geopolymer composition, microstructure and mechanical properties. Colloids Surfaces A Physicochem Eng Asp 269:47–58. https://doi.org/10.1016/j.colsurfa.2005.06.060

Duxson P, Provis JL, Lukey GC, van Deventer JSJ (2007) The role of inorganic polymer technology in the development of "green concrete." Cem Concr Res 37:1590–1597. https://doi.org/10.1016/j.cemconres.2007.08.018

Feldman RF, Sereda PJ (1964) Sorption of water on compacts of bottle-hydrated cement: I. The sorption and length-change isotherms. J Appl Chem 14:87–93. https://doi.org/10.1002/jctb.5010140206

Gallé C (2001) Effect of drying on cement-based materials pore structure as identified by mercury intrusion porosimetry. Cem Concr Res 31:1467–1477. https://doi.org/10.1016/S0008-8846(01)00594-4

Hardjito D et al. (2004) On the development of fly ash-based geopolymer concrete. ACI Mat J 101(6):467–472
Habert G, Ouellet-Plamondon C (2016) Recent update on the environmental impact of geopolymers. RILEM Tech Lett 1:17. https://doi.org/10.21809/rilemtechlett.v1.6
Hagymassy J, Brunauer S, Mikhail RS (1969) Pore structure analysis by water vapor adsorption: I. t-Curves for water vapor. J Colloid Interface Sci 29:485–491. https://doi.org/10.1016/0021-9797(69)90132-5
Hansen W (1987) Drying shrinkage mechanisms in Portland cement paste. J Am Ceram Soc 70:323–328. https://doi.org/10.1111/j.1151-2916.1987.tb05002.x
Huang Y (2020) Influence of calcium bentonite addition on the compressive strength, efflorescence extent and drying shrinkage of fly-ash based geopolymer mortar. Trans Indian Ceram Soc 79:77–82. https://doi.org/10.1080/0371750X.2020.1719206
Hobbs DW (1978) Expansion and shrinkage of oversulphated portland cements. Cement Concrete Res 8(2): 211–222
Khan I, Xu T, Castel A et al (2019) Risk of early age cracking in geopolymer concrete due to restrained shrinkage. Constr Build Mater 229:116840. https://doi.org/10.1016/j.conbuildmat.2019.116840
Kong DLY, Sanjayan JG, Sagoe-Crentsil K (2007) Comparative performance of geopolymers made with metakaolin and fly ash after exposure to elevated temperatures. Cem Concr Res 37:1583–1589. https://doi.org/10.1016/j.cemconres.2007.08.021
Kriven WM, Bell JL, Gordon M (2003) Microstructure and microchemistry of fully-reacted geopolymers and geopolymer matrix composites. Ceram Trans 153:227–250. https://doi.org/10.1002/9781118406892.ch15
Kuenzel C, Vandeperre LJ, Donatello S et al (2012) Ambient temperature drying shrinkage and cracking in metakaolin-based geopolymers. J Am Ceram Soc 95:3270–3277. https://doi.org/10.1111/j.1551-2916.2012.05380.x
Langmuir I (1918) The adsorption of gases on plane surfaces of glass, mica and platinum. J Am Chem Soc 40:1361–1403. https://doi.org/10.1021/ja02242a004
Li Z, Zhang S, Zuo Y et al (2019) Chemical deformation of metakaolin based geopolymer. Cem Concr Res 120:108–118. https://doi.org/10.1016/j.cemconres.2019.03.017
Ling Y, Wang K, Fu C (2019) Shrinkage behavior of fly ash based geopolymer pastes with and without shrinkage reducing admixture. Cem Concr Compos 98:74–82. https://doi.org/10.1016/j.cemconcomp.2019.02.007
Ma H (2014) Mercury intrusion porosimetry in concrete technology: tips in measurement, pore structure parameter acquisition and application. J porous Mater 21:207–215
Ma Y, Ye G (2015) The shrinkage of alkali activated fly ash. Cem Concr Res 68:75–82. https://doi.org/10.1016/j.cemconres.2014.10.024
MacKenzie JK (1950) The elastic constants of a solid containing spherical holes. Proc Phys Soc Sect B 63:2–11. https://doi.org/10.1088/0370-1301/63/1/302
Maitland CF, Buckley CE, Oconnor BH et al (2011) Characterization of the pore structure of metakaolin-derived geopolymers by neutron scattering and electron microscopy. J Appl Crystallogr 44:697–707. https://doi.org/10.1107/S0021889811021078
Mastali M, Kinnunen P, Dalvand A et al (2018) Drying shrinkage in alkali-activated binders—a critical review. Constr Build Mater 190:533–550. https://doi.org/10.1016/j.conbuildmat.2018.09.125
Mobili A, Belli A, Giosuè C et al (2016) Metakaolin and fly ash alkali-activated mortars compared with cementitious mortars at the same strength class. Cem Concr Res 88:198–210. https://doi.org/10.1016/j.cemconres.2016.07.004
Mosale Vijayakumar R (2014) Evaluating shrinkage of fly ash—slag geopolymers. University of Illinois at Urbana-Champaign
Nastic M, Bentz EC, Kwon OS et al (2019) Shrinkage and creep strains of concrete exposed to low relative humidity and high temperature environments. Nucl Eng Des 352:110154. https://doi.org/10.1016/j.nucengdes.2019.110154

Novais RM, Ascensão G, Ferreira N et al (2018) Influence of water and aluminium powder content on the properties of waste-containing geopolymer foams. Ceram Int 44:6242–6249. https://doi.org/10.1016/j.ceramint.2018.01.009

Panda B, Tan MJ (2018) Experimental study on mix proportion and fresh properties of fly ash based geopolymer for 3D concrete printing. Ceram Int 44:10258–10265. https://doi.org/10.1016/j.ceramint.2018.03.031

Pinson MB, Masoero E, Bonnaud PA et al (2015) Hysteresis from multiscale porosity: modeling water sorption and shrinkage in cement paste. Phys Rev Appl 3:064009. https://doi.org/10.1103/PhysRevApplied.3.064009

Punurai W, Kroehong W, Saptamongkol A, Chindaprasirt P (2018) Mechanical properties, microstructure and drying shrinkage of hybrid fly ash-basalt fiber geopolymer paste. Constr Build Mater 186:62–70. https://doi.org/10.1016/j.conbuildmat.2018.07.115

Rahier H, Wastiels J, Bisemans M et al (2007) Reaction mechanism, kinetics and high temperature transformations of geopolymers. J Mater Sci 42:2982–2996. https://doi.org/10.1007/s10853-006-0568-8

Rangan BV et al. (2006) Heat-cured, low-calcium, fly ash-based geopolymer concrete. Indian Concrete J 80(6): 47–52

Sadat MR, Bringuier S, Asaduzzaman A et al (2016) A molecular dynamics study of the role of molecular water on the structure and mechanics of amorphous geopolymer binders. J Chem Phys 145:134706. https://doi.org/10.1063/1.4964301

Si R, Dai Q, Guo S, Wang J (2020) Mechanical property, nanopore structure and drying shrinkage of metakaolin-based geopolymer with waste glass powder. J Clean Prod 242:118502. https://doi.org/10.1016/j.jclepro.2019.118502

Thokchom S, Ghosh P, Ghosh S (2009) Effect of water absorption, porosity and sorptivity on durability of geopolymer mortars. J Eng Appl Sci 4:28–32

Thokchom S, Ghosh P, Ghosh S (2010) Performance of fly ash based geopolymer mortars in sulphate solution. J Eng Sci Technol Rev 3:36–40. https://doi.org/10.25103/jestr.031.07

van Jaarsveld JGS, van Deventer JSJ, Lukey GC (2002) The effect of composition and temperature on the properties of fly ash-and kaolinite-based geopolymers. Chem Eng J 89:63–73. https://doi.org/10.1016/S1385-8947(02)00025-6

White CE, Provis JL, Llobet A et al (2011) Evolution of local structure in geopolymer gels: an in situ neutron pair distribution function analysis. J Am Ceram Soc 94:3532–3539. https://doi.org/10.1111/j.1551-2916.2011.04515.x

Wittmann FH (1973) Interaction of hardened cement paste and water. J Am Ceram Soc 56:409–415. https://doi.org/10.1111/j.1151-2916.1973.tb12711.x

Xiang J, Liu L, Cui X et al (2019) Effect of fuller-fine sand on rheological, drying shrinkage, and microstructural properties of metakaolin-based geopolymer grouting materials. Cem Concr Compos 104:103381. https://doi.org/10.1016/j.cemconcomp.2019.103381

Yang T, Zhu H, Zhang Z (2017) Influence of fly ash on the pore structure and shrinkage characteristics of metakaolin-based geopolymer pastes and mortars. Constr Build Mater 153:284–293. https://doi.org/10.1016/j.conbuildmat.2017.05.067

Ye H, Radlińska A (2016) A review and comparative study of existing shrinkage prediction models for Portland and non-Portland cementitious materials. Adv Mater Sci Eng 2016:1–13. https://doi.org/10.1155/2016/2418219

Yousefi Oderji S, Chen B, Ahmad MR, Shah SFA (2019) Fresh and hardened properties of one-part fly ash-based geopolymer binders cured at room temperature: effect of slag and alkali activators. J Clean Prod 225:1–10. https://doi.org/10.1016/j.jclepro.2019.03.290

Zheng L, Wang W, Shi Y (2010) The effects of alkaline dosage and Si/Al ratio on the immobilization of heavy metals in municipal solid waste incineration fly ash-based geopolyme. Chemosphere 79:665–671. https://doi.org/10.1016/j.chemosphere.2010.02.018

(1832) An essay on the cohesion of fluids. Abstr Pap Print Philos Trans R Soc London 1:171–172. https://doi.org/10.1098/rspl.1800.0095

Chapter 5
Sulfate Corrosion of MKG

Abstract The durability of geopolymers in sulfate-rich environments is a critical factor for their utilization in marine and salty area. This chapter explores the impact of the Si/Al ratio, a significant compositional parameter of MKG, on its resilience when exposed to sodium and magnesium sulfate solutions. Additionally, this study thoroughly investigates the stability of MKG mortars' chemical, physical, and mechanical properties over the course of 180 days by evaluating a number of factors, including appearance, mass, dimensions, strength, microstructure, and chemical composition, while subjected to the immersion effects of both sodium sulfate and magnesium sulfate solutions.

5.1 Introduction

The potential detrimental effects of sulfate attacks pose a significant threat to the durability and service life of structures, particularly in industrial, marine, and coastal areas known for their high sulfate content (Whittaker and Black 2015; Li et al. 2020). Conventional OPC concretes typically exhibit degradation when subjected to sulfate attack due to intricate physico-chemical processes. These processes include the crystallization of gypsum and ettringite crystals within narrow pores, as well as the softening of material matrices caused by the decomposition and leaching of calcium silicate hydrate (C–S–H) gels and calcium hydroxide (CH) crystals (Hekal et al. 2002; Whittaker and Black 2015; Zhutovsky and Douglas Hooton 2017). Sulfate corrosion, being a severe degradation factor, can result in significant expansion and compromise the integrity of OPC concrete (Mehta 1983). Therefore, the research community has shown considerable interest in evaluating the sulfate durability of geopolymers.

Previous studies have established a resemblance between the deterioration mechanisms of geopolymers with high calcium content and conventional OPC. The degradation of these geopolymers is attributed to the formation of gypsum and ettringite, leading to volume expansion (Bakharev 2005; Chindaprasirt et al. 2013). In contrast,

D. Yan et al., *Metakaolin-Based Geopolymers*,
https://doi.org/10.1007/978-981-97-0652-5_5

geopolymers with low calcium content, such as those derived from metakaolin or low-calcium fly ash, exhibit a different pattern of degradation. These geopolymers demonstrate enhanced performance in sulfate environments due to their limited calcium content in the precursors. Bakharev (2005) observed that geopolymers produced from class F fly ash exhibited less severe deterioration compared to OPC counterparts. Furthermore, the type of activator used, which varies in silicate content and alkali cation, also significantly influences the response to sulfate attacks. El-Sayed et al. (2011) conducted a study showing that an appropriate mixture ratio of sodium hydroxide and sodium silicate as an activator leads to improved sulfate resistance of geopolymers. Recent research by Elyamany et al. (2018) suggests that the addition of silica fume enhances the resistance of geopolymer mortars to magnesium sulfate. These investigations indicate that increasing the silicate content in geopolymers may yield beneficial effects on their ability to withstand sulfate corrosion. Notably, previous research has demonstrated that a lower Ca/Si ratio in OPC results in increased resistance under sulfate attack due to the decelerated decalcification of C–S–H gel (Kunther et al. 2015). However, the underlying mechanism responsible for the change in resistance of geopolymers remains uncertain.

The response and underlying mechanisms of geopolymer concretes to sulfate attacks can be substantially altered due to significant disparities in their chemical composition and microstructure. To understand the physico-chemical interactions between sulfate solutions and geopolymer matrices, several experimental investigations have been carried out up to this point. In general, geopolymer materials made from different minerals, such as fly ash (or bottom ash), slag, and metakaolin, show exceptional resistance to sulfate exposure (Bakharev 2005; Sata et al. 2012; Komljenović et al. 2013; Duan et al. 2016; Rashidian-Dezfouli and Rangaraju 2017; Salami et al. 2017). The addition of nano additives such as nano silica (Çevik et al. 2018) or bio-additives like Terminalia chebula and natural sugars (Karthik et al. 2017) further enhances the sulfate resistance of these material matrices. The effectiveness of this resistance is intrinsically linked to the specific geopolymer mixtures employed (Khatib and Wild 1998; Yusuf 2015; Elyamany et al. 2018). Similar to OPC systems, the calcium content plays a crucial role in influencing the durability of geopolymer materials against sulfate attacks. For instance, Chindaprasirt et al. (2013) discovered that the primary degradation mechanism for high-calcium fly ash-based geopolymer materials under sulfate exposure is still attributed to gypsum crystallization and decomposition of the C–S–H phase. Consequently, employing a low-calcium geopolymer matrix may prove effective in mitigating sulfate-induced damage. Indeed, studies have demonstrated that partial replacement of fly ash with low-calcium metakaolin significantly improves the mechanical properties and densification of the geopolymer matrix, thereby enhancing its durability against sulfate actions (Nuaklong et al. 2018). These examples highlight the potential of low-calcium minerals, such as metakaolin, as promising raw materials for the synthesis of long-lasting geopolymer concretes (Kwasny et al. 2018).

In engineering applications, it has been demonstrated by Zhang et al. (2010) that utilizing a 10% granulated blast furnace slag-coated MKG over a marine concrete structure effectively safeguards against erosion caused by marine environments. The stability of MKG or fly-ash-based geopolymer cements in harsh conditions also holds promise for constructing innovative structures for transporting and storing domestic sewage and industrial wastewater (Montes and Allouche 2012). These examples illustrate the wide-ranging prospects for applying MKG materials in buildings and infrastructure subjected to challenging environments.

Considering the widespread erosion of buildings and infrastructure caused by concentrated sulfate solutions, particularly sodium sulfate (Na_2SO_4) and magnesium sulfate ($MgSO_4$), in many regions of China, including the northwest and coastal areas, MKG materials offer considerable potential to prolong the service life of local structures and reduce maintenance costs. While the positive influence of MKG materials on mechanical properties against attacks from sodium sulfate and magnesium sulfate solutions has been extensively documented, debates persist regarding their physico-chemical behavior, as the aluminate-silicate hydrates in MKG may react with magnesium sulfate based on thermodynamic principles (Walling and Provis 2016). This issue has been partially addressed by Ismail et al. (2013a, p. 361), who noted that the "sulfate attack" process on geopolymer binders is significantly influenced by the accompanying cation. Consequently, the application of geopolymer materials in sulfate environments should be approached with caution until fundamental questions are thoroughly investigated.

One crucial aspect pertains to the physical and chemical stability of the MKG system exposed to sulfate solutions containing different cations. The mechanisms underlying these interactions, to the best knowledge of the present authors, have not been fully comprehended yet.

This chapter presents a systematic investigation into the influence of silicate content, specifically the Si/Al ratio, on the ion exchange process, micromorphology, and stability of MKG. The study combines various characterizations including compressive strength, pH value, ion concentration, X-ray diffraction (XRD), Fourier transform infrared spectroscopy (FTIR), and scanning electron microscopy (SEM) to gain a thorough understanding of the coupling effect between the Si/Al ratio and sulfate type on the stability of MKG. Moreover, the chemical–physical–mechanical stability of MKG mortars under sulfate attacks is thoroughly examined.

To investigate the influential factors, three mixtures and two sulfate environments (5 wt% magnesium sulfate and sodium sulfate solutions) are designed in relation to the chemical composition of the materials and the environmental conditions. To assess changes in appearance, mass, dimensions, strength, and material microstructure, experimental tests are carried out. Special attention is given to investigating the interactions between cations (Na^+ and Mg^{2+}) and the active minerals present in the MKG materials, which have received limited previous attention. The findings from this study enhance our understanding of the chemical–physical–mechanical mechanisms governing the response to sulfate exposure. Furthermore, these findings help bridge the gap between laboratory tests and field applications of MKG materials in future engineering projects.

5.2 The Influence of Si/Al Ratio on Sulfate Durability of Metakaolin-Based Geopolymer

5.2.1 Experimental Programs

5.2.1.1 Materials

(1) Metakaolin

Metakaolin, selected for its high reactivity, served as the precursor material for geopolymer synthesis in this study. The metakaolin powder employed was subjected to characterization using X-ray fluorescence analysis (XRF-1800) to determine its chemical composition. Additionally, the phase structure was evaluated through X-ray diffraction (XRD) using the Ultima IV method. Table 5.1 presents the chemical composition, while Fig. 5.1 displays the XRD pattern. The experimental findings indicate that the main phase of the metakaolin powder is amorphous, with additional crystal phases observed including anatase (TiO_2) and quartz (SiO_2). It is worth noting that there might be some impurities present in the commercial product. Furthermore, employing a laser particle size analyzer (Coulter LS-230), the average particle size of the powder was determined to be approximately 5.91 μm.

(2) Alkali-Activator

Sodium hydroxide pellets and sodium silicate solution were used to make the alkali activator. The sodium silicate solution had a SiO_2/Na_2O ratio of 3.07, and its detailed chemical composition can be found in Table 5.2. Sodium hydroxide with an analytical purity level exceeding 96% was used. The required amount of sodium hydroxide was dissolved in the sodium silicate solution along with extra water as needed based on the compositions stated in Table 5.3. Subsequently, the solution was sealed and left in the ambient environment for 24 h prior to commencing the synthesis process.

(3) Sulfates

Anhydrous sodium sulfate (SS) and magnesium sulfate (MS) pellets with analytical purity levels exceeding 98% were employed to prepare the immersion solutions.

Table 5.1 Chemical composition of metakaolin powder

Components	Al_2O_3	SiO_2	K_2O	Na_2O	CaO	TiO_2	Fe_2O_3	LOI
Contents (wt%)	39.68	57.26	0.21	0.27	0.04	1.78	0.43	0.33

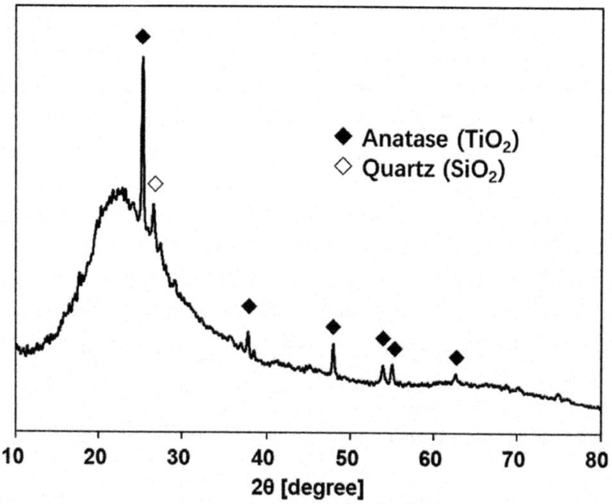

Fig. 5.1 XRD pattern of metakaolin powder. Reproduced from [The influence of Si/Al ratio on sulfate durability of metakaolin-based geopolymer] by [Shikun Chen] with permission from [Construction and Building Materials]

Table 5.2 Chemical composition of sodium silicate solution

Components	Al_2O_3	SiO_2	K_2O	Na_2O	Fe_2O_3	LOI
Content (wt%)	0.054	18.39	1.25	5.97	0.016	74.32

Table 5.3 Mixture compositions

Mixture composition	MKG-1	MKG-2
Metakaolin (g/L)	863	805
Sodium silicate solution (g/L)	1139	1405
Sodium hydroxide (g/L)	145	98
Additional water (g/L)	160	0
Si/Al (mol/mol)	2.00	2.25
Na/Al (mol/mol)	1	1
Water/solid ratio (g/g)	0.7	0.7

5.2.1.2 Mixture Synthesis

Based on earlier studies by Yan et al. (2016), which showed excellent mechanical performance, two MKG mixes with differing Si/Al ratios (2.00 and 2.25) were created. The composition details of the mixtures can be found in Table 5.3. Consistent with the research conducted by Hung et al. (2013), a water/solid ratio of 0.7 was chosen for all mixtures. It was observed that these MKG compositions exhibited a satisfactory balance between workability and strength, indicating their potential application in corrosive protection coatings, such as those used for concrete.

To synthesize the mixture, an auto mixer was utilized to combine metakaolin powder and alkali activator. Initially, they were gently mixed for 30 s and then vigorously mixed for 1 min. Subsequently, the resulting mixture was poured into plastic molds measuring 40 mm × 40 mm × 40 mm. To eliminate any trapped air bubbles, the filled molds were vibrated for 2 min. After being finished, the molds were covered with plastic film and exposed to room temperature and high humidity for two days to develop the necessary strength for demolding. On the third day, the samples were demolded and further cured under the same conditions for an additional 3 days prior to conducting tests, thereby enhancing the stability of their strength.

5.2.1.3 Sulfate Corrosion Test

In the accelerated sulfate corrosion test, highly concentrated sulfate solutions were employed. The experimental procedure followed the guidelines outlined in the National Standard GB/T 749-2008 (GB/T 749-2008 2008). To prepare the test solution, SS and MS pellets were dissolved in deionized water at a concentration of 10 wt%. The volume ratio of the specimen to the solution was approximately 0.05. The specimens were then submerged in plastic tanks containing sulfate solutions for a total of 180 days. The tanks were securely sealed and kept in the laboratory throughout the corrosion test. For the sake of identification, the samples were labeled using the notation consisting of the composition reference (MKG-I or MKG-II) followed by the immersion solution designation (SS or MS).

In order to investigate the chemical exchange occurring during immersion, periodic measurements of pH value and ion concentrations (Na^+, Mg^{2+}) were conducted throughout the corrosion test. The pH value was determined using a digital pH meter (pH Testr30), while the ion concentrations were analyzed using an atomic absorption spectrometer (ICE3500). Prior to immersion, the pH value and ion concentrations of the prepared solutions were measured. Following immersion, the pH value and ion concentrations were sampled at regular intervals during the test. Each sampling involved collecting approximately 10 mL of the immersion solution. Firstly, the pH value was measured, and then the sampled solution was filtered to eliminate impurities. Subsequently, the filtered solution was diluted 1300 times with deionized water. Finally, the samples of the diluted solution's sodium and magnesium ion concentrations were described and recorded for examination.

5.2.1.4 Mechanical Test

The compressive strength of the immersed specimens was evaluated at specific time intervals, namely 0, 30, 60, 90, and 180 days. Following the guidelines provided by the National Standard GB/T-17671-1999 (GB/T-17671-1999 1999), three specimens were selected from the immersion tank for each mixture. The surfaces of the specimens were carefully cleaned using a cloth. Subsequently, they were subjected to crushing using a universal mechanical test machine (Instron 8802) at a constant speed of 2.4 kN/s. The compressive strength was determined using the following equation:

$$f_c = \frac{F}{A} \tag{5.1}$$

where (F) represents the recorded peak force and (A) denotes the cross-sectional area of the specimen. In this study, the cross-sectional area was equal to 1600 mm^2 for the cubic specimens measuring 40 mm × 40 mm × 40 mm.

5.2.1.5 SEM

The morphology of the immersed specimens at the microscale was examined using a scanning electron microscope (SEM). Fragments ranging in size from 0.5 to 1 cm were chosen from the crushed specimens and subjected to drying at 60 °C for 24 h. Subsequently, manual fracturing was performed on the dried fragments, which were then observed using a FEG-650 scanning electron microscope. The observations and imaging were carried out on relatively smooth regions of the unpolished surface, maintaining a working distance of 11.7–14.5 mm and an accelerating voltage of 20 kV. Additionally, the local elemental composition was simultaneously analyzed using an energy dispersive spectrometer (EDS) during the observation process.

5.2.1.6 XRD

The crystal phases present in the immersed specimens were analyzed using X-ray diffraction (XRD). Surface samples from within a depth of 1 mm were collected and subsequently ground into powder, passing through an 80 μm sieve. The powdered samples were then dried at a temperature of 60 °C for 24 h. X-ray polycrystalline diffraction analysis was performed using an Ultima IV diffractometer, with a scanning range spanning from 10° to 80°.

5.2.1.7 FTIR

In addition to XRD analysis, the same samples used in the previous test were subjected to characterization using the Fourier-transform infrared spectroscopy (FTIR) technique. An AVATAR 370 infrared spectrometer was employed for this purpose, with a scanning range spanning from 400 to 4000 cm^{-1}.

Fig. 5.2 Appearance of **a**, **b** unexposed specimens and immersed specimens in **c**, **d** sodium sulfate solution and **e**, **f** magnesium sulfate solution for 180 days. Reproduced from [The influence of Si/Al ratio on sulfate durability of metakaolin-based geopolymer] by [Shikun Chen] with permission from [Construction and Building Materials]

5.2.2 Results and Discussions

5.2.2.1 Appearance

The appearance of the immersed specimens at 180 days is depicted in Fig. 5.2. It can be observed that there is no discernible change in the appearance of any of the specimen following immersion in sulfate solutions. This indicates that macro-scale MKG exhibits excellent stability in sulfate environments. Unlike high-calcium geopolymers and ordinary Portland cement (OPC), which typically experience significant deteriorations caused by sulfate corrosion due to the formation of expanding deposits such as gypsum and ettringite (Bakharev 2005; Thokchom et al. 2010; Chindaprasirt et al. 2013), these detrimental deposits are absent in MKG due to its lower calcium content. This characteristic aligns with previous studies on MKG (Duan et al. 2016) as well as low-calcium fly ash-based geopolymer (Bakharev 2005).

5.2.2.2 Compressive Strength

Figure 5.3 presents the compressive strength of the specimens throughout the corrosion test. As depicted in Fig. 5.3a, there is no significant change observed in the strength of MKG-1 and MKG-2 specimens during the entire 180-day immersion in sodium sulfate solution. In fact, the strengths at 180 days are slightly higher compared to the initial strengths before immersion, with an increase of 7.8% for MKG-1 and 6.6% for MKG-2. This rising tendency was also seen in a prior research (Bakharev

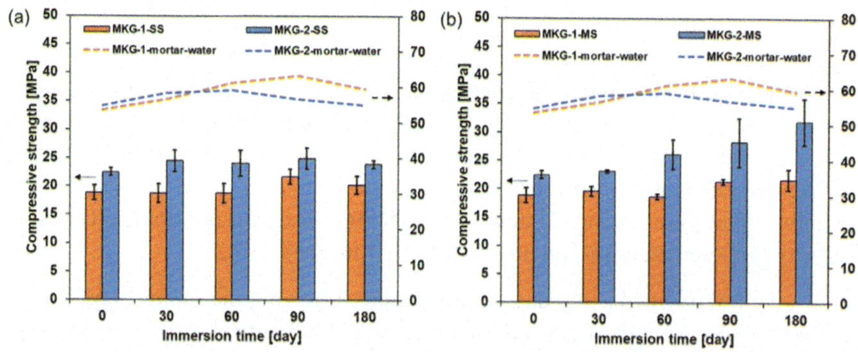

Fig. 5.3 Compressive strength of specimens immersed in **a** sodium sulfate solution and **b** magnesium sulfate solution. The mortar strengths (dash lines) immersed in water are also shown for reference. Reproduced from [The influence of Si/Al ratio on sulfate durability of metakaolin-based geopolymer] by [Shikun Chen] with permission from [Construction and Building Materials]

2005), and it has been partially explained by the ongoing cure of the sample. This conclusion was also reached by comparing the test findings to the strengths of mortar specimens submerged in water (with a comparable binder makeup), which are displayed as referenced dashed lines in Fig. 5.3. However, a recent study by Guo and Curtis (2015) revealed that fiber-reinforced MKG degrades to different degrees when submerged in sodium sulfate solution than when not exposed. This tendency runs counter to the current study's conclusions in certain ways. The difference may be caused in part by the various curing times before to the corrosion test (2 days in (Guo et al. 2020) and 5 days in present study). Indeed, it is important to consider other factors that can influence the results, such as the source materials used in the MKG and the inclusion of fibers. The specific properties and characteristics of these materials can have an impact on the overall behavior and performance of the specimens during the corrosion test. Therefore, it is essential to take into account these additional factors when analyzing and interpreting the findings.

Conversely, the trend of strength change in MKG when immersed in magnesium sulfate solution differs from that observed in sodium sulfate solution. In fact, the strength enhancements are more pronounced in magnesium sulfate solution, with the Si/Al ratio exerting a more significant influence. After 180 days of immersion, the strength of MKG-1 increases by 15.1% in magnesium sulfate solution. In the same period, the strength of MKG-2 experiences a remarkable rise of 42.0%, nearly three times higher than that of MKG-1. Notably, the strength enhancement is particularly prominent in the mixture with a higher silicate content (MKG-2). Further analysis integrating various characterizations will be conducted to explore the mechanisms underlying this notable strength improvement.

5.2.2.3 pH Value

The pH values of immersion solutions are displayed in Fig. 5.4. Both the initial pH of the sodium sulfate solution and the first pH of the magnesium sulfate solution are consistent with the results of the prior investigation (Thokchom et al. 2010).

The pH values exhibit diverse changes in the two types of sulfate solutions. In sodium sulfate solution, the pH trends for both mixtures are quite similar. The entire 180-day immersion period can be divided into four major stages:

(1) Stage I, 0 day to 11 ± 6 days. Immediately following the immersion of MKG specimens, the pH values rapidly rise. The pH levels peak between one to two weeks. Peak pH levels are around 10.6 0.1, and there is very little difference between the two mixes.
(2) Stage II, 11 ± 6 days to 42 ± 10 days. After the peaks, the pH readings go through a somewhat constant phase. The pH readings barely slightly fall throughout this time. However, their rates quickly rise. This time frame is around one month long.
(3) Stage III, 42 ± 10 days to 118 ± 17 days. The pH readings reach a point where they are steadily dropping. As pH levels decrease, they eventually reach zero. This period lasts for two to three months.
(4) Stage IV, after 118 ± 17 days. The pH readings finally reach a steady condition. The pH readings vary between 9.1 and 0.1. Additionally, the ultimate pH values for various mixes show minimal variation.

However, the pH values in magnesium sulfate solution exhibit different variations compared to sodium sulfate solution. The immersion period in magnesium sulfate solution can be divided into three distinct stages:

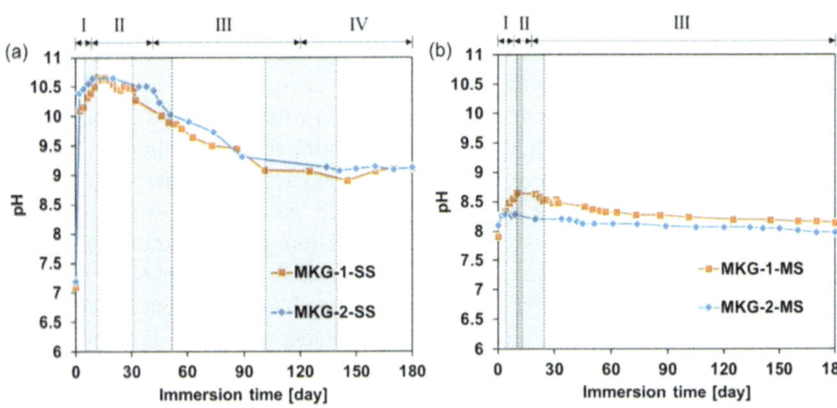

Fig. 5.4 pH values in **a** sodium sulfate solution and **b** magnesium sulfate solution. Reproduced from [The influence of Si/Al ratio on sulfate durability of metakaolin-based geopolymer] by [Shikun Chen] with permission from [Construction and Building Materials]

(1) Stage I, 0 day to 8 ± 4 days. Similar to sodium sulfate solution, the first rising stage is present. This phase lasts one to two weeks. The greatest pH value is just 8.7, which is significantly lower than sodium sulfate solution's (10.6 + / − 0.1) value. Low Si/Al ratio MKG-1 has this pH value.
(2) Stage II, 8 ± 4 days to 16 ± 9 days. The second stage in magnesium sulfate solution is shorter compared with that in sodium sulfate solution. It closely follows the stage I and ends up in only 1–2 weeks.
(3) Stage III, after 16 ± 9 days. The pH levels all enter a long-lasting degradation stage after roughly a month. pH readings change at a fairly slow rate (approximately 0.1/month). In comparison to MKG-1 (8.1), the MKG-2 has a significantly lower ultimate pH value (8.0).

5.2.2.4 Ion Concentrations

Figure 5.5 illustrates the Na^+ ion concentration in diluted sodium sulfate immersion solution. Throughout the immersion test, the Na^+ ion concentration remains relatively constant around 22 mg/L. Unlike previous research findings (Bakharev 2005) where a significant rise in Na^+ ion concentration was observed, this study indicates a different trend. This discrepancy can be attributed to the high concentration of the immersion solution, which limits the diffusion of Na^+ ions from the specimen into the surrounding environment.

Figure 5.6 displays the concentrations of Na^+ and Mg^{2+} ions in the magnesium sulfate immersion solution. The Na^+ ion concentration exhibits a rapid increase within the first week, followed by a gradual increase at a rate of approximately 0.24 mg/(L·month) for both mixtures. In contrast, the Mg^{2+} ion concentration experiences a rapid decrease during the initial week, followed by a gradual decline over subsequent days. This reduction in Mg^{2+} concentration aligns with previous research findings (Bakharev 2005). The decreasing rates are approximately 0.34 mg/(L·month) for MKG-1 and 0.59 mg/(L·month) for MKG-2. Notably, it is evident

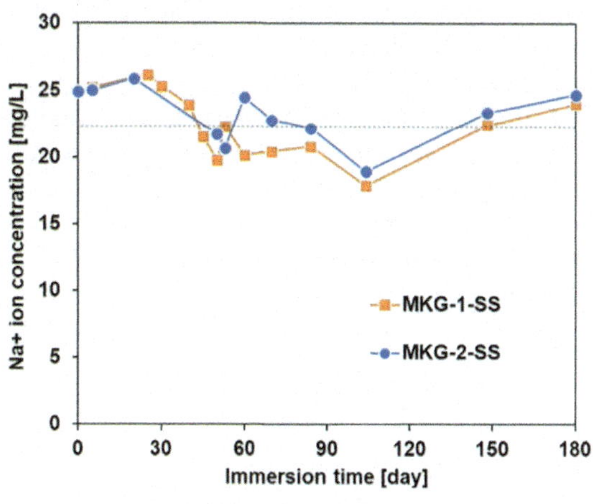

Fig. 5.5 Na^+ ion concentrations in sodium sulfate immersion solution. Reproduced from [The influence of Si/Al ratio on sulfate durability of metakaolin-based geopolymer] by [Shikun Chen] with permission from [Construction and Building Materials]

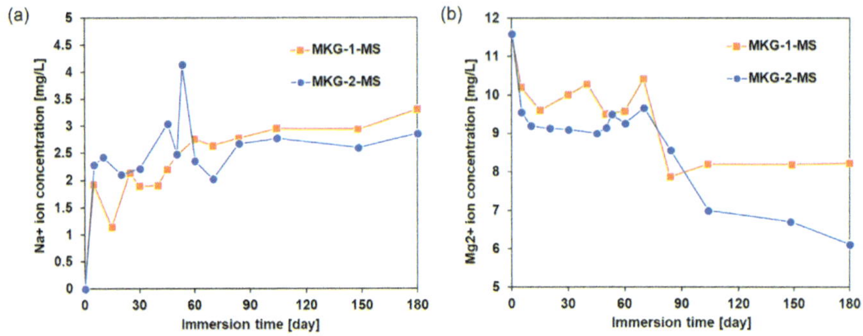

Fig. 5.6 Ion concentrations in magnesium sulfate solution: **a** Na^{+} ion; **b** Mg^{2+} ion. Reproduced from [The influence of Si/Al ratio on sulfate durability of metakaolin-based geopolymer] by [Shikun Chen] with permission from [Construction and Building Materials]

that the reduction in Mg^{2+} ion concentration is higher for the mixture with a higher silicate content (MKG-2).

5.2.2.5 XRD

Figure 5.7 depicts the X-ray diffraction (XRD) patterns of MKG-1 and MKG-2 before immersion. The patterns exhibit a high degree of similarity between the two mixtures. A prominent feature observed in the XRD patterns of both MKGs is a broad hump centered at 26°–28°, indicating the presence of an amorphous aluminosilicate gel within the MKG samples (Ozer and Soyer-Uzun 2015). Additionally, there are discernible peaks corresponding to impurities present in the raw materials, such as anatase (TiO_2) and pyrophyllite ($Al_2Si_4O_{10}(OH)_2$), as shown in Fig. 5.1.

Figure 5.8a shows the XRD patterns of samples immersed in sodium sulfate solution. It is discovered that even after 180 days of immersion, MKG's amorphous characteristic remains unchanged. However, additional peaks that resemble thenardite were detected in these samples. It suggests that samples' surfaces have been permeated by sulfate.

Figure 5.8b shows the XRD patterns of samples immersed in magnesium sulfate solution. The amorphous characteristic does not alter significantly either. The crystal peaks associated to Na_2SO_4 are not exhibited in comparison to Fig. 5.8a. Instead, the MKG-1 sample shows a minor peak at 27.8°. It corresponds to one of the principal peaks of the mineral caminite, a kind of magnesium sulfate hydrate. However, the MKG-2 sample lacks this peak.

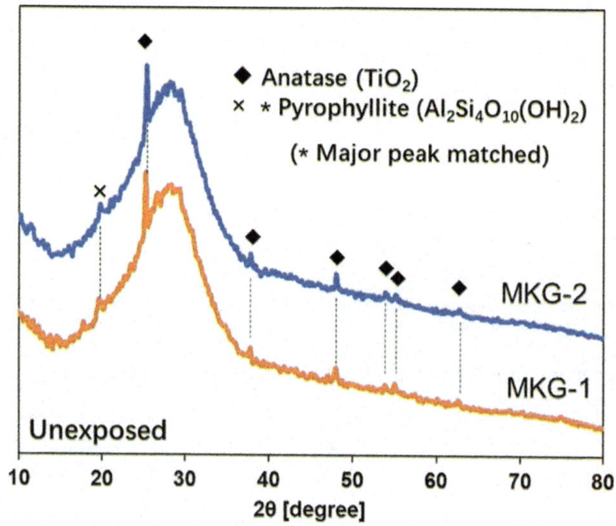

Fig. 5.7 XRD patterns of unexposed MKG samples. Reproduced from [The influence of Si/Al ratio on sulfate durability of metakaolin-based geopolymer] by [Shikun Chen] with permission from [Construction and Building Materials]

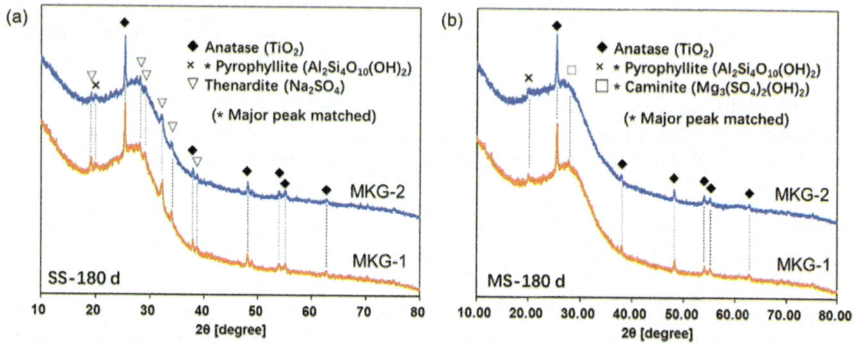

Fig. 5.8 XRD patterns of specimens immersed in **a** sodium sulfate solution and **b** magnesium sulfate solution for 180 days. Reproduced from [The influence of Si/Al ratio on sulfate durability of metakaolin-based geopolymer] by [Shikun Chen] with permission from [Construction and Building Materials]

5.2.2.6 FTIR

The FTIR patterns in sodium and magnesium sulfate solutions are shown in Figs. 5.9 and 5.10, respectively. Table 5.4 provides a summary of the absorbance peaks in FTIR patterns along with their explanations.

As shown in Fig. 5.9, the specimens immersed in sodium sulfate solution shows pattern features as:

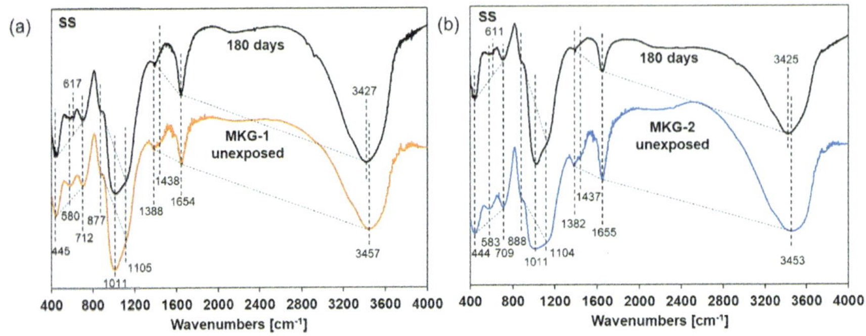

Fig. 5.9 FTIR patterns of **a** MKG-1 and **b** MKG-2 immersed in sodium sulfate solution

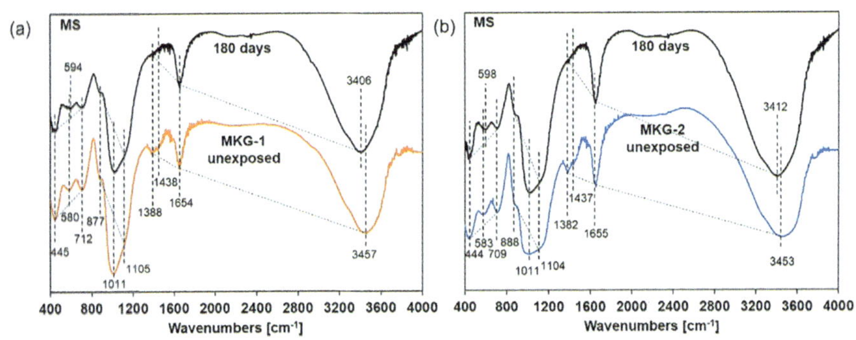

Fig. 5.10 FTIR patterns of **a** MKG-1 and **b** MKG-2 immersed in magnesium sulfate solution. Reproduced from [The influence of Si/Al ratio on sulfate durability of metakaolin-based geopolymer] by [Shikun Chen] with permission from [Construction and Building Materials]

Table 5.4 Absorbance peaks in FTIR patterns

Peak position $[cm^{-1}]$		Description
MKG-1	MKG-2	
445	444	Bending of Si–O bonds (Bakharev 2005; He et al. 2016)
580	583	Bending of Si–O–Al bonds (He et al. 2016)
712	709	Symmetrical stretching of Si–O bonds (Bakharev 2005)
877	888	Stretching of Si–OH bonds (Bakharev 2005)
1011	1011	Asymmetrical stretching of Si–O–Si and Si–O–Al bonds (He et al. 2016)
1105	1104	Stretching of SiO_4 units (Tchakouté and Rüscher 2017)
1388	1382	Stretching of O–C–O bonds in $CO_3{}^{2-}$ ions (Cyr and Pouhet 2016)
1438	1437	Bending of O–C–O bonds in $CO_3{}^{2-}$ ions (Hasnaoui et al. 2019)
1654	1655	Bending of H–O–H bonds (Bakharev 2005)
3457	3453	Stretching of (OH)– groups (Bakharev 2005)

(1) In samples that have not been exposed to light, the absorbance peaks at 3453–3457 cm^{-1} shift to a lower value (3427 cm^{-1} for MKG-1 and 3425 cm^{-1} for MKG-2). At the same time, the strength of the peak for MKG-2 at 1655 cm^{-1} marginally declines while that for MKG-1 at 1654 cm^{-1} remains same.
(2) The carbonation of MKG during the curing process is shown by the peaks at 1382–1388 cm^{-1} and 1437–1438 cm^{-1}. The formation of sodium bicarbonate from the extra sodium in geopolymers reacting with atmospheric CO^2 has previously been documented (Ren et al. 2017). Compared with the peak at 1654 and 1655 cm^{-1}, the intensity of peaks at 1382 and 1388 cm^{-1} after immersion decreases. At the same time, those at 1437 and 1438 cm^{-1} disappear. These modifications show that carbonation products decrease when exposed to sodium sulphate solution. When immersed, these goods could disintegrate into the surrounding fluid.
(3) The intensity of the peak at 877 cm^{-1} for MKG-1 does not vary after immersion in comparison to the shoulder at 1105 cm^{-1}, which might be caused by the slop of the inclined auxiliary line connecting the two peaks. At the same time, the intensity of peak at 1011 cm^{-1} slightly decreases after immersion. This indicates a depolymerization may happen to the Si–O–Al and Si–O–Si networks. For MKG-2, there is a distinct pattern. The peak intensity at 888 cm^{-1} for this combination dramatically reduces, but the peak intensity at 1011 cm^{-1} increases. These modifications point to an improvement in the polymerization of Si–OH groups into Si–O–Si or Si–O–Al networks rather than the networks' depolymerization.
(4) As indicated by the auxiliary lines connecting the peaks at 444–445 cm^{-1} and 709–712 cm^{-1}, the intensity of peaks at 444–445 cm^{-1} increases while that at 580–583 cm^{-1} decreases. There are also shifts of the peak at 580–583 cm^{-1} (into 617 cm^{-1} for MKG-1 and 611 cm^{-1} for MKG-2). Together, these modifications suggest that the aluminosilicate gel networks may undergo reconfiguration. Si–O–Al bond connections might be disrupted, whilst Si–O–Si bond connections could possibly be strengthened.

As shown in Fig. 5.10, the specimens immersed in magnesium sulfate solution shows pattern features as:

(1) The peak shifts at 3453–3457 cm^{-1} are also shown in magnesium sulfate solution. The peak at 3457 cm^{-1} shifts to 3406 cm^{-1} while that at 3453 cm^{-1} shifts to 3412 cm^{-1}. When submerged in magnesium sulfate solution, the peak at 1655 cm^{-1} for MKG-2's intensity decreases more noticeably.
(2) The peaks at 1382–1388 cm^{-1} and 1437–1438 cm^{-1} show similar trends as those in sodium sulfate solution. The removal of carbonates, however, is more important. After 180 days of immersion, the peaks at those locations vanish. The magnesium sulfate solution has a greater dissolving impact on carbonates in binders, according to this.
(3) The changing trends of peaks at 877–888 cm^{-1}, 1011 cm^{-1} and 1104–1105 cm^{-1} are also similar to those shown in sodium sulfate solution. The intensity fluctuations, nevertheless, are less pronounced than in sodium sulfate solution. This

suggests that the linkages of the gel networks may be slightly more affected by the magnesium sulfate solution.

(4) For MKG-1 and MKG-2, the peaks at 444–445 cm^{-1} and 709–712 cm^{-1} have varying patterns. Peaks at MKG-1's peak intensity of 445 cm^{-1} and MKG-2's peak intensity of 444 cm^{-1} are in comparison to peaks at 709 and 712 cm^{-1}, respectively, and both are somewhat less intense. However, the difference is small. The peaks at 580–583 cm^{-1} shifts to a higher value (594 cm^{-1} for MKG-1 and 598 cm^{-1} for MKG-2) which is similar to the trend happens in sodium sulfate solution. Peaks at 594–598 cm^{-1} and 580–583 cm^{-1} have similar intensities. These changes confirm the trends at 877–1105 cm^{-1} which indicate a slighter influence of magnesium sulfate solution on the structure of the gel networks.

5.2.2.7 Micromorphology

The micromorphology of samples submerged in sodium sulfate solution for 180 days is displayed in Fig. 5.11. It demonstrates that the submerged specimens contain many sodium sulfate crystals with a size range of 3–5 m. Due to sodium sulfate's high solubility, drying the pore solution during sample preparation may cause these crystals to develop. Besides that, it shows no significant deterioration of the matrix at microscale. Otherwise, the matrix even become densified partially due to the crystal formation. However, there are more densified gels present without the crystals. It indicates the densification may be attributed to the continuous polymerization of the binder (Bakharev 2005; Guo et al. 2020) rather than crystal formation. The continuous polymerization and densification could also be responsible for the increased strength (as shown in Fig. 5.3).

Figure 5.12 shows the micromorphology of samples immersed in magnesium sulfate solution for 180 days. From MKG-1 contains some magnesium sulfate crystals, as can be observed in Fig. 5.12a. These crystals are a bit bigger in size (about 6–8 m) than sodium sulfate crystals (3–5 m). They could dry out during processing, much as how sodium sulfate crystals do. However, there are far fewer crystals than in Fig. 5.11. These magnesium sulfate crystals are not shown in MKG-2. Instead, there are a great number of flower-like phases among the matrix shown in Fig. 5.12b, which is similar to the brucite $Mg(OH)_2$ (Gomez-Villalba et al. 2018; Maltseva et al. 2019). However, these phases have a higher silicate and magnesium content. Thus, it may also be the magnesium silicate hydrate (M–S–H) gels, $(MgO)_x - SiO_2 - (H_2O)_y$ (Brew and Glasser 2005; Chiang et al. 2014). To better identify the phases in the future, further work is required. Near the micropores and fissures, these gels coexisted with the aluminosilicate gels. They add more overpasses for these flaws. Previous studies have also demonstrated the emergence of these gel phases, which lead to the matrix's densification (Thokchom et al. 2010).

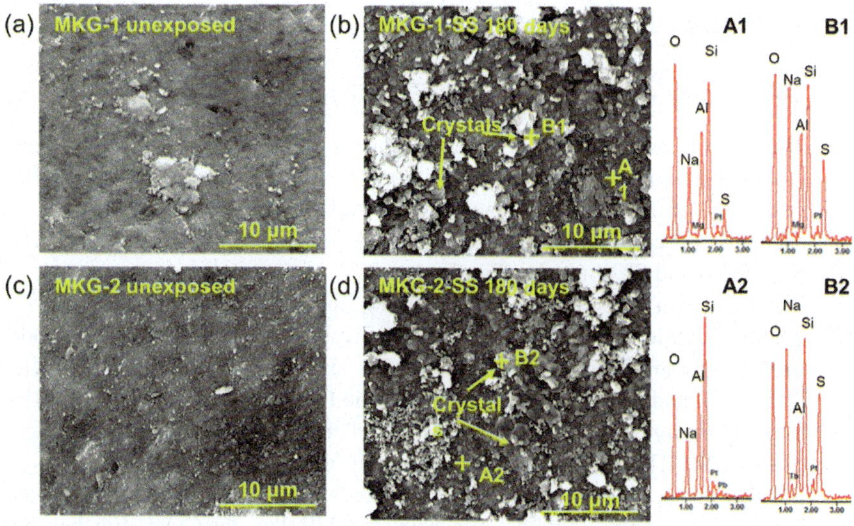

Fig. 5.11 Micromorphology of **a**, **b** MKG-1 and **c**, **d** MKG-2 before and after immersion in sodium sulfate solution for 180 days. Reproduced from [The influence of Si/Al ratio on sulfate durability of metakaolin-based geopolymer] by [Shikun Chen] with permission from [Construction and Building Materials]

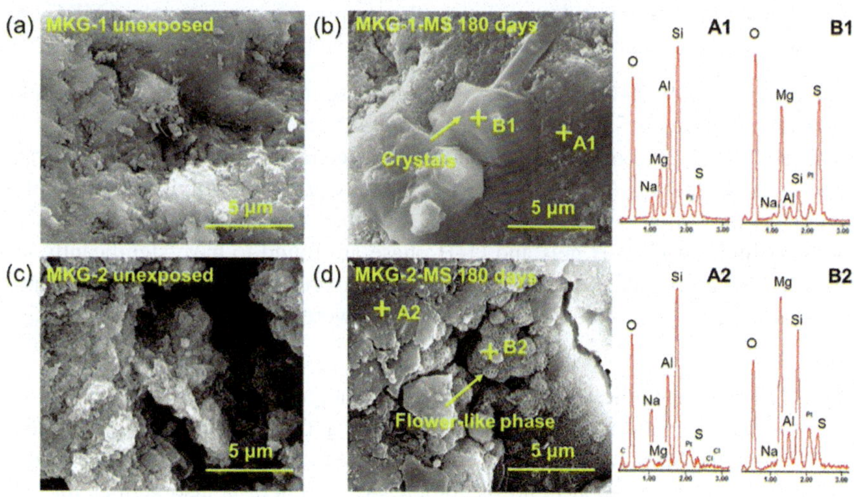

Fig. 5.12 Micromorphology of **a**, **b** MKG-1 and **c**, **d** MKG-2 before and after immersion in magnesium sulfate solution. Reproduced from [The influence of Si/Al ratio on sulfate durability of metakaolin-based geopolymer] by [Shikun Chen] with permission from [Construction and Building Materials]

5.2.3 Further Discussions

Combined the different experiments, some important facts could be recognized:

(1) The ion exchange between MKG and sulfate solution is influenced by the cation type. The amount of Na^+ that diffuses from the specimen into the sodium sulfate solution is little. The exchange of OH^- and $SO_4{}^{2-}$ ions dominates the ion exchange process, which is also indicated from pH value (Fig. 5.4a), XRD (Fig. 5.8a) and SEM results (Fig. 5.11).

(2) Meanwhile, in magnesium sulfate solution, the exchange of Na^+ and Mg^{2+} is dominant while that of OH^- and $SO_4{}^{2-}$ is restrained. These can be concluded from the trends in ion concentrations (Fig. 5.6), pH value (Fig. 5.4b), XRD (Fig. 5.8b) and SEM results (Fig. 5.12). The OH^- and $SO4^{2-}$ bonds in MKG may be strengthened by the production of magnesium silicate hydrate gels (Fig. 5.12b). Thus, it restricts the movement of these ions even further.

(3) Besides these, there is also a different trend in dissolving of the carbonates between two sulfate solutions. According to FTIR data, the $CO_3{}^{2-}$ ion dissolves more significantly in magnesium sulfate solution than in sodium sulfate solution (Fig. 5.9 and Fig. 5.10). However, the influence of $CO_3{}^{2-}$ dissolving was not shown in other results. Thus, its potential influences and underlying mechanisms need further studies.

(4) Due to immersion in sulfate solution, the aluminosilicate network of MKG may alter, and this change is more pronounced in sodium sulfate solution.

The variation of Si–O–Si and Si–O–Al response in FTIR (Fig. 5.9) indicates the connections between Si sites in networks are enhanced while that of Al sites are weakened. The previously reported increase in mean chain length and Si/Al ratio support these findings (Bakharev 2005). This network connection change is not as serious as those that occurred with OPC, though. After sodium sulfate corrosion, the OPC binder's C–S–H gels and O–H phase are found to have significantly lost strength (Bhutta et al. 2014). But for MKGs in present study, these remarkable changes, especially for the change in strength, were not founded after 180 days' immersion.

(5) Si/Al ratio has little effect on materials submerged in sodium silicate solution. It can be seen from the pH value (Fig. 5.4a) and ion concentrations (Fig. 5.5) results, the Si/Al ratio does not have a substantial impact on the ion exchange process. The variation trend of compressive strength is also similar for both mixtures with different Si/Al ratios. Only minor changes are present in the intensity of the peaks of XRD (Fig. 5.8a) and FTIR (Fig. 5.9). It may be inferred that the corrosion of MKG in a sodium sulfate environment is not significantly influenced by the Si/Al ratio.

The impact of the Si/Al ratio on samples submerged in magnesium silicate solution is apparent. The high silicate content mixture (MKG-2) has great strength improvement after immersion in magnesium sulfate solution (Fig. 5.3b). From clues in ion concentrations (Fig. 5.6b), the migration of Mg2^{+} is more significant for this mixture. At the same time, the dissolving of OH^- (Fig. 5.4b) and penetration of SO_4^{2-} (Fig. 5.12) are more constrained for MKG-2. These alterations all point to a high silicate concentration. In a solution of magnesium sulfate, MKG is more reactive. This reactivity might lead to the development of a new binding phase and an increase in the strength following immersion.

Beyond these findings, it should be noted that the trends found in this chapter is limited to MKG which has very low calcium content. The calcium in geopolymers may lead to much more complex results (Chotetanorm et al. 2013). Furthermore, the number of samples we had prevented us from precisely establishing a precise mathematical relationship between the Si/Al ratio and the sulfate durability of MKG at this time. Future improvements still require a lot of work. However, the results of this study offer some fundamental hints about how geopolymers are subject to sulfate corrosion. Additional research is still required to determine how other elements (such as calcium concentration, carbonate dissolving, etc.) affect the outcome.

5.3 Chemical–Physical–Mechanical Stability of MKG Mortars Under Sulfate Attacks

5.3.1 Notations

L current length of specimen
li initial length of the specimen
m current mass of the specimen
mi initial mass of the specimen
w/b water-to-binder ratio

5.3.2 Materials and Methods

5.3.2.1 Materials and Sample Preparation

A commercial metakaolin powder (Metamax, BASF Co.) was used as the aluminosilicate source of geopolymers. The particle size distribution of the metakaolin powder was measured by using a laser particle size analyser (LS-230, Coulter). The 90% passing particle size was 13.59 m, while the mean particle size was 5.91 m. The metakaolin powder's packing density was found to be 421.8 kg/m^3. As the fine aggregate in the mortars, fine sands with a fineness modulus of 1.75 and density of 2510 kg/m^3 were utilized. The alkaline activator was made using pellets of sodium hydroxide and liquid sodium silicate. The molar ratio of silicon dioxide (SiO_2) to metal oxide

Table 5.5 Chemical and physical characteristics of the MK powder and liquid sodium silicate

Materials	Al_2O_3	SiO_2	K_2O	Na_2O	CaO	TiO_2	Fe_2O_3	LOI
Metakaolin	39.68	57.26	0.21	0.27	0.04	1.78	0.43	0.34
Liquid sodium silicate	0.054	18.39	1.25	5.97	–	–	0.016	74.32

Reproduced from [Chemical–physical–mechanical stability of MKG mortars under sulfate attacks] by [Dongming Yan] with permission from [Advances in Cement Research]
Note Al_2O_3, aluminium oxide; SiO_2, silicon dioxide; K_2O, potassium oxide; Na_2O, sodium oxide; CaO, calcium oxide; TiO_2, titanium dioxide; Fe_2O_3, iron oxide; LOI, loss on ignition

[M_2O = sodium oxide (Na_2O) or potassium oxide (K_2O)] in the liquid sodium silicate was 2.87 and the pelleted sodium hydroxide was analytical reagent level with a purity of 96% (Shanghai Chemical Reagent, China). The alkali activator solutions were initially created using a correct mixture of the sodium hydroxide pellets, liquid sodium silicate, and water at 20 1 °C. The activator solutions for the geopolymer synthesis were easily made after being permitted to settle for 24 h. Table 5.5 displays the chemical make-up of sodium silicate liquid and metakaolin powder.

5.3.2.2 Mixture Design

Considering the possible influences of material composition—that is, the silicon dioxide/aluminum oxide (Al_2O_3) (Si/Al) ratio and sodium oxide/aluminum oxide (Na/Al) ratio (or sodium oxide/silicon dioxide (Na/Si) ratio, alternatively)—on the product, microstructure and mechanical properties of the geopolymers (Duxson et al. 2007a; Rowles and O'Connor 2009; Guo and Curtis 2015; He et al. 2016; Yan et al. 2016; Alanazi et al. 2017), here three mixes were designed with different Si/Al ratios and Na/Al ratios. A review of the literature found that Si/Al ratios between 2.0 and 6.0 and Na/Al ratios between 0.8 and 2.0 are necessary for the composition of geopolymers with uniform material structures and high mechanical characteristics (Duxson et al. 2007b; Rowles and O'Connor 2009; He et al. 2016; Yan et al. 2016). As a result, in this investigation, MKG mortars with high mechanical characteristics but potential for varying responses in sulfate environments were created using Si/Al ratios of 4 and 4.5 and Na/Al ratios of 1.0 and 0.8. Therefore, the geopolymers were labelled as mortar-Si/Al-Na/Al (Table 5.6). According to the experience of the current authors, a water-to-binder (w/b) ratio of 0.7 and a sand-to-binder ratio of 3.0 would be suitable to control the fluidity of the MKG slurries and to produce satisfactory mechanical properties in the hardened MKG mortars. When opposed to geopolymers manufactured of fly ash, metakaolin geopolymers often require a substantially greater water content to obtain optimum workability. It is a frequent trade-off between workability and material qualities to raise the w/b ratio for producing high-quality geopolymers, even if the high water content may cause higher shrinkage (White et al. 2013). In Table 5.6, the precise mixing ratios are displayed.

Table 5.6 Mixture proportions of the geopolymer mortars (volume in total 5.1 l)

Specimen ID	M4-1	M4-0.8	M4.5-1
MK powder: g	1438	1483	1341
Water glass: g	1899	1957	2342
Sodium hydroxide: g	242	157	163
Water: g	267	248	–
Sand: g	6915	6915	6915
Silicon dioxide/aluminium oxide	4	4	4.5
Sodium oxide/aluminium oxide	1	0.8	1
Sodium oxide/silicon dioxide	0.25	0.2	0.22
Sodium oxide/($SiO_2 + Al_2O_3$)	0.2	0.16	0.18
Water/binder	0.7	0.7	0.7
Sand/binder	3	3	3
Compressive strength: MPa	53.471	54.334	54.744
Flexural strength: MPa	2.471	2.241	2.398
Porosity: %	17.75	21.19	19.01

Reproduced from [Chemical–physical–mechanical stability of MKG mortars under sulfate attacks] by [Dongming Yan] with permission from [Advances in Cement Research]

5.3.2.3 Specimen Preparation

Fresh MKG mortars were made by first pouring the rapidly settled alkali activator solution into a Hobart mixing bowl, then gradually adding the carefully weighed metakaolin powder and water while stirring for three minutes to create MKG paste slurries. After that, sands were added to the paste slurries, and the MKG mortar was homogenized for another 3 min of churning. The mortar slurries were then poured into cuboid molds measuring 40 by 40 by 160 mm. To release the air bubbles trapped in the mortars, a vibration operation was applied to the full moulds. The specimen surfaces were coated with plastic sheets to prevent any water loss that would lead to drying shrinkage of the mortar specimens.

Due to the relatively high chemical activity of metakaolin, a standard room-temperature is adequate to achieve the desired strengths as opposed to steam curing at a high temperature, which is utilized for some binders with low chemical activity (such as fly ash) (Alanazi et al. 2017). To support the strength growth in this investigation, the MKG specimens and the molds were cured in a typical curing environment (20 ± 1 °C, 90% relative humidity) for 2 days to support the strength gain. Then all mortars were demoulded and stored in the same chamber for another 3 days. Following that, the mix design was evaluated using tests of the compressive and flexural strengths. It appears that the compressive (53–55 MPa) and flexural (2.2–2.5 MPa) strengths of all three combinations were comparable. Because the specimens are substantially smaller than those needed for the conventional compression testing, the compressive strengths might be overstated. According to Table 5.6, the porosities of M4-1, M4-0.8, and M4.5-1 were 17.75%, 21.19%, and 19.01%, respectively. After curing, all additional specimens were ready for immersion testing.

5.3.2.4 Testing Programme

According to Chinese standards, two separate solutions containing 5 weight percent sodium sulfate (SS) and 5 weight percent magnesium sulfate (MS) were applied to the manufactured MKG mortars (SPC, 2008). In order for the solutions to cover each specimen, the solid–liquid ratio was set at 0.5. Every two weeks, the solutions were changed to maintain the salt concentration. As a guide, certain mortars were submerged in distilled water. The specimens were marked with the following information for the durability tests: material-solution-time. For example, M4-1-SS-60 represents the M4-1 mortar exposed to 5 wt% sodium sulfate solution for 60 days.

The experimental arrangement is shown in Fig. 5.13. The visual appearance, mass, and dimensions of the specimens were evaluated after immersion for a certain number of days in order to evaluate the physical characteristics of the MKG mortars submerged in the sulfate solutions. An electronic balance was used to record the original mass (mi) and the current mass (m) for the mass test. The relative change in mass was then determined using the formula (m mi)/mi 100%. The length test was conducted by using a comparator, and the length change before (li) and after the exposures (l) was evaluated by (l − li)/li × 100%.

The mechanical resistance of the MKG mortars against the sulfate solutions' immersing activities was evaluated using tests on compressive and flexural strengths. An Instron 8802 fully working test apparatus was used to do the mechanical tests. In the flexural tests, three cuboid mortars were loaded at the speed of 3 kN/min until they broke into six half-cuboid pieces. The compressive strength of the six mortars was then evaluated at a rate of 144 kN/min.

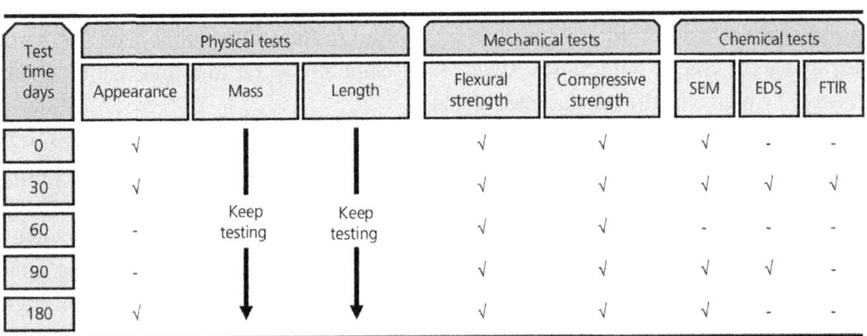

Fig. 5.13 Experimental arrangements used in the present study. Reproduced from [Chemical–physical–mechanical stability of MKG mortars under sulfate attacks] by [Dongming Yan] with permission from [Advances in Cement Research]

A Quanta FEG650 field emission environmental scanning electron microscope with an energy-dispersive spectrometer (SEM–EDS) was used to examine the microstructure of the MKG mortars. The microstructure tests required small mortar slices, which were created. The samples were pre-dried at 75 °C for 24 h, and the bottom surfaces were polished to adhere to the test bench, before being put into the frame of the SEM–EDS equipment. With 30 keV of accelerating voltage and a 13.4 mm working distance, SEM–EDS experiments were performed.

After the sulfate assaults, X-ray diffraction (XRD) examinations were also carried out using an X'Pert Pro Panalytical X-ray diffractometer at a rate of 5°/min at 5°–80° to analyze the crystals in the MKG mortars. A Fourier transform infrared spectroscopy (FTIR) test was conducted in an Avatar 370 FTIR spectrometer to analyse the surface deposits on the specimens immersed in the magnesium sulfate solution at early age (30 days). The spectra between 4000 and 400 cm^1 were captured. The identical pre-drying method (75 °C for 24 h) was carried out before the testing on the XRD and FTIR samples, which were milled to a thin size less than 150 m.

5.3.3 Results and Discussion

5.3.3.1 Visual Appearance

Figure 5.14 depicts the surface alterations of the MKG specimens that had been immersed in the sodium sulfate and magnesium sulfate solutions at different ages. As seen in the image, the specimens in every case had essentially no physical deterioration, such as fractures and debris. However, after being exposed to the magnesium sulfate solution for 30 days, two groups of the specimens—namely, M4-1-MS-30 and M4-0.8-MS-30—showed white 'rime' on the surfaces (Fig. 5.2g and h), but M4.5-1-MS-30 did not (Fig. 5.14i). A similar phenomenon was also reported in a previous research study by Valencia Saavedra et al. (2016), who found a white crystalline layer on fly-ash-based and slag-based geopolymer specimens immersed in 5 wt% magnesium sulfate solution after 270 days and took it as salt deposits. White deposits on the surfaces of geopolymer materials made from slag and fuel ash and immersed in magnesium sulfate solution were also observed by Yusuf (2015). The white "rime" appears to diminish as the amount of water glass in the activator solutions increases, according to a thorough analysis of Fig. 5.14g–i (Table 5.6). The white 'rime' could be a mixture of magnesium hydroxide ($Mg(OH)_2$), magnesium sulfate and magnesium carbonate ($MgCO_3$) crystals, magnesium silicate hydrates (M–S–H) and probably magnesium silicate aluminate hydrates (M–S–Al–H) (see the section later entitled 'Further discussion'). The white "rime," nevertheless, vanished after 180 days of soaking. Even yet, it appeared that the MGK specimens were unaffected by the emergence and removal of the surface deposits.

Fig. 5.14 Selected visual images of the MKG specimens immersed in different sulfate solutions for 30 days, shown in **a**, **b**, **c**, **g**, **h** and **i** and 180 days, shown in **d**, **e**, **f**, **j**, **k** and **l**: **a** M4-1-SS-30; **b** M4-0.8-SS-30; **c** M4.5-1-SS-30; **d** M4-1-SS-180; **e** M4-0.8-SS-180; **f** M4.5-1-SS-180; **g** M4-1-MS-30; **h** M4-0.8-MS-30; **i** M4.5-1-MS-30; **j** M4-1-MS-180; **k** M4-0.8-MS-180; **l** M4.5-1-MS-180. Reproduced from [Chemical–physical–mechanical stability of MKG mortars under sulfate attacks] by [Dongming Yan] with permission from [Advances in Cement Research]

5.3.3.2 Change in Mass

Figure 5.15 shows the mass changes of the MKG mortars submerged in the sodium sulfate and magnesium sulfate solutions. Apparently, a consistently increased mass gain can be observed for all mixes under the actions of the sodium sulfate solution (Fig. 5.15a). Particularly, when the time of immersion was extended up to 24 weeks, the mass values of the M4.5-1 and M4-1 specimens immersed in the sodium sulfate solution increased gradually and similarly, but those of the M4-0.8 specimens were somewhat higher. For the purpose of comparison, the data of OPC mortars (Biricik et al. 2000) are plotted in Fig. 5.15a as well. In the sodium sulfate solution-immersed MKG and OPC mortars, there were no discernible changes in weight growth. In the immersion tests, the weight gains of both the MKG and OPC specimens can be ascribed to the ingress of the external solution to the open pores. All of the specimens submerged in the sodium sulfate solution experienced weight increases that were less than 0.75% (Fig. 5.15a). It is reasonable to say at this time that the MKG mortars had a good degree of mass stability when subjected to the action of the sodium sulfate solution.

The mass changes of the MKG mortars were significantly different when the immersion solution was replaced by magnesium sulfate (Fig. 5.15b). Roughly, all curves exhibited a rising stage followed by a declining and a consistently varying (or

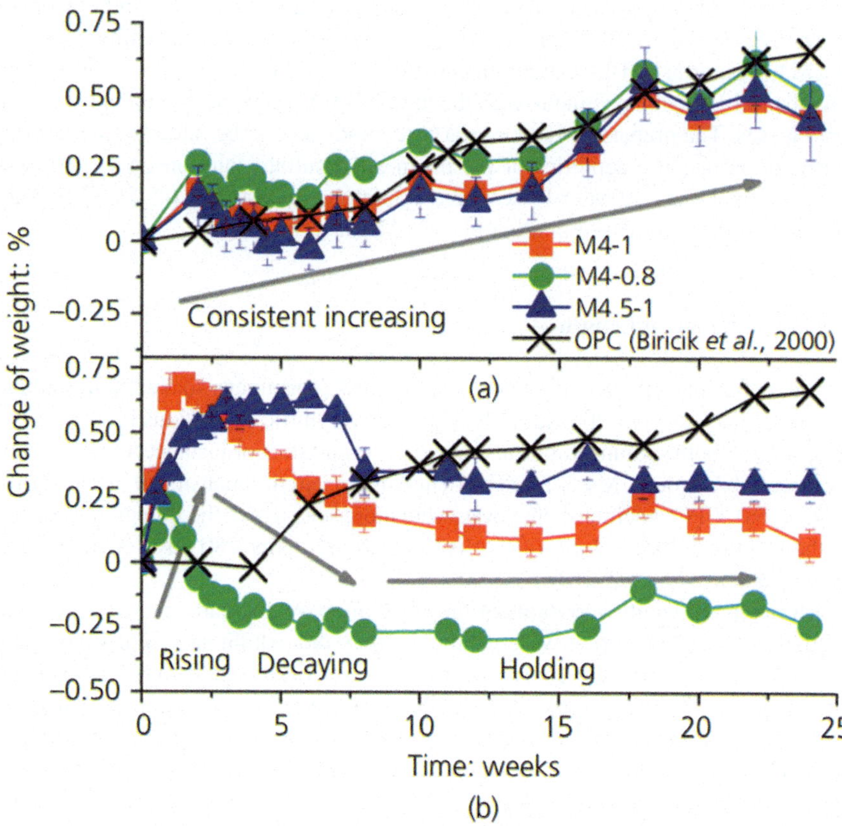

Fig. 5.15 Changes in mass of the MKG and OPC specimens immersed in **a** sodium sulfate and **b** magnesium sulfate solutions. Reproduced from [Chemical–physical–mechanical stability of MKG mortars under sulfate attacks] by [Dongming Yan] with permission from [Advances in Cement Research]

holding) stage. These outcomes differed noticeably from OPC specimens exposed to the same environmental factors, which showed no changes in mass in the first five weeks then rose dramatically and monotonously to about 0.7% after 180 days (Biricik et al. 2000). The entry of the solution and the creation of the deposits, as seen in Fig. 5.14g and h, might be the source of the first fast mass increases of the MKG specimens immersed in the magnesium sulfate solution in the first few weeks. At this point, M4.5-1-MS and M4-1-MS both had greater mass increases than M4-0.8-MS. The leaching of the material matrix was the cause of the weight losses during the latter immersion time. M4-0.8-MS lost mass by roughly 0.25% after 180 days of exposure, but M4.5-1-MS and M4-1-MS kept their positive mass increases (Fig. 5.15b). The rise and reduction in mass of the MKG mortars immersed in the magnesium sulfate solution were also reported by Yusuf (2015).

The material microstructure is closely related to the mass changes. For instance, the insufficient alkali content (e.g. low Na/Al ratios) can reduce the microstructure compactness of geopolymer materials (Duxson et al. 2007b; Rowles and O'Connor 2009; Yan et al. 2016). Additionally, studies conducted by the current authors showed that M4-0.8 mortar had the maximum porosity (21.19%; Table 5.6). This may explain why the M4-0.8-MS specimens with the lowest Na/Al ratio showed the most severe mass losses. The prepared MKG mortars were indicated to be rather stable in mass under the immersion activities of the magnesium sulfate solution up to 180 days since the final mass gains or losses were extremely modest (i.e., 0.5%). This point of view has been discussed in other places (Albitar et al. 2017).

5.3.3.3 Change in Length

Figure 5.16 depicts the length changes of the MKG specimens soaked in the sulfate solutions. The MKG mortars initially displayed very little expansions in response to the effects of both sodium sulfate and magnesium sulfate. For instance, the biggest of the three MKG groups, M4.5-1-SS, had a maximum length expansion of just 0.0325% (Fig. 5.16a). OPC specimens, however, were more sensitive to the immersing actions of the sodium sulfate solution—that is, up to 0.17% after 180 days of immersion (Valencia Saavedra et al. 2016).

Although the length variations of the MKG mortars that had been soaked in the magnesium sulfate solution were negligible, they did exhibit several distinct traits (Fig. 5.16b). M4-1-MS specimens specifically shrank somewhat, and it appears that the maximum shrinking took place at the same time as the largest weight gain (Fig. 5.15b). This behavior could be the result of a delicate equilibrium between the leaching of soluble minerals, which causes rapid early shrinkage, and the development of surface deposits, which simultaneously heavily and quickly increase the mass. Note that this case may only occur for the specimens with high contents of dissolvable minerals, such as the M4-1 specimens with the high content of sodium hydroxide (NaOH) (Table 5.6). However, those physico-chemical actions had no obvious influences on the final lengths. For instance, both M4-1-MS and M4.5-1-MS only shrank by around 0.006% and M4-0.8 expanded by 0.03% at 180 days.

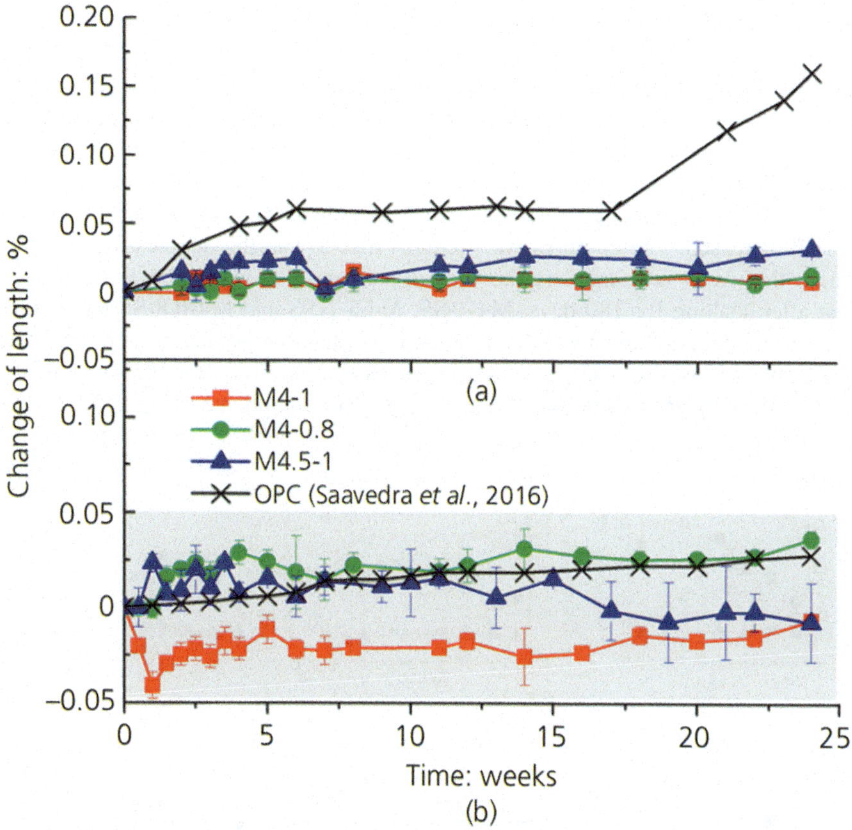

Fig. 5.16 Change in length of the MKG and OPC mortars in **a** sodium sulfate and **b** magnesium sulfate solutions. Reproduced from [Chemical–physical–mechanical stability of MKG mortars under sulfate attacks] by [Dongming Yan] with permission from [Advances in Cement Research]

The length changes obtained for the MKG mortars are close to those of fly-ash-slag geopolymer concrete after aging for 360 days in magnesium sulfate (final expansion of 0.04%) (Valencia Saavedra et al. 2016). The authors also reported the stable, constant dimensions of OPC mortars submerged in magnesium sulfate solution (Valencia Saavedra et al. 2016) (Fig. 5.16b). Similar outcomes were recorded for geopolymer materials made from fly ash that were subjected to sulfate solutions for 720 days (Skvára et al. 2005). The findings of Fig. 5.16 and those from the literature (Skvára et al. 2005; Valencia Saavedra et al. 2016) show that the kinds of concrete binder and sulfate have no impact on how the dimensions of concrete alter when immersed.

5.3.3.4 Mechanical Properties

Figure 5.17 depicts the flexural and compressive strengths of the MKG mortars under sulfate assault. During the whole immersing time, the flexural strengths of the MKG mortars fluctuated around 2.5 MPa in a constrained range. Similar outcomes were reported for MKG materials subjected to sodium sulfate attack (Palomo et al. 1999).

The MKG mortars' compressive strengths, however, changed as the soaking period increased. The compressive strengths rose to varying degrees for the specimens that were submerged in the sodium sulfate solution. Figure 5.17a–c show that after soaking for 180 days, M4-1-SS, M4.5-1-SS and M4-0.8-SS had gained in compressive strength by 22%, 17% and 7%, respectively. However, when the MKG mortars were submerged in the magnesium sulfate solution, the development of compressive strength became significantly more complicated (Fig. 5.5d–f); that

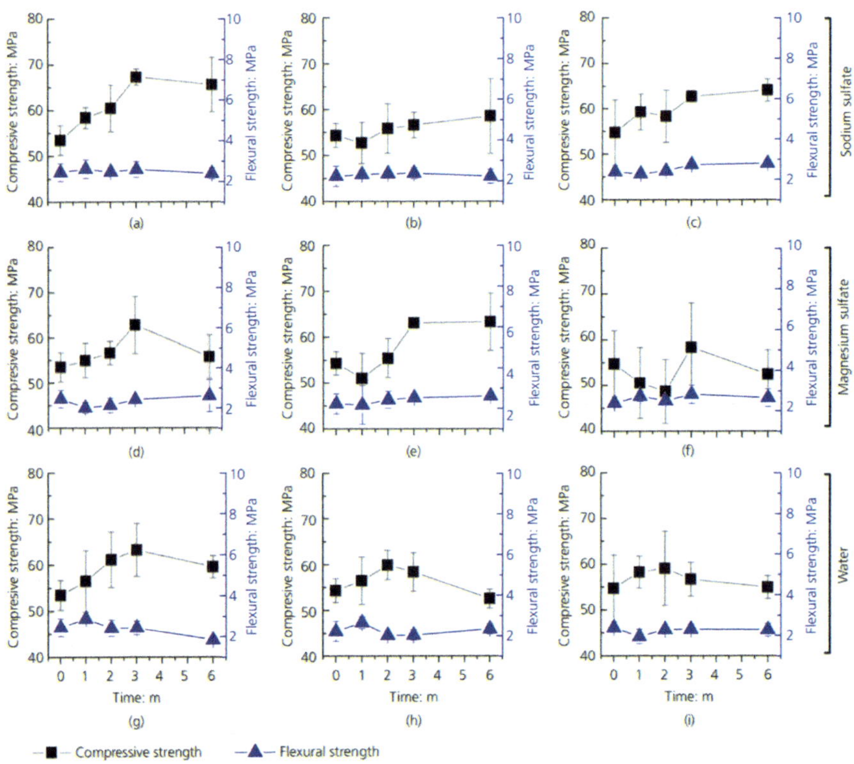

Fig. 5.17 Evolutions of compressive and flexural strengths of the MKG mortars immersed in sodium sulfate solution, shown in **a**, **b** and **c**, magnesium sulfate solution, shown in **d**, **e** and **f** and water, shown in **g**, **h** and **i**: **a** M4-1-SS; **b** M4-0.8-SS; **c** M4.5-1-SS; **d** M4-1-MS; **e** M4-0.8-MS; **f** M4.5-1-MS; **g** M4-1-H2O; **h** M4-0.8-H_2O; **i** M4.5-1-H_2O. Reproduced from [Chemical–physical–mechanical stability of MKG mortars under sulfate attacks] by [Dongming Yan] with permission from [Advances in Cement Research]

is, the compressive strengths for all mixtures altered in an erratic manner with the amount of time that they were submerged. The results were in line with the experimental data elsewhere (Fernandez-Jimenez et al. 2007; Khater 2013; Baščarević et al. 2014). Evidently, the reference mortar's compressive strengths grew while it was submerged in water due to the ongoing geopolymerization process, and subsequently they declined as a result of likely mineral leaching (Fig. 5.17g–i).

According to the findings illustrated in Fig. 5.17, it can be inferred that the impact of sulfate on the mechanical properties of MKG mortars, during immersion tests lasting up to 180 days, is not more detrimental than that of pure water. Nonetheless, when subjected to the influence of magnesium sulfate, the strength of MKG mortars exhibited a downward trend as the curing period extended, suggesting the possibility of strength deterioration during prolonged immersion exposure. As a matter of fact, Valencia Saavedra et al. (2016) reported a strength decrease by 33% for fly-ash-slag geopolymer concrete immersed in magnesium sulfate solution for 360 days. Thus, more thorough research is needed to understand the long-term mechanical stability of geopolymer materials due to the prolonged immersion period.

5.3.3.5 Micro Morphology

The SEM method was used to examine the microstructure of MKG specimens that had been submerged in sulfate solutions for varying lengths of time. Figures 5.18 and 5.19 are exemplary representations of this investigation. The samples were taken from the specimens' surface layers, which ranged in depth from 0 to 1 mm.

According to Fig. 5.18a, b, e, f, i, and j, all three groups of the MKG mortars immersed in the sodium sulfate solution for 30 days showed a relatively loose and porous microstructure without obvious crystals on the surfaces. In the matrix of M4-1-SS-30, there were cracking meshes with short, linked cracks (Fig. 5.18a), and in the two other mortars (Fig. 5.18e and i), there were porous surfaces with unevenly dispersed particles. At higher magnifications, the material heterogeneities may be seen (Fig. 5.18b, f, and j). Due to drying degradation, cracked meshes were also noted in the interior matrix (Kuenzel et al. 2012). Similar drying-induced microcracks were observed on the matrices of the fly ash and slag geopolymer materials at ambient and elevated temperatures (Çelikten et al. 2019). Severe drying may also alter geopolymer microstructure (Ismail et al. 2013a), but this is outside the scope of the present study.

The changes in micro morphology of the exposed geopolymer mortar surfaces under the immersion action of 5 wt% sodium sulfate for 180 days are illustrated in Fig. 5.18c, d, g, h, k and l. Although the material matrices were more denser and more uniform and had some significant flaws and/or spaces, the mortar surfaces became significantly smoother. Figure 5.17a–c illustrate the highly homogeneous microstructures and densely compacted matrices of the MKG mortars 180 days after immersion, which could be the result of continuous geopolymerization.

The magnesium sulfate solution was used as the exposure environment, and this affected how the MKG mortars' microstructure appeared. For instance, the surfaces

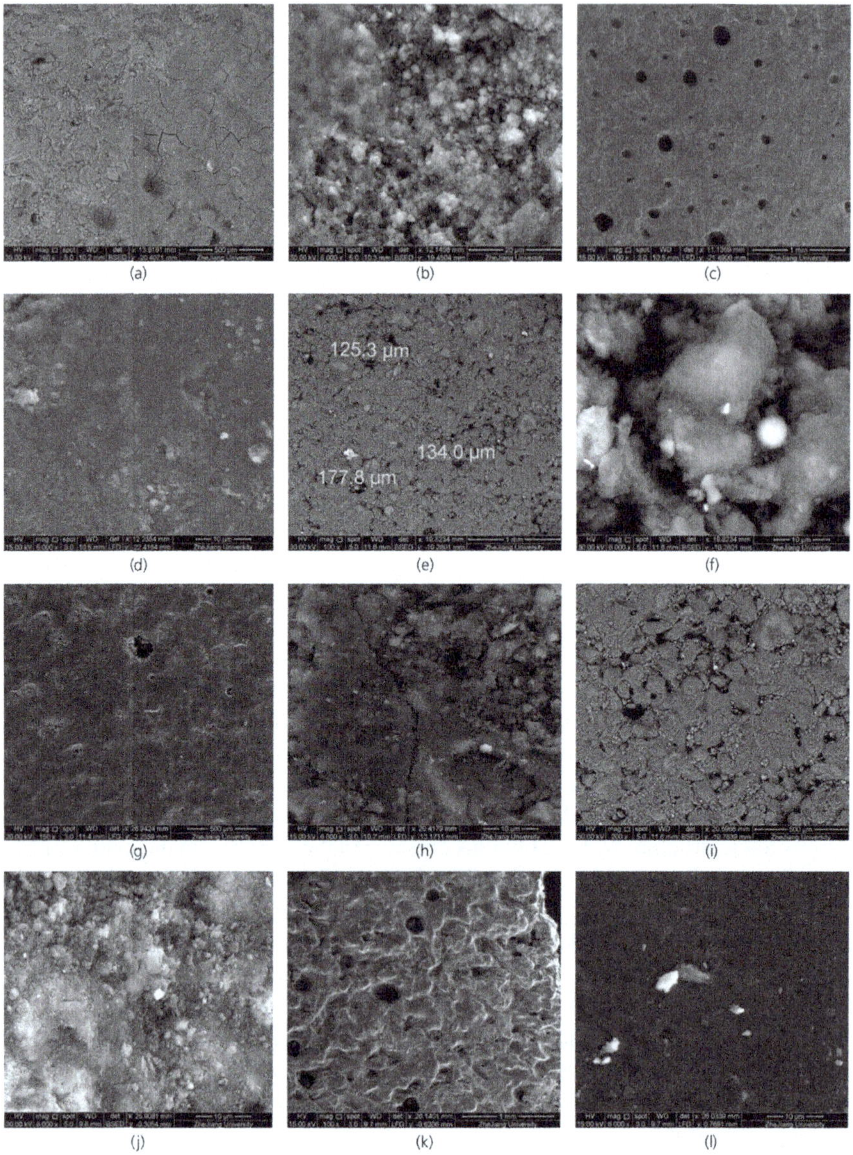

Fig. 5.18 Typical SEM images of the MKG mortars immersed in sodium sulfate solution: M4-1-SS-30 at **a** low and **b** high magnifications; M4-1-SS-180 at **c** low and **d** high magnifications; M4-0.8-SS-30 at **e** low and **f** high magnifications; M4-0.8-SS-180 at **g** low and **h** high magnifications; M4.5-1-SS-30 at **i** low and **j** high magnifications; and M4.5-1-SS-180 at **k** low and **l** high magnifications. Reproduced from [Chemical–physical–mechanical stability of MKG mortars under sulfate attacks] by [Dongming Yan] with permission from [Advances in Cement Research]

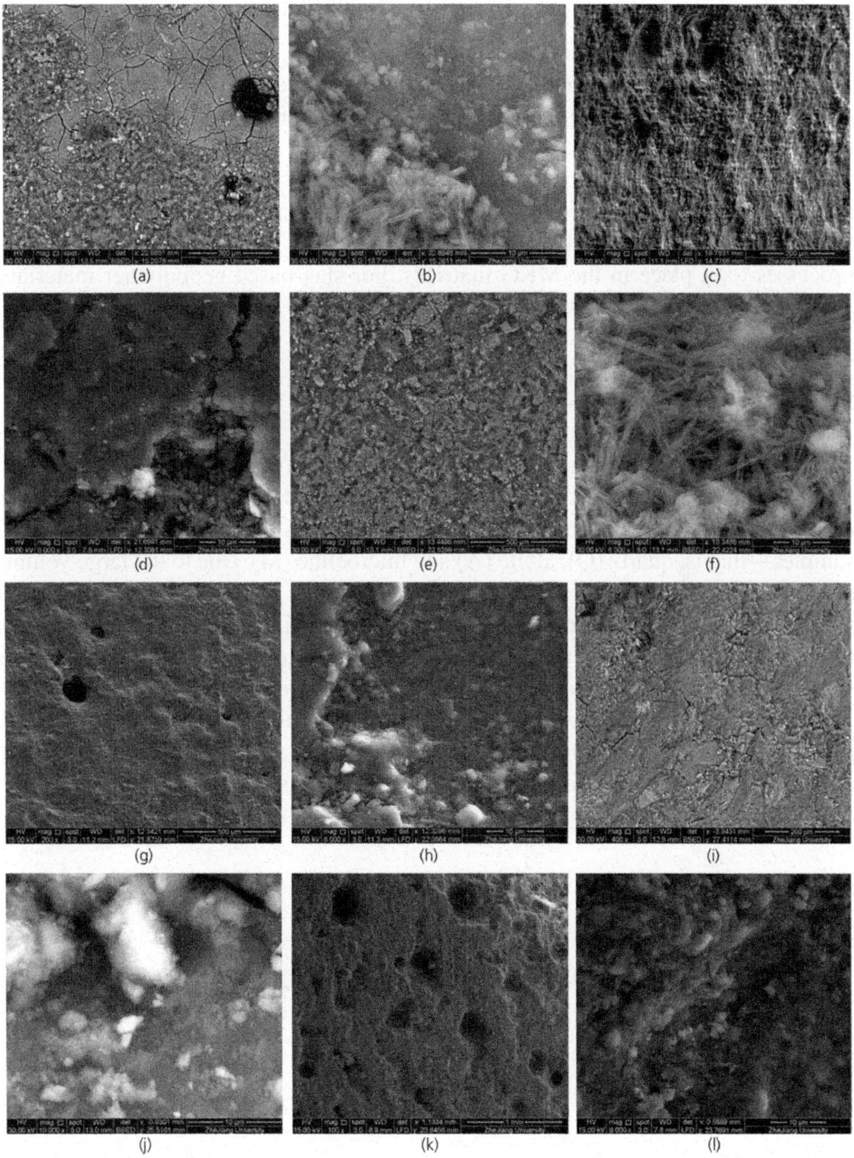

Fig. 5.19 Typical SEM images of the MKG mortars immersed in magnesium sulfate solution: M4-1-MS-30 at **a** low and **b** high magnifications; M4-1-MS-180 at **c** low and **d** high magnifications; M4-0.8-MS-30 at **e** low and **f** high magnifications; M4-0.8-MS-180 at **g** low and **h** high magnifications; M4.5-1-MS-30 at **i** low and **j** high magnifications; and M4.5-1-MS-180 at **k** low and **l** high magnifications. Reproduced from [Chemical–physical–mechanical stability of MKG mortars under sulfate attacks] by [Dongming Yan] with permission from [Advances in Cement Research]

of M4-1-MS-30 and M4-0.8-MS-30 displayed fasciculate crystals, whereas M4.5-1-MS-30's surfaces were comparatively clean (Fig. 5.19a, b, e, f, i and j). Those crystals were the white 'rime' shown in Fig. 5.14g and h. When the exposure age reached 180 days, again, the crystals vanished and the microstructure remained dense (Fig. 5.19c, d, g, h, k and l).

The MKG mortars' microstructure changed, but it is unlikely that it degraded as a result of the immersion actions of the sodium sulfate and magnesium sulfate solutions, according to the SEM analyses (Figs. 5.18 and 5.19), because no degradation processes took place in the MKG matrices. For slag-based geopolymer materials, similar findings were reached (Karakoç et al. 2016).

5.3.3.6 XRD Results

The XRD tests were conducted on superficial samples (0–1 mm depth) of the MKG specimens with and without the actions of sulfate, and the main outcomes are shown in Fig. 5.20. At first look, the main crystals were almost the same for all the mortar samples—that is, quartz (Q), albite (A) and microcline (M). Due to the large volume sand fractions, the peaks of quartz nearly completely dominated the entire XRD spectrum. The remaining crystals of A and M were minor, which in turn indicated that most of the MKG hydrates were in the amorphous state. This viewpoint was supported by the experimental results reported by Hawa et al. (2013).

Comparing the XRD curves before and after the actions of immersion, the peak positions did not change. This indicates that during the sulfate solutions' immersing operations, all crystal phases formed from the source materials were stable. Additionally, prolonging the immersion time had no influences on the XRD patterns of

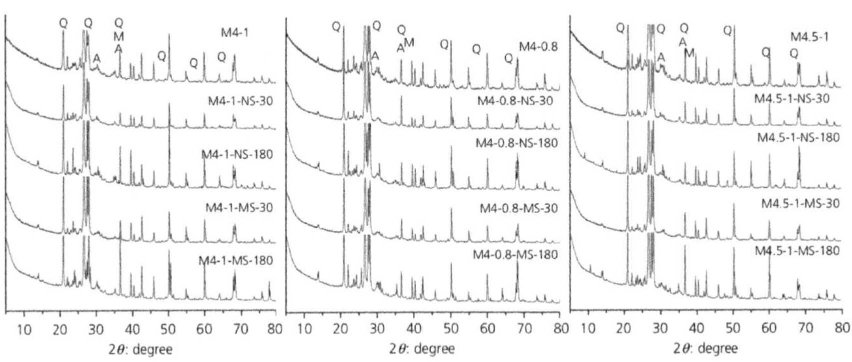

Fig. 5.20 XRD results of the MKG mortars immersed in the sulfate solutions for different exposure periods: **a** M4-1; **b** M4-0.8; **c** M4.5-1 (Q = quartz (silicon dioxide); A = albite (Na(Si3Al)O8) and M = microcline (KAlSi3O8)). Reproduced from [Chemical–physical–mechanical stability of MKG mortars under sulfate attacks] by [Dongming Yan] with permission from [Advances in Cement Research]

the MKG mortars (Fig. 5.8a–c). The expected magnesium hydroxide, which was supposed to account for the white deposits shown in Fig. 5.14g and h, however, could not be identified from the XRD curves. This can be as a result of the low magnesium hydroxide concentration and the variations in sample collection.

In Fig. 5.20, no obvious zeolites or zeolite-like crystals were probed, although geopolymers may be viewed as an amorphous type of zeolites (Provis et al. 2005). Indeed, the degree of crystallinity in the geopolymer microstructure is highly dependent on product formulation and synthesis conditions. It was reported that the crystalline structure can hardly form when the Si/Al ratio is higher than 2:1 (Wan et al. 2017). In addition, heating conditions are more likely to be used in the synthesis of zeolites (Provis et al. 2005). Both the mix formulations and the synthesising temperature in the study of the current authors did not promote the development of zeolites.

The XRD results (Fig. 5.8a–c) showed that the MKG mortars would retain their stability in mass, dimensions, microstructure, and mechanical properties after being submerged in the sodium sulfate and magnesium sulfate solutions. This is because sodium sulfate and calcium sulfate ($CaSO_4$), which typically generate large crystallization pressures on the pore wall, do not typically produce massive harmful crystals.

5.3.4 *Further Discussion*

5.3.4.1 Microstructure of Surface Deposits

The microstructure of the superficial deposits shown in Fig. 5.14g and h was investigated by the SEM–EDS technique (Fig. 5.9a–c). According to analysis, the white "rime" is composed of druse-like crystals with a predominance of Mg, O, and S (Fig. 5.21a) that are covered by an amorphous phase with a predominance of C, O, Na, Mg, Al, and Si (Fig. 5.21b). In the form of M–S–H and M–S–Al–H gels containing Mg, Al, Si, and O, the element distributions of the surface deposits were different from those of the tightly compacted geopolymer matrix (Bakharev 2005; Brew and Glasser 2005; Zhang et al. 2011; Chindaprasirt et al. 2013; Ismail et al. 2013b; Walling et al. 2015).

The chemical interactions of the deposits were measured by the FTIR measurement (Fig. 5.21d). According to Palomo et al. (1999), Guo et al. (2019), the large bands at 3418 cm^{-1} and 1650 cm^{1} were caused by the free water (or capillary water) and the chemically linked water, respectively. Additionally, the symmetric and asymmetric vibration of calcite was said to be the cause of the bands at 2927 cm^{-1} (Yusuf 2015). Magnesium carbonate was present as evidenced by the O–C–O bond's vibration at the bands of 1430–1490 cm^{-1} (Guo et al. 2019). According to Walling et al. (2015), the tiny absorption peak at 1052 cm^{1} indicated the development of M–S–H (Walling et al. 2015). The vibration of O–Si–O characterised at the band 1049 cm^{-1} (Yusuf 2015), which although it is minor in Fig. 5.21b, can still be

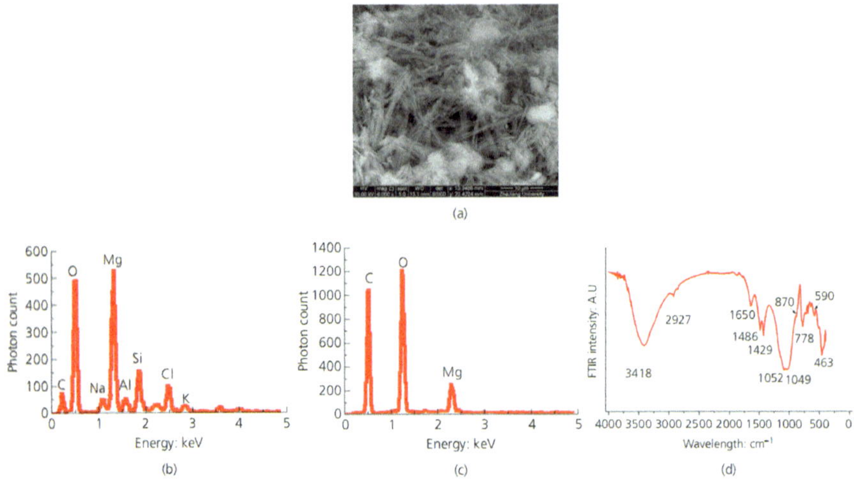

Fig. 5.21 **a** Typical SEM picture results of superficial localities for M4-1-MS-30; the EDS results of **b** an amorphous phase and **c** crystals; and **d** FTIR spectra of the white 'rime' on the same sample. Reproduced from [Chemical–physical–mechanical stability of MKG mortars under sulfate attacks] by [Dongming Yan] with permission from [Advances in Cement Research]

observed. According to Walling and Provis (2016), the vibration of Si–OH was indicated by the absorption peak about 870 cm^{-1}. According to Walling and Provis (2016), the combination of HO–Mg was responsible for the significant absorption peak at 778 cm^{-1}. The many absorption peaks that occurred in the FTIR curve between 590 and 700 cm^{-1} may have been brought on by the distorted tetrahedron shape of SO_4^{2} (Walling and Provis 2016). The absorption peaks in the range of 460–465 cm^{-1} were mainly due to the bending and stretching of Si–O and the connection of Al–O (He et al. 2016).

This white "rime" may represent a combination of magnesium hydroxide, magnesium sulfate, magnesium carbonate, and M–S(–Al)–H, according to the SEM–EDS and FTIR data. The formation of magnesium hydroxide was governed by the reaction (Walling and Provis 2016): $MgSO_4 + 2OH^- \rightarrow Mg(OH)_2 \downarrow + SO_4^{2-}$. Magnesium hydroxide and magnesium sulfate were carbonated to produce magnesium carbonate. The M–S(–Al)–H phase was a product of the geopolymerising reaction. M–S(–Al)–H gels were created as a result of the reaction between the produced magnesium hydroxide crystals and the dissolved SiO_4^{2-} (or AlO_3^{3-}) (Brew and Glasser 2005; Walling et al. 2015). In this regime, the surface deposits disappeared with immersion time (Fig. 5.2g and h).

5.3.4.2 The Role of Mg

Under the immersion action of magnesium sulfate, the exchanges between the cations (mainly Mg^{2+} and Na^+) might take place progressively from the superficial to inner areas. The authors used EDS analysis to quantify the concentrations of Mg and Na from the outer to the interior regions in order to gain a better grasp of the roles that the cations performed in the immersion activities. Figure 5.22 displays the distribution profiles of Mg and Na in the M4-1-MS-90 and M4.5-1-MS-90 specimens. It is obvious that as the probed locations went from the exposed surfaces to the interior regions, the intensity of Mg gradually decreased while that of Na gradually rose. This could be because while the specimens were continuously submerged in the magnesium sulfate solution, Mg eventually replaced the Na that was originally linked with polymerized silica-aluminate minerals. Similar observations have been documented elsewhere (Ismail et al. 2013a; Walling et al. 2015).

The Mg and Na profile discrepancies between the M4-1 and M4.5-1 MKG mortars were also investigated. Apparently, the penetration depth of Mg was around 1 mm for the M4-1 material (Fig. 5.22a), but increased to over 4 mm for the M4.5-1 one (Fig. 5.22b). The shorter invasion distance of Mg into the M4-1 material may be due to its denser microstructure with the lower porosity (Table 5.6) and the filling effect of the superficial deposits (mainly magnesium hydroxide) that could block the open channels for Mg penetration. The saturation of the Mg bindings to the superficially polymerized silica-aluminate minerals resulted in essentially identical Mg concentrations at the specimen surfaces for the two mortars (15.74% for M4.1 vs. 15.41% for M4.5-1) for both mortars. The M4.5-1 material's larger mass gains after 180 days of immersion were consistent with the increased Mg penetration depth (Fig. 5.15b). A research by Ismail et al. (2013a) also reported considerable changes in the silicate gel bonding environment of geopolymers submerged in magnesium sulfate. However, it is still hard to draw any conclusions from the current studies on how the penetrated Mg would alter the nanostructure of MKG gels and the engineering consequences. Considering the possible data bias of EDX tests induced by surface roughness, more tough evidence is required from further rigorous investigations.

5.4 Conclusions

We investigated the ion exchange procedure, MKG structures, and physical alterations in concentrated sodium and magnesium sulfate environments in this chapter. By combining the experimental results of compressive strength, pH value, ion concentrations, XRD, FTIR, and SEM, several significant conclusions can be inferred:

(1) The cation of the sulfate has an impact on the ion exchange between MKG and the sulfate solution. In sodium sulfate solution, the ion exchange process is dominated by the interchange of OH^- and SO^{2-}, whereas in magnesium sulfate solution, it is dominated by the exchange of Na^+ and Mg^{2+}. The sulfate type would also affect the migration of other ions (such CO^{2-}).

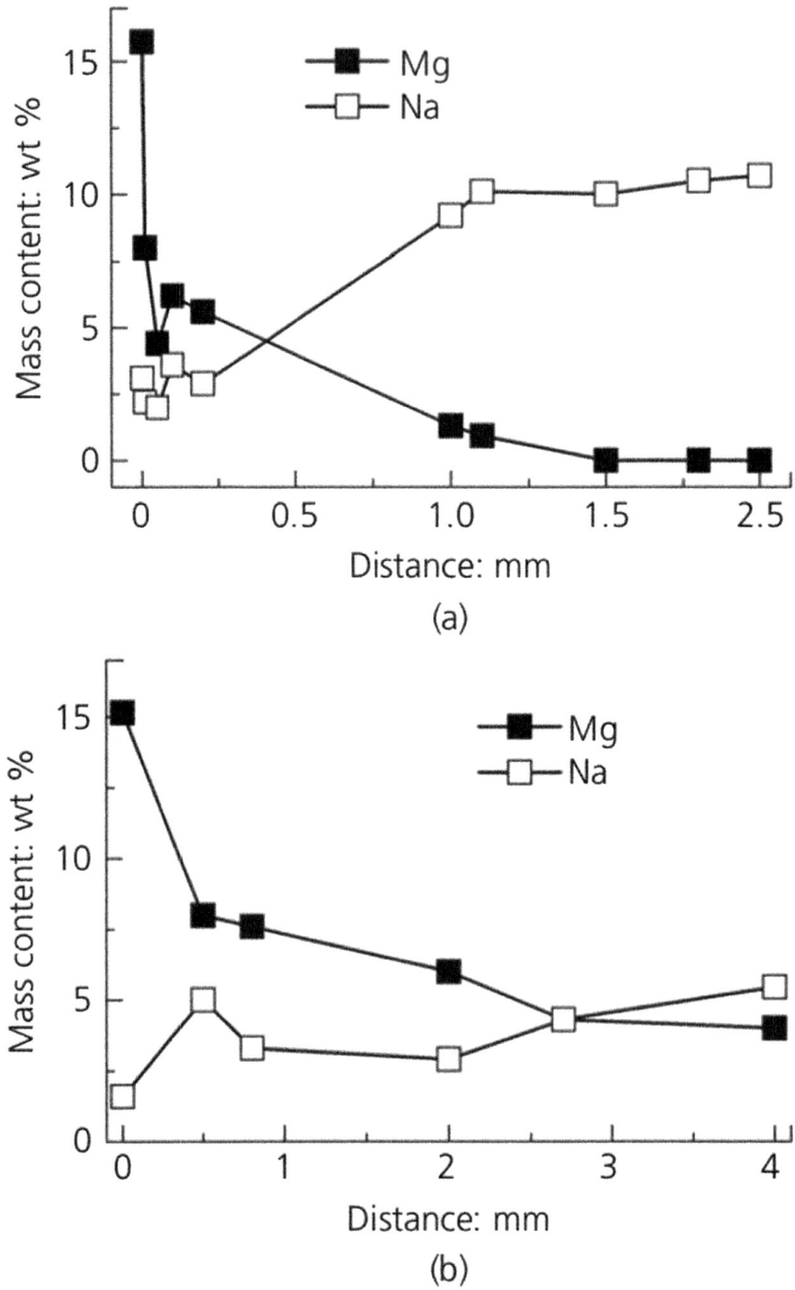

Fig. 5.22 Contents of Na and Mg in the geopolymer mortars immersed in magnesium sulfate solution probed by EDS: **a** M4-1-MS-90 and **b** M4.5–1-MS-90. Reproduced from [Chemical–physical–mechanical stability of MKG mortars under sulfate attacks] by [Dongming Yan] with permission from [Advances in Cement Research]

(2) The immersion in sulfate solution may alter the aluminosilicate network of MKG, and this alteration was more pronounced in sodium sulfate solution. When compared to changes that occurred in OPC, this change in gel structure is minor.
(3) The impact of Si/Al ratio on samples immersed in sodium silicate solution was found to be negligible. The overall trend observed in various test results was similar for both mixtures with different Si/Al ratios, with only minor differences evident. Therefore, the Si/Al ratio is not considered a significant factor in the corrosion of MKG in a sodium sulfate environment.
(4) In the case of samples immersed in magnesium silicate solution, the influence of Si/Al ratio was found to be significant. The mixture with a high silicate content demonstrated notable changes in micromorphology and strength following immersion. These changes suggest a higher reactivity of MKG with a high silicate content in magnesium sulfate solution. Consequently, an increase in the Si/Al ratio resulted in improved mechanical performance.
(5) The results in this chapter, however, are only applicable to MKG with a very low calcium concentration. Additional research is still required to determine how other elements (such as calcium concentration, carbonate dissolving, etc.) affect the outcome.
(6) The MKG mortars' length, mass, and strength did not change significantly during the course of their exposure to the sodium sulfate and magnesium sulfate solutions. Under the effects of immersion, the MKG mortars' chemical composition and surface microstructure displayed various features.
(7) During the initial immersion of MKG mortars in magnesium sulfate solution, a layer of deposits consisting of magnesium hydroxide, magnesium sulfate, magnesium carbonate, and M–S(–Al)–H formed on the surfaces of the M4-1-MS and M4-0.8-MS mortars. This occurrence was attributed to the relatively high alkali content present in the material matrix. However, with further prolonged exposure, the superficial deposits gradually disappeared due to the continuous consumption of magnesium hydroxide.
(8) The gradual replacement of sodium (Na) by magnesium (Mg) in the geopolymer matrices resulted in changes to the distribution of Mg near the surfaces of the specimens. Currently, these Mg–Na exchanges appear to have minor effects on the material and mechanical properties of MKG mortars when immersed in magnesium sulfate solution. However, it is important to focus on the long-term implications of microstructural and chemical alterations in MKG materials under sulfate environments. These aspects require further attention and investigation.
(9) In general, the MKG mortars demonstrated no deterioration in terms of mass, dimensions, and strength when subjected to immersion in sodium sulfate and magnesium sulfate solutions. The favorable physical and mechanical stability observed in the MKG materials aligns with findings reported for other geopolymer materials in existing literature. Nevertheless, the chemical stability

of MKG materials is highly influenced by their composition and the interactions between active minerals and environmental conditions. Because of this, engineering applications of geopolymer-based structures and infrastructure in sulfate settings should be carefully considered.

References

Alanazi H, Yang M, Zhang D, Gao Z (2017) Early strength and durability of metakaolin-based geopolymer concrete. Mag Concr Res 69:46–54. https://doi.org/10.1680/jmacr.16.00118

Albitar M, Mohamed Ali MS, Visintin P, Drechsler M (2017) Durability evaluation of geopolymer and conventional concretes. Constr Build Mater 136:374–385. https://doi.org/10.1016/j.conbuildmat.2017.01.056

Bakharev T (2005) Durability of geopolymer materials in sodium and magnesium sulfate solutions. Cem Concr Res 35:1233–1246. https://doi.org/10.1016/j.cemconres.2004.09.002

Baščarević Z, Komljenović M, Miladinović Z et al (2014) Impact of sodium sulfate solution on mechanical properties and structure of fly ash based geopolymers. Mater Struct Constr 48:683–697. https://doi.org/10.1617/s11527-014-0325-4

Bhutta MAR, Hussin WM, Azreen M, Tahir MM (2014) Sulphate resistance of geopolymer concrete prepared from blended waste fuel ash. J Mater Civ Eng 26. https://doi.org/10.1061/(ASCE)MT.1943-5533.0001030

Biricik H, Aköz F, Türker F, Berktay I (2000) Resistance to magnesium sulfate and sodium sulfate attack of mortars containing wheat straw ash. Cem Concr Res 30:1189–1197. https://doi.org/10.1016/S0008-8846(00)00314-8

Brew DRM, Glasser FP (2005) Synthesis and characterisation of magnesium silicate hydrate gels. Cem Concr Res 35:85–98. https://doi.org/10.1016/j.cemconres.2004.06.022

Çelikten S, Sarıdemir M, Özgür Deneme İ (2019) Mechanical and microstructural properties of alkali-activated slag and slag + fly ash mortars exposed to high temperature. Constr Build Mater 217:50–61. https://doi.org/10.1016/j.conbuildmat.2019.05.055

Çevik A, Alzeebaree R, Humur G et al (2018) Effect of nano-silica on the chemical durability and mechanical performance of fly ash based geopolymer concrete. Ceram Int 44:12253–12264. https://doi.org/10.1016/j.ceramint.2018.04.009

Chiang W-S, Ferraro G, Fratini E et al (2014) Multiscale structure of calcium-and magnesium-silicate-hydrate gels. J Mater Chem A 2:12991–12998

Chindaprasirt P, Rattanasak U, Taebuanhuad S (2013) Resistance to acid and sulfate solutions of microwave-assisted high calcium fly ash geopolymer. Mater Struct Constr 46:375–381. https://doi.org/10.1617/s11527-012-9907-1

Chotetanorm C, Chindaprasirt P, Sata V et al (2013) High-calcium bottom ash geopolymer: sorptivity, pore size, and resistance to sodium sulfate attack. J Mater Civ Eng 25:105–111. https://doi.org/10.1061/(ASCE)MT.1943-5533.0000560

Cyr M, Pouhet R (2016) Carbonation in the pore solution of metakaolin-based geopolymer. Cem Concr Res 88:227–235. https://doi.org/10.1016/j.cemconres.2016.05.008

Duan P, Yan C, Zhou W (2016) Influence of partial replacement of fly ash by metakaolin on mechanical properties and microstructure of fly ash geopolymer paste exposed to sulfate attack. Ceram Int 42:3504–3517. https://doi.org/10.1016/j.ceramint.2015.10.154

Duxson P, Mallicoat SW, Lukey GC et al (2007a) The effect of alkali and Si/Al ratio on the development of mechanical properties of metakaolin-based geopolymers. Colloids Surfaces A Physicochem Eng Asp 292:8–20. https://doi.org/10.1016/j.colsurfa.2006.05.044

Duxson P, Provis JL, Lukey GC, van Deventer JSJ (2007b) The role of inorganic polymer technology in the development of "green concrete." Cem Concr Res 37:1590–1597. https://doi.org/10.1016/j.cemconres.2007.08.018

El-Sayed HA, El-Enein SAA, Khate HM, Hasanein SA (2011) Resistance of alkali activated water-cooled slaggeopolymer to sulphate attack. Ceram—Silikaty 55:153–160

Elyamany HE, Abd Elmoaty AEM, Elshaboury AM (2018) Magnesium sulfate resistance of geopolymer mortar. Constr Build Mater 184:111–127. https://doi.org/10.1016/j.conbuildmat.2018.06.212

Fernandez-Jimenez A, García-Lodeiro I, Palomo A (2007) Durability of alkali-activated fly ash cementitious materials. J Mater Sci 42:3055–3065. https://doi.org/10.1007/s10853-006-0584-8

GB/T-17671–1999 (1999) Method of testing cements: Determination of strength

GB/T 749-2008 (2008) Test method for determining the capability of resisting sulfate corrode of cement. 151–156

Gomez-Villalba LS, Sierra-Fernandez A, Quintana P et al (2018) Correlation between microstructure and cathodoluminescence properties of $Mg(OH)_2$ (brucite) nanoparticles: effect of synthesis method. CrystEngComm 20:5632–5640. https://doi.org/10.1039/c8ce00942b

Guo Y, Curtis JS (2015) Discrete element method simulations for complex granular flows. Annu Rev Fluid Mech 47:21–46. https://doi.org/10.1146/annurev-fluid-010814-014644

Guo Z, Wang Y, Hou P et al (2019) Comparison study on the sulfate attack resistivity of cement-based materials modified with nano SiO_2 and conventional SCMs: mechanical strength and volume stability. Constr Build Mater 211:556–570. https://doi.org/10.1016/j.conbuildmat.2019.03.235

Guo L, Wu Y, Xu F et al (2020) Sulfate resistance of hybrid fiber reinforced metakaolin geopolymer composites. Compos Part B Eng 183. https://doi.org/10.1016/j.compositesb.2019.107689

Hasnaoui A, Ghorbel E, Wardeh G (2019) Optimization approach of granulated blast furnace slag and metakaolin based geopolymer mortars. Constr Build Mater 198:10–26. https://doi.org/10.1016/j.conbuildmat.2018.11.251

Hawa A, Tonnayopas D, Prachasaree W, Taneerananon P (2013) Investigating the effects of oil palm ash in metakaolin based geopolymer. Ceram—Silikaty 57:319–327

He P, Wang M, Fu S et al (2016) Effects of Si/Al ratio on the structure and properties of metakaolin based geopolymer. Ceram Int 42:14416–14422. https://doi.org/10.1016/j.ceramint.2016.06.033

Hekal EE, Kishar E, Mostafa H (2002) Magnesium sulfate attack on hardened blended cement pastes under different circumstances. Cem Concr Res 32:1421–1427. https://doi.org/10.1016/S0008-8846(02)00801-3

Hung TC, Huang JS, Wang YW, Fan YC (2013) Microstructure and properties of metakaolin-based inorganic polymer foams. J Mater Sci 48:7446–7455. https://doi.org/10.1007/s10853-013-7559-3

Ismail I, Bernal SA, Provis JL et al (2013a) Influence of fly ash on the water and chloride permeability of alkali-activated slag mortars and concretes. Constr Build Mater 48:1187–1201. https://doi.org/10.1016/j.conbuildmat.2013.07.106

Ismail I, Bernal SA, Provis JL et al (2013b) Microstructural changes in alkali activated fly ash/slag geopolymers with sulfate exposure. Mater Struct Constr 46:361–373. https://doi.org/10.1617/s11527-012-9906-2

Karakoç MB, Türkmen I, Maraş MM et al (2016) Sulfate resistance of ferrochrome slag based geopolymer concrete. Ceram Int 42:1254–1260. https://doi.org/10.1016/j.ceramint.2015.09.058

Karthik A, Sudalaimani K, Vijayakumar CT (2017) Durability study on coal fly ash-blast furnace slag geopolymer concretes with bio-additives. Ceram Int 43:11935–11943. https://doi.org/10.1016/j.ceramint.2017.06.042

Khater HM (2013) Effect of cement kiln dust on geopolymer composition and its resistance to sulfate attack. Green Mater 1:36–46. https://doi.org/10.1680/gmat.12.00003

Khatib JM, Wild S (1998) Sulphate resistance of metakaolin mortar. Cem Concr Res 28:83–92. https://doi.org/10.1016/S0008-8846(97)00210-X

Komljenović M, Baščarević Z, Marjanović N, Nikolić V (2013) External sulfate attack on alkali-activated slag. Constr Build Mater 49:31–39. https://doi.org/10.1016/j.conbuildmat.2013.08.013

Kuenzel C, Vandeperre LJ, Donatello S et al (2012) Ambient temperature drying shrinkage and cracking in metakaolin-based geopolymers. J Am Ceram Soc 95:3270–3277. https://doi.org/10.1111/j.1551-2916.2012.05380.x

Kunther W, Lothenbach B, Skibsted J (2015) Influence of the Ca/Si ratio of the C–S–H phase on the interaction with sulfate ions and its impact on the ettringite crystallization pressure. Cem Concr Res 69:37–49. https://doi.org/10.1016/j.cemconres.2014.12.002

Kwasny J, Aiken TA, Soutsos MN et al (2018) Sulfate and acid resistance of lithomarge-based geopolymer mortars. Constr Build Mater 166:537–553. https://doi.org/10.1016/j.conbuildmat.2018.01.129

Li J, Shi M, Zhao G (2020) Degradation and life prediction model for piles due to crystallisation attack in sulfate saline area. Adv Cem Res 32:181–195. https://doi.org/10.1680/jadcr.18.00147

Maltseva A, Shkirskiy V, Lefèvre G, Volovitch P (2019) Effect of pH on $Mg(OH)_2$ film evolution on corroding Mg by in situ kinetic Raman mapping (KRM). Corros Sci 153:272–282. https://doi.org/10.1016/j.corsci.2019.03.024

Mehta PK (1983) Mechanism of sulfate attack on portland cement concrete—another look. Cem Concr Res 13:401–406. https://doi.org/10.1016/0008-8846(83)90040-6

Montes C, Allouche EN (2012) Evaluation of the potential of geopolymer mortar in the rehabilitation of buried infrastructure. Struct Infrastruct Eng 8:89–98. https://doi.org/10.1080/15732470903329314

Nuaklong P, Sata V, Chindaprasirt P (2018) Properties of metakaolin-high calcium fly ash geopolymer concrete containing recycled aggregate from crushed concrete specimens. Constr Build Mater 161:365–373. https://doi.org/10.1016/j.conbuildmat.2017.11.152

Ozer I, Soyer-Uzun S (2015) Relations between the structural characteristics and compressive strength in metakaolin based geopolymers with different molar Si/Al ratios. Ceram Int 41:10192–10198. https://doi.org/10.1016/j.ceramint.2015.04.125

Palomo A, Blanco-Varela MT, Granizo ML et al (1999) Chemical stability of cementitious materials based on metakaolin. Cem Concr Res 29:997–1004. https://doi.org/10.1016/S0008-8846(99)00074-5

Provis JL, Lukey GC, van Deventer JSJ (2005) Do geopolymers actually contain nanocrystalline zeolites? A reexamination of existing results. Chem Mater 17:3075–3085

Rashidian-Dezfouli H, Rangaraju PR (2017) A comparative study on the durability of geopolymers produced with ground glass fiber, fly ash, and glass-powder in sodium sulfate solution. Constr Build Mater 153:996–1009. https://doi.org/10.1016/j.conbuildmat.2017.07.139

Ren D, Yan C, Duan P et al (2017) Durability performances of wollastonite, tremolite and basalt fiber-reinforced metakaolin geopolymer composites under sulfate and chloride attack. Constr Build Mater 134:56–66. https://doi.org/10.1016/j.conbuildmat.2016.12.103

Rowles MR, O'Connor BH (2009) Chemical and structural microanalysis of aluminosilicate geopolymers synthesized by sodium silicate activation of metakaolinite. J Am Ceram Soc 92:2354–2361. https://doi.org/10.1111/j.1551-2916.2009.03191.x

Salami BA, Megat Johari MA, Ahmad ZA, Maslehuddin M (2017) Durability performance of palm oil fuel ash-based engineered alkaline-activated cementitious composite (POFA-EACC) mortar in sulfate environment. Constr Build Mater 131:229–244. https://doi.org/10.1016/j.conbuildmat.2016.11.048

Sata V, Sathonsaowaphak A, Chindaprasirt P (2012) Resistance of lignite bottom ash geopolymer mortar to sulfate and sulfuric acid attack. Cem Concr Compos 34:700–708. https://doi.org/10.1016/j.cemconcomp.2012.01.010

Škvára F, Jílek T, Kopecký L (2005) Geopolymer materials based on fly ash. Ceram—Silikaty 49:195–204

Tchakouté HK, Rüscher CH (2017) Mechanical and microstructural properties of metakaolin-based geopolymer cements from sodium waterglass and phosphoric acid solution as hardeners: a comparative study. Appl Clay Sci 140:81–87. https://doi.org/10.1016/j.clay.2017.02.002

Thokchom S, Ghosh P, Ghosh S (2010) Performance of fly ash based geopolymer mortars in sulphate solution. J Eng Sci Technol Rev 3:36–40. https://doi.org/10.25103/jestr.031.07

Valencia Saavedra WG, Angulo DE, Mejía de Gutiérrez R (2016) Fly ash slag geopolymer concrete: resistance to sodium and magnesium sulfate attack. J Mater Civ Eng 28. https://doi.org/10.1061/(ASCE)MT.1943-5533.0001618

Walling SA, Kinoshita H, Bernal SA et al (2015) Structure and properties of binder gels formed in the system $Mg(OH)_2$-SiO_2-H_2O for immobilisation of Magnox sludge. Dalt Trans 44:8126–8137. https://doi.org/10.1039/c5dt00877h

Walling SA, Provis JL (2016) Magnesia-based cements: a journey of 150 years, and cements for the future? Chem Rev 116:4170–4204. https://doi.org/10.1021/acs.chemrev.5b00463

Wan Q, Rao F, Song S et al (2017) Geopolymerization reaction, microstructure and simulation of metakaolin-based geopolymers at extended Si/Al ratios. Cem Concr Compos 79:45–52. https://doi.org/10.1016/j.cemconcomp.2017.01.014

White CE, Page K, Henson NJ, Provis JL (2013) In situ synchrotron X-ray pair distribution function analysis of the early stages of gel formation in metakaolin-based geopolymers. Appl Clay Sci 73:17–25. https://doi.org/10.1016/j.clay.2012.09.009

Whittaker M, Black L (2015) Current knowledge of external sulfate attack. Adv Cem Res 27:532–545. https://doi.org/10.1680/adcr.14.00089

Yan D, Chen S, Zeng Q et al (2016) Correlating the elastic properties of metakaolin-based geopolymer with its composition. Mater Des 95:306–318. https://doi.org/10.1016/j.matdes.2016.01.107

Yusuf MO (2015) Performance of slag blended alkaline activated palm oil fuel ash mortar in sulfate environments. Constr Build Mater 98:417–424. https://doi.org/10.1016/j.conbuildmat.2015.07.012

Zhang Z, Yao X, Zhu H (2010) Potential application of geopolymers as protection coatings for marine concrete I. Basic properties. Appl Clay Sci 49:1–6. https://doi.org/10.1016/j.clay.2010.01.014

Zhang T, Cheseman CR, Vandeperre LJ (2011) Development of low pH cement systems forming magnesium silicate hydrate (M-S-H). Cem Concr Res 41:439–442. https://doi.org/10.1016/j.cemconres.2011.01.016

Zhutovsky S, Douglas Hooton R (2017) Experimental study on physical sulfate salt attack. Mater Struct Constr 50. https://doi.org/10.1617/s11527-016-0936-z

Chapter 6
Heat Resistance of MKG

Abstract This chapter investigated the degradation process of metakaolin-based geopolymers under high temperatures of up to 1000 °C. The study focused on examining changes in the microstructure, phase assemblage, and mechanical properties of the samples. The results indicated that the type of activator—$Na_2SiO_3 + NaOH$ (Na/Na) and $Na_2SiO_3 + KOH$ (Na/K) solution—influenced the degradation mechanisms of geopolymer mortars at high temperatures. Geopolymers based on (Na, K) exhibited superior thermal resistance above 200 °C, revealing higher compressive strength, lower porosity, and reduced cracking tendency compared to those based on (Na, Na). Mortars based on (Na, Na) experienced more mass loss, leading to significant drying shrinkage, and further crack development at 200 °C. At temperatures exceeding 200 °C, crack development and material property degradation resulted in the decline of the mechanical properties of geopolymers. In contrast to their (Na-Na) counterparts, (Na–K)-based geopolymers showed improved chemical stability and did not develop new crystalline phases over 1000 °C. Higher temperatures (1000 °C) caused geopolymers to sinter significantly, resulting in a thick and homogenous matrix and bettering the specimens' mechanical characteristics. Overall, the results point to the critical role that Na^+ and K^+'s mutual encouraged effects play in the development of cracks, sintering, and new crystallization in geopolymers at high temperatures.

6.1 Introduction

In high-temperature environments, reinforced concrete structures using hardened ordinary Portland cement (OPC) tend to undergo phase decomposition leading to significant performance deterioration (EAPFP 1987; Ali et al. 2001; Phan and Carino 2004). Consequently, substitutes to OPC, such as alkali-activated materials and high-temperature resistant geopolymers, have gained attention due to their superior properties (Davidovits 1991; Duxson et al. 2007a; Komnitsas and Zaharaki 2007; Benito et al. 2013; Thomas et al. 2016). In comparison to OPC, geopolymers demonstrate high early-age strength, excellent chemical durability, and enhanced heat resistance (Papakonstantinou et al. 2001; Mohd Salahuddin et al. 2015; Huiskes et al. 2016).

D. Yan et al., *Metakaolin-Based Geopolymers*,
https://doi.org/10.1007/978-981-97-0652-5_6

Additionally, geopolymers are environmentally friendly due to low energy consumption and reduced greenhouse gas emissions during production (Davidovits 1993). Geopolymers have found innovative applications in precast geopolymer concretes and 3D printing mortars (Panda et al. 2019, 2020; Kastiukas et al. 2020).

Geopolymers are more resistant to degradation at high temperatures than OPC due to the superior thermal stability of their gel matrix, which lacks portlandite (Barbosa and MacKenzie 2003a, b; Lahoti et al. 2018). Fly ash and metakaolin are the most commonly used precursors for geopolymer synthesis. Fly ash-based geopolymers have been found to introduce additional interconnected pores, leading to improved volumetric stability compared to their metakaolin-based counterparts (Kong et al. 2007; Zhang et al. 2014; Lahoti et al. 2018). The high-temperature resistance of metakaolin-based geopolymers has been studied by various researchers who examined parameters such as aluminosilicate precursors and alkaline activators (Duxson et al. 2006a, 2007a; Kong et al. 2007). However, the effect of high temperatures on the mechanical properties of geopolymers appears to be contradictory. Some studies have shown an increase in strength for metakaolin-based geopolymers after exposure to a temperature of 1000 °C (Bernal et al. 2011; Kuenzel et al. 2013), while others observed adverse effects (Kong et al. 2007).

The type of alkaline cations (such as Na, K, Li, or a combination thereof) is a crucial factor affecting the thermal resistance of metakaolin or fly ash-based geopolymers (Duxson et al. 2006b, 2007c; Lahoti et al. 2018). According to Lahoti et al. (2018), when fly ash-based geopolymers were subjected to high temperatures, the strength of the material was significantly impacted by the type of alkali cations employed (i.e., Na and K). Additionally, alkali cations played a role in the thermal shrinkage and phase composition of metakaolin-based geopolymers at elevated temperatures (Duxson et al. 2006b, 2007c). Uncertainty persists regarding the effect of different alkali cations on the mechanical characteristics and fracture formation of metakaolin-based geopolymers at high temperatures. Furthermore, the methods by which geopolymers containing various kinds of activators degrade after being exposed to a wide range of temperatures remain poorly known.

The primary objective of this chapter is to examine the correlation between mechanical properties, pore/crack structures, and the chemical compositions of geopolymers under high-temperature conditions. Specifically, the impact of alkali cation types on the mechanical properties and microstructure of metakaolin-based geopolymers exposed to elevated temperatures was scrutinized. Mass loss, compression, and bending experiments were carried out on geopolymer specimens after exposure to various temperatures to assess its macroscopic behavior. Additionally, high temperature geopolymer microstructural and compositional changes were examined.

Table 6.1 Oxide composition of raw metakaolin

Oxide	SiO_2	Al_2O_3	TiO_2	Fe_2O_3	Na_2O	K_2O	CaO	LOI
Mass content (%)	57.46	39.81	1.79	0.43	0.27	0.21	0.04	0.34

Reproduced from [Effects of Activator Types on Degradation Mechanisms of Metakaolin Geopolymer Mortars Exposed to High Temperature] by [Yajun Zhang] with permission from [Journal of Materials in Civil Engineering]

Table 6.2 Oxide composition of sodium silicate solution

Oxide	SiO_2	Na_2O	H_2O
Mass content (%)	26.00	8.20	65.80

Reproduced from [Effects of Activator Types on Degradation Mechanisms of Metakaolin Geopolymer Mortars Exposed to High Temperature] by [Yajun Zhang] with permission from [Journal of Materials in Civil Engineering]

6.2 Materials and Methods

6.2.1 Materials

This chapter utilized commercially available metakaolin powder (Metamax, BASF) as the solid aluminosilicate precursor. Table 6.1 provides the composition of the metakaolin powder as measured by X-ray fluorescence (XRF) spectrometry. The metakaolin powder had an average particle size of 5.89 m. In this work, sodium hydroxide (NaOH) and potassium hydroxide (KOH) were dissolved in a sodium silicate solution of commercial quality to create two distinct activator solutions. NaOH and KOH had purity levels of 96% and 88%, respectively. The chemical composition of the sodium silicate solution is provided in Table 6.2, indicating that the solution consisted of 26% SiO_2 by weight, 8.20% Na_2O by weight, and 65.8% H_2O by weight. Geopolymer mortars were produced by utilizing sand with a fineness modulus of 1.75 and a density of 2510 kg/m^3 as aggregate.

6.2.2 Preparation of Specimens

The compositional ratios of geopolymer paste, expressed in molar ratio, can be found in Table 6.3. The resulting alkaline activators were cooled and sealed at room temperature for a day, then mixed with metakaolin powder for five minutes. The mixture was then hermetically sealed for three days at room temperature before demolding and subjected to moist curing for up to seven days. To prevent furnace damage, all specimens were dried in an oven at 60 °C for one day.

Geopolymer paste samples were molded into molds measuring 40 mm by 40 mm by 40 mm for X-ray diffraction (XRD) and thermogravimetric analysis (TGA).

Table 6.3 Composition proportion [molar ratio: SiO_2/Al_2O_3, $Na_2O(K_2O)$/Al_2O_3, H_2O/Al_2O_3] of metakaolin-based geopolymer

Chemical	SiO_2/Al_2O_3	$Na_2O(K_2O)/Al_2O_3$	H_2O/Al_2O_3
Molar ratio	4.00	1.00	14.76

Reproduced from [Effects of Activator Types on Degradation Mechanisms of Metakaolin Geopolymer Mortars Exposed to High Temperature] by [Yajun Zhang] with permission from [Journal of Materials in Civil Engineering]

Accordingly, (Na-Na)-based activator and (Na–K)-based activator, respectively, were produced as GP-NaNa and GP-NaK. Geopolymer mortar, GM-NaNa, and GM-NaK with an aggregate-to-geopolymer paste mass ratio of 3.0 were also prepared using the same alkaline activators as in the paste preparation.

6.3 Experimental

6.3.1 Exposure to High Temperatures

The cured specimens were subjected to different temperatures (200, 400, 600, 800, and 1000 °C) using a heating rate of 5 °C/min starting from room temperature. Each target temperature was maintained for 1 h after the furnace had reached it, before turning off the furnace and allowing the specimens to cool down naturally.

6.3.2 Mechanical Properties

In order to assess the reduction in strength of geopolymers following exposure to high temperatures, a universal testing machine (Instron 8802, INSTRON-Division of ITW, Norwood, Massachusetts) was used to conduct flexural and compressive tests. The flexural and compressive strengths were measured using prisms of size 40 × 40 × 160 mm and cubes of size 40 × 40 × 40 mm respectively, in accordance with the GB/T 17671–1999 standard test method (National Standard of People's Republic of China 1999). Six samples were examined for each mixture to guarantee repeatability, and the findings are displayed in Figs. 6.1 and 6.2 with their means and error bars. It should be mentioned that each geopolymer mortar prism was put through compressive testing with displacement control at a loading speed of 1 mm/min after undergoing a three-point bending test with a span length of 100 mm and a loading rate of 3 kN/min.

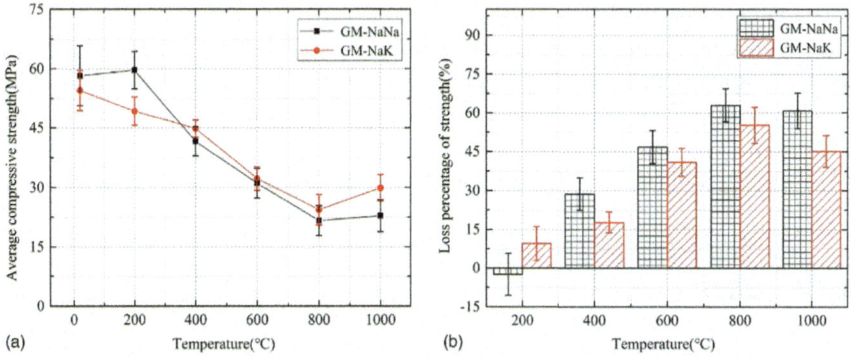

Fig. 6.1 **a** Average compressive strength; and **b** loss percentage of compressive strength of geopolymer mortars containing different activators at various elevated temperatures. Reproduced from [Effects of Activator Types on Degradation Mechanisms of Metakaolin Geopolymer Mortars Exposed to High Temperature] by [Yajun Zhang] with permission from [Journal of Materials in Civil Engineering]

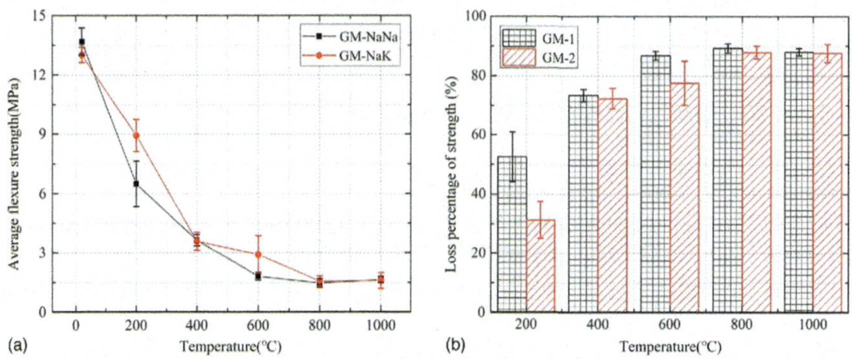

Fig. 6.2 **a** Residual flexural strength; and **b** loss percentage of flexural strength of geopolymer mortars containing different activators at various elevated temperatures. Reproduced from [Effects of Activator Types on Degradation Mechanisms of Metakaolin Geopolymer Mortars Exposed to High Temperature] by [Yajun Zhang] with permission from [Journal of Materials in Civil Engineering]

6.3.3 X-Ray Computed Microtomography

Specimens were subjected to X-ray computed microtomography (μCT) scanning both at room temperature and after exposure to temperatures ranging from 200 to 1000 °C. Raw μCT data were collected from 2D X-ray projections captured at various angles, and the linear attenuation coefficients of different specimen components were recorded. In general, the densities of different components are positively correlated with their linear attenuation coefficients. The interior architecture of the specimens were detected by gray threshold segmentation (Otsu's method) on a series of 2D cross sectional pictures using the maximum between-class variance approach.

Accordingly, morphological variables including volume and location were examined. With a tube voltage of 80 kV and a tube current of 120 μA, a Nikon XT H 320 (Nikon Corporation, Tokyo) equipment was employed for the μCT scanning of the study. Each specimen was scanned for 15 min, capturing 1600 2D X-ray projections. The voxel size of the resulting 3D digital model was $3.5 \times 3.5 \times 3.5$ μm. Finally, VGStudio was utilized for analyzing the scanning results.

6.3.4 XRD

An X'Pert Pro (PANalytical B.V., Almelo, Netherlands) apparatus with Ni-filtered CuK radiation was used to perform the X-ray diffraction (XRD) examination. The tube current and generator voltage were both set to 40 mA and 40 kV, respectively. With a step size of 0.02° and a collecting duration of 0.2 s each step, the 2 range was between 5° and 80°. In order to learn more about how geopolymers degrade at high temperatures, specimens that had been heated to 400, 800, and 1000 °C were ground, and their phases were determined using XRD analysis.

6.3.5 Scanning Electron Microscopy

Subsequent to conducting compressive tests, remaining specimens were used for microstructural analysis without polishing. The crushed specimens were analyzed at 20 kV through an FEI Quanta FEG650 (FEI Corporation, Hillsboro, Oregon) field emission environmental scanning electron microscope, both before and after being exposed to elevated temperatures.

6.3.6 Thermogravimetric Analysis

Geopolymer specimens containing NaOH or KOH were ground using a 120-mesh sieve. Prior to testing, the specimens were not pre-dried, in order to preserve their original condition. Thermal gravimetric analysis (TGA) and differential scanning calorimetry (DSC) were performed using a NETZSCH STA 449 F3 (NETZSCH Corporation, Selb, Deutschland) instrument, with the specimens being heated from 25 to 1200 °C in air at a constant heating rate of 5 °C/min.

6.4 Results and Discussion

6.4.1 *Mechanical Properties: Compressive Strength*

GM-NaNa and GM-NaK specimens' compressive strength and strength loss % after being subjected to both room temperature and increased temperature are shown in Fig. 6.1. It was clear from a comparison of the compressive strengths of GM-NaNa and GM-NaK that the type of alkali cations had little bearing on the strength development at different temperatures. At room temperature, there was no significant difference between the compressive strengths of GM-NaNa (58.2 MPa) and GM-NaK (54.6 MPa) specimens. However, when exposed to 200 °C, a minor increase in the compressive strength of GM-NaNa specimens was observed while the GM-NaK specimens experienced a slight decrease. Nonetheless, this alteration can be considered negligible given the error bars. From 200 to 800 °C, the compressive strength of both GM-NaNa and GM-NaK groups reduced gradually as a result of material degradation, sintering, crystallization, internal structural changes such as pore structure and cracks. Prior studies have indicated that decreased compressive strength is attributed to differential thermal expansion of aggregate and geopolymer binder, along with crack formation (Kong and Sanjayan 2008; Pan et al. 2014). The compressive strength of GM-NaNa and GM-NaK specimens both somewhat increased after exposure to 1000 °C, however GM-NaK specimens' average compressive strength was much greater than that of GM-NaNa specimens.

6.4.2 *Flexural Strength*

The geopolymer mortars' flexural strength and strength loss % with regard to temperature are shown in Fig. 6.2. The GM-NaNa group's average flexural strength at room temperature (13.70 MPa) was comparable to the GM-NaK group's (i.e., 13 MPa). Nevertheless, there was a decrease in the average flexural strength of both groups with increasing exposure temperature.

Once the exposure temperature reached 200 °C, comparable reductions in average flexural strength were observed between the GM-NaNa and GM-NaK groups. The average flexural strength of GM-NaNa specimens decreased by more than half compared to specimens at room temperature, whereas a 31.30% decrease in the average flexural strength of GM-NaK was evident at temperatures above 200 °C. According to Lahoti et al. (2018), maximum thermal deformation of geopolymers using different types of alkali cation occurred within the range of 150–300 °C, and that the major shrinkage of (Na–K)-based geopolymer was less significant compared to its (Na-Na)-based counterpart, suggesting that the GM-NaNa group may have considerably more cracks than the GM-NaK group at this stage. The average flexural strengths of the GM-NaNa and GM-NaK groups were close, reaching 3.64 (3.60), 1.47 (1.57), and 1.64 (1.60) MPa at 400 °C, 800 °C, and 1000 °C, respectively. Experimental results indicate that degradation of flexural strength of both GM-NaNa and GM-NaK groups was more significant compared to their compressive strength before reaching 200 °C. This shows that, compared to the compressive strength, the flexural strength of the geopolymer mortar was more vulnerable to microstructural changes before reaching 200 °C. The process of strength deterioration below 800 °C would be very different from that beyond 800 °C, as shown by the fact that there was only a modest drop in flexural strength after reaching 400 °C and notably after 800 °C.

6.4.3 Microstructural Analysis with μCT: Cracking Patterns

Pores that have diameters larger than 100 μm and low sphericity (sphericity ratio of surface area of equal volume sphere to actual surface area less than 0.3) can be regarded as cracks. The volume of cracks is a crucial factor in studying the crack propagation process, and therefore, 3D digital models of the cracks were developed. Figures 6.3 and 6.4 depict the evolution and progression of cracks in specimens after being subjected to high temperatures.

Figure 6.3 revealed that the GM-NaNa group exhibited internal cracking at 200 °C, with three isolated cracks having a total volume of 92.50 mm^3, which was consistent with the notable drop in flexural strength. As a result, the flexural strength of geopolymer mortars was significantly impacted by crack development. However, when the temperature approached 200 °C, the compressive strength rose, demonstrating that tiny, evenly spaced fractures were not affected by the specimens' compressive strength during this time. These fissures continued to spread and partially linked as the temperature rose. At 400 °C, the strength was significantly reduced and the existing fractures reached the surface with a volume of 228.80 mm^3. The specimen developed new fractures in several locations as the exposure temperature was raised to 800°C and 1000°C, and the pre-existing cracks widened. Furthermore, the growth rate of crack volume accelerated, with values reaching 334.60 and 654.68 mm^3 at 800 °C and 1000 °C, respectively.

Figures 6.1 and 6.2 indicate that above 800 °C, there was no decrease in compressive and flexural strengths observed with an increase in crack volume. The inherent characteristics of the mortar's matrix material and internal structure, including cracks and pores, are what essentially define the material's compressive strength. Due to the possibility of forming nepheline or sintering at 1000 °C (Kljajević et al. 2017; Lahoti et al. 2018), it is plausible to enhance the mechanical properties of geopolymers through phase change and sintering mechanisms in specimens.

The crack propagation laws in the GM-NaK group were akin to those observed in the GM-NaNa group. However, as depicted in Fig. 6.4, the rates of propagation varied between the two groups. The first cracks with a volume of 44.80 mm^3 emerged in the GM-NaK group at 200 °C. Subsequently, these cracks extended to the surface at 600 °C and continued to grow upon exposure to temperatures of 800 and 1000 °C.

The strength analysis of the GM-NaNa and GM-NaK groups revealed that cracks played a crucial role in the development of strength levels between 200 and 800 °C,

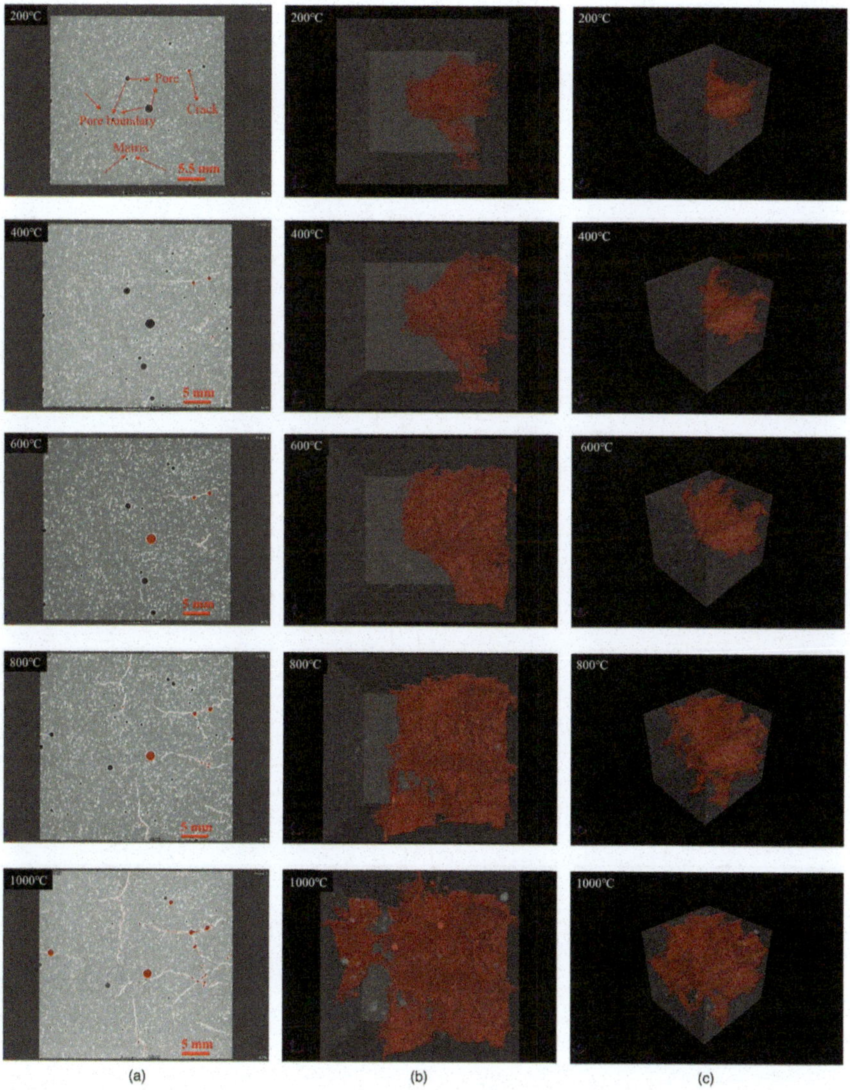

Fig. 6.3 CT microtomography of cracks in geopolymer mortars of GM-NaNa after different elevated temperature exposure: **a** unsegmented 2D slice; **b** segmented crack; and **c** 3D crack visualization. Reproduced from [Effects of Activator Types on Degradation Mechanisms of Metakaolin Geopolymer Mortars Exposed to High Temperature] by [Yajun Zhang] with permission from [Journal of Materials in Civil Engineering]

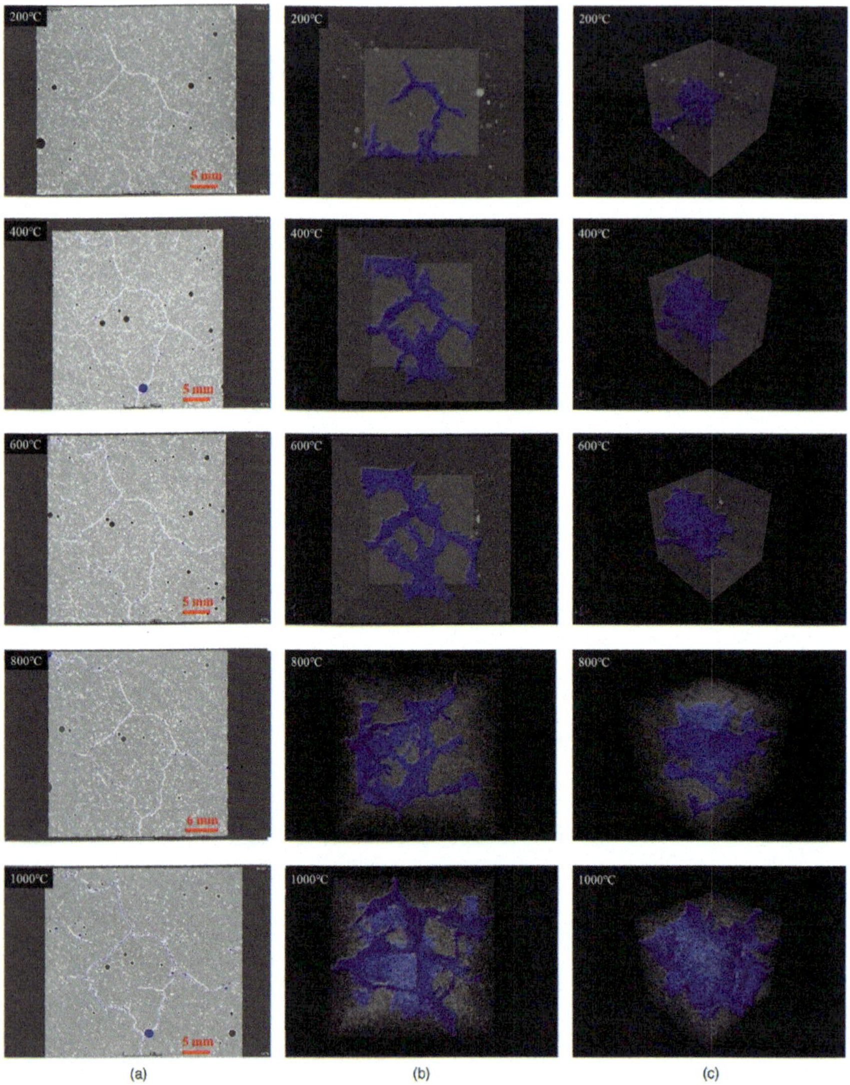

Fig. 6.4 CT microtomography of cracks in geopolymer mortars of GM-NaK after different elevated temperature exposure: **a** unsegmented 2D slice; **b** segmented crack; and **c** 3D crack visualization. Reproduced from [Effects of Activator Types on Degradation Mechanisms of Metakaolin Geopolymer Mortars Exposed to High Temperature] by [Yajun Zhang] with permission from [Journal of Materials in Civil Engineering]

with crack development resulting in a reduction in specimen strength. Cracks continued to widen at 800 °C, although phase change or sintering took over as the main determinants of high temperature strength levels. There are several processes over 800 °C (such as sintering and crystallization) that can take the place of the mechanism of strength deterioration below 800 °C (internal structure such as pore structure and fracture. The results highlight that geopolymer mortar specimens with different types of alkali activators exhibited distinct crack development patterns. In geopolymer mortar containing (Na-Na)-based activators, cracks can emerge early and propagate rapidly towards the surface. The formation of cracks in the GM-NaK group was less substantial than in the GM-NaNa group. This observation could be attributed to the inferior volume stability of geopolymers containing (Na-Na)-based activators (Duxson et al. 2006b; Lahoti et al. 2018). Prior research has shown that NaK geopolymers manifest less thermal shrinkage compared to Na geopolymers (Duxson et al. 2006b; Lahoti et al. 2018). As the GM-NaNa and GM-NaK groups had the same aggregate, one could expect fewer cracks and better volume stability in the GM-NaK group.

6.4.4 Porosity Analysis

Porosity was evaluated based on the μCT data for further analysis (Fig. 6.5). Pores exceeding 3.5 μm in size, including significant cracks as per voxel size, were considered. Subsequently, sample images of cubes were obtained via μCT, and pores of varying sizes at room temperature were represented using different colors. The porosity of the GM-NaNa group was found to be 1.27 times greater than that of the GM-NaK group at room temperature. This could be attributed to the thicker and more paste-like slurry texture of the GM-NaNa group during specimen preparation, which made it more challenging to eliminate air and excess water during the vibration process, thereby reducing compactness and increasing porosity. Additionally, Na^+ pairs displayed smaller silicate monomers compared to K^+ pairing with larger silicate oligomers (Xu and Deventer 2002). This disparity may account for the differences in the microstructure of specimens containing various alkali activators.

Specimens experience water loss at temperatures ranging between room temperature and 200 °C (Kamseu et al. 2010; Kuenzel et al. 2013), leading to an increased porosity of the specimens. Additionally, the expansion of water compresses pore walls and triggers microcrack formation. The drying process of two specimens prior to high-temperature exposure limited their free water content, rendering it challenging to detect microcracks efficiently via μCT. The findings indicated that the porosity of these two specimens remained almost constant between ambient temperature and 200 °C. It is worth noting that the CT possesses a high resolution (i.e., 3.5 μm) and the low density of water makes its color appear similar to air on the CT image. Therefore, the increase in porosity can be attributed to pore expansion and crack development. Furthermore, when the specimens were subjected to high temperatures of 400, 600, 800, and 1000 °C, both types of mortars demonstrated a

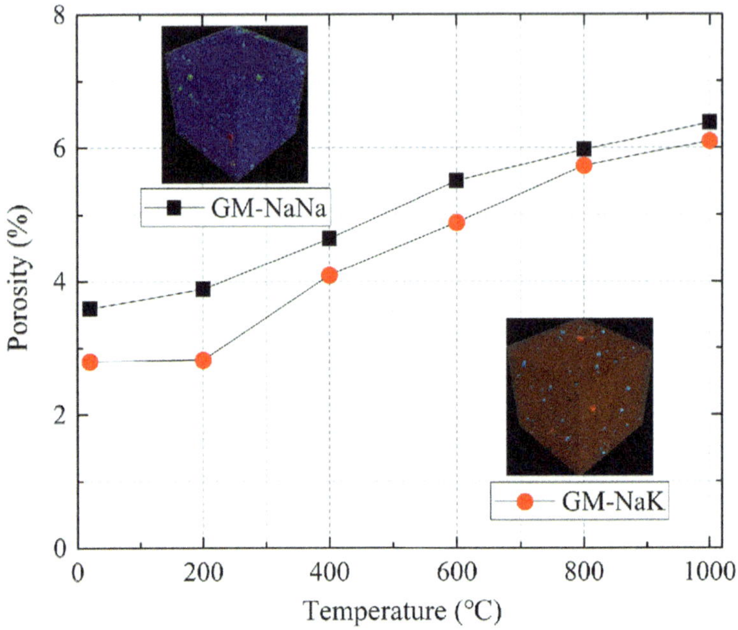

Fig. 6.5 Porosities of geopolymer mortars containing different activators at different temperatures. Reproduced from [Effects of Activator Types on Degradation Mechanisms of Metakaolin Geopolymer Mortars Exposed to High Temperature] by [Yajun Zhang] with permission from [Journal of Materials in Civil Engineering]

continuous increase in porosity. The GM-NaK group exhibited lower porosity than the GM-NaNa group, indicating superior volume stability of geopolymers containing K^+ at high temperatures (Sindhunata et al. 2006).

6.4.5 *Pore-Size Distribution*

The pore-size distribution of the GM-NaNa and GM-NaK groups is shown in Fig. 6.6. The number of pores with a diameter of 5.77 μm, which had the highest count in both GM-NaNa and GM-NaK, were 14,018 and 6518, respectively, under ambient temperature. The pore-size distribution underwent changes with increasing heat exposure. With more heat exposure, the distribution of pore sizes changed. In specimens heated to 200 °C, the number of tiny holes (i.e., 5.77 m) was significantly reduced in the GM-NaNa group. This phenomenon suggests that exposure to high temperatures may catalyze the geopolymerization of unreacted metakaolin particles. After being exposed to a temperature of 200 °C, the GM-NaK group showed only minor variations in its pore structure, which was consistent with the specimens' stable porosity.

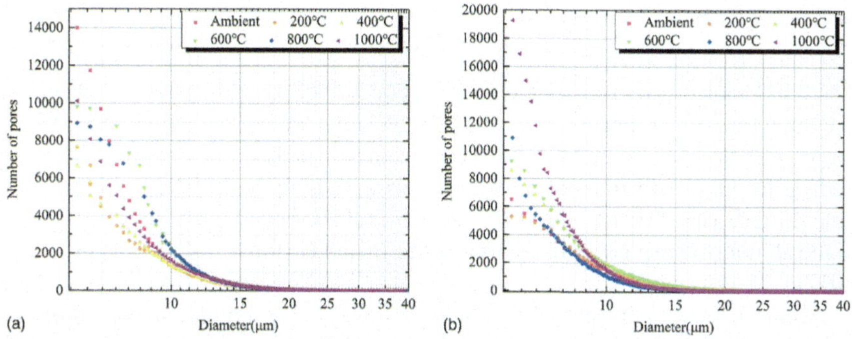

Fig. 6.6 Pore-size distribution of geopolymer mortars containing different activators at different temperatures: **a** GM-NaNa; and **b** GM-NaK. Reproduced from [Effects of Activator Types on Degradation Mechanisms of Metakaolin Geopolymer Mortars Exposed to High Temperature] by [Yajun Zhang] with permission from [Journal of Materials in Civil Engineering]

From temperatures of 200 up to 600 °C, the small pore-size distribution of the GM-NaNa group demonstrated no evident changes due to evaporation of the pore water before reaching 200 °C. Nonetheless, upon exposure to temperatures of 600 and 800 °C, the microstructure of the GM-NaNa group underwent dramatic changes, and the number of pores increased significantly compared to that of specimens exposed to 400 °C. This change may not be attributable to water in specimens (including free water and bound water), as both evaporate at 20–250 °C (Abdulkareem et al. 2014; Zhang et al. 2016), and dehydroxylation of the gel occurs at 250–600 °C (Duxson et al. 2007b; Kuenzel et al. 2012). We believe that partial sintering was a crucial contributor to the variations in microstructure and pore size in these specimens. The number of some pores (such as those with diameters of 6.60, 6.95, 7.56, and 8.54 μm) declined above 1000 °C. At this temperature, geopolymers can undergo crystallization or sintering, leading to changes in the pore structure of the specimens (Kuenzel et al. 2013). In the GM-NaK group, changes in pore size were mostly prominent above 800 and 1000 °C due to sintering and melting processes.

At exposure temperatures of 600, 800, and 1000 °C, there are noticeable differences between the pore architectures of GM-NaNa and GM-NaK specimens. Specifically, the pore size of the GM-NaNa group underwent the most substantial increase at 600 °C, while that for the GM-NaK group experienced a remarkable rise at 800 °C. This phenomenon points to a difference in their sintering temperatures. Due to the presence of K, the enhanced Al–O bonds made it more challenging for the GM-NaK group to achieve density (Duxson et al. 2007c; Lahoti et al. 2018). The result indicates that sintering is not an effortless process for the GM-NaK group. The number of holes changed differently in the GM-NaNa and GM-NaK groups after exposure to 1000 °C, indicating substantial variations in the internal structure of geopolymer mortars containing various activators.

6.4.6 XRD

The XRD analysis of raw metakaolin and geopolymer paste based on metakaolin is shown in Fig. 6.7. Anatase (TiO_2) and quartz (SiO_2) are present in tiny quantities in the amorphous phase of metakaolin, which has a halo in the 2 range of 15°–30°. Upon adding the alkali activators to metakaolin during geopolymerization, the amorphous halo shifted to the 2θ range of 22°–35°, which could be attributed to potential changes in the aluminosilicate network environment (Ye and Radlińska 2016; Kljajević et al. 2017).

At an increased exposure temperature of 800 °C, an amorphous structure was detected in the GP-NaNa group (Fig. 6.7a), similar to that observed at room temperature. However, the peak moved to a lower 2 range of 20°–25°, and the amorphous halo enlarged (18°–35°). The structure almost completely disappeared at 1000 °C, and new peaks appeared at 10.22°, 26.98°, 27.50°, 29.16°, 34.74°, 30.86°, 39.24°, 43.52°, 59.28°, 68.52°, 65.52°, 71.86°, and 74.22°. Subsequently, the geopolymer was annealed and converted to nepheline.

After being exposed to high temperatures, the amorphous structure of the GP-NaK group displayed a characteristic change process. The height of the halo peak shrank but the amorphous halo's breadth and position virtually stayed unaltered. This observation suggests a nanostructural change such as gel decomposition. Unlike the GP-NaK group, a new crystallization peak emerged in the GP-NaNa group. The results indicate that (Na–K)-based geopolymers provided improved phase stability over their (Na-Na)-based counterparts without forming new crystal phases upon thermal exposure. Free alkalis can lower the crystallization temperature of nepheline (Krivenko and Kovalchuk 2007; Rickard et al. 2012; Lahoti et al. 2018). Consequently, the X-ray diffraction (XRD) patterns of the GP-NaK group showed no emergence of new phases, implying that crystalline phases, such as nepheline, were challenging to

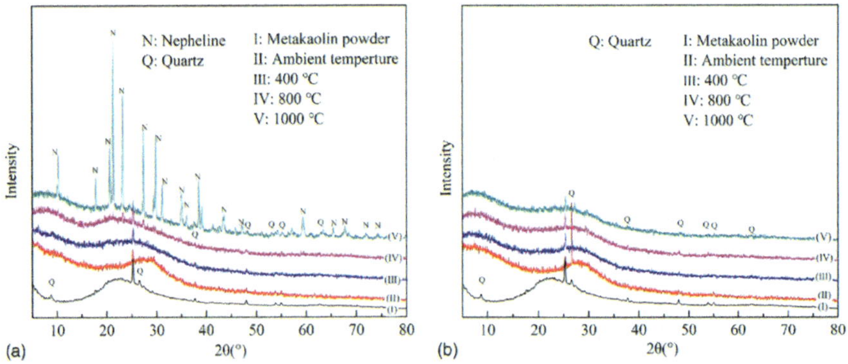

Fig. 6.7 XRD results of metakaolin powder and geopolymer pastes containing different activators at various elevated temperatures: **a** GP-NaNa; and **b** GP-NaK. Reproduced from [Effects of Activator Types on Degradation Mechanisms of Metakaolin Geopolymer Mortars Exposed to High Temperature] by [Yajun Zhang] with permission from [Journal of Materials in Civil Engineering]

form due to the deficit of free alkalis. In geopolymers, Na^+ is prone to form $Si(OH)_4$, while K^+ can hasten the polycondensation reaction between $Si(OH)_4$ and $Al(OH)_4$, contributing to the growth of Si–O–Al bonds (Yachao 2014). It may be concluded that Na^+ and K^+'s mutually encouraged actions resulted in greater geopolymerization, preventing excessive free alkalis from being seen after being exposed to high temperatures.

6.4.7 SEM Observation

Figures 6.8 and 6.9 present SEM images of the GM-NaNa and GM-NaK groups before and after exposure to high temperatures at 200, 400, 600, 800, and 1000 °C. The images were considered representative microstructures. They demonstrate that at room temperature, the microstructural differences between the GM-NaNa and GM-NaK groups are negligible. Less residual metakaolin particles were seen as layered phases in the GM-NaK group, indicating that the interaction between Na^+ and K^+ facilitated geopolymerization.

Up to 600 °C, the microstructures of the GM-NaNa and GM-NaK groups exhibited no apparent changes. At 600 °C, the matrix in the GM-NaNa specimens partially sintered and melted, with the phenomenon of melted material becoming more evident at 800°C. In contrast, the sintering onset temperature of the GM-NaK specimens was

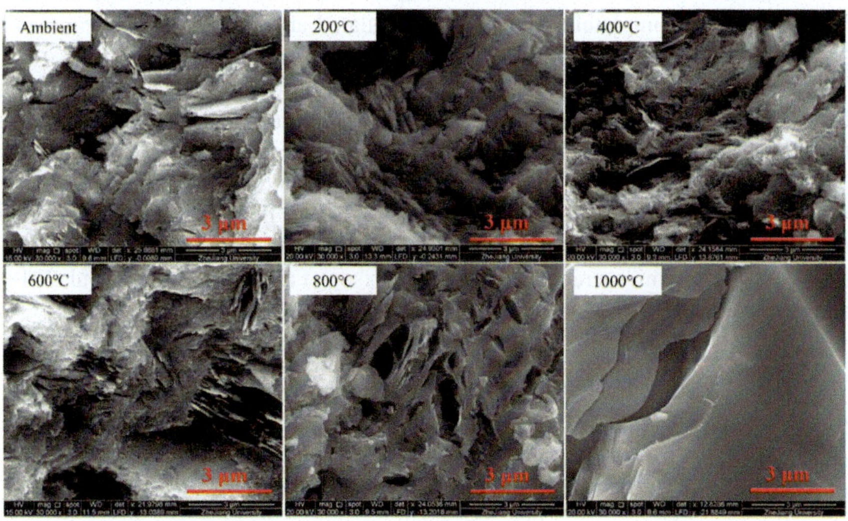

Fig. 6.8 SEM micrographs of geopolymer mortars of GM-NaNa at various elevated temperatures. Reproduced from [Effects of Activator Types on Degradation Mechanisms of Metakaolin Geopolymer Mortars Exposed to High Temperature] by [Yajun Zhang] with permission from [Journal of Materials in Civil Engineering]

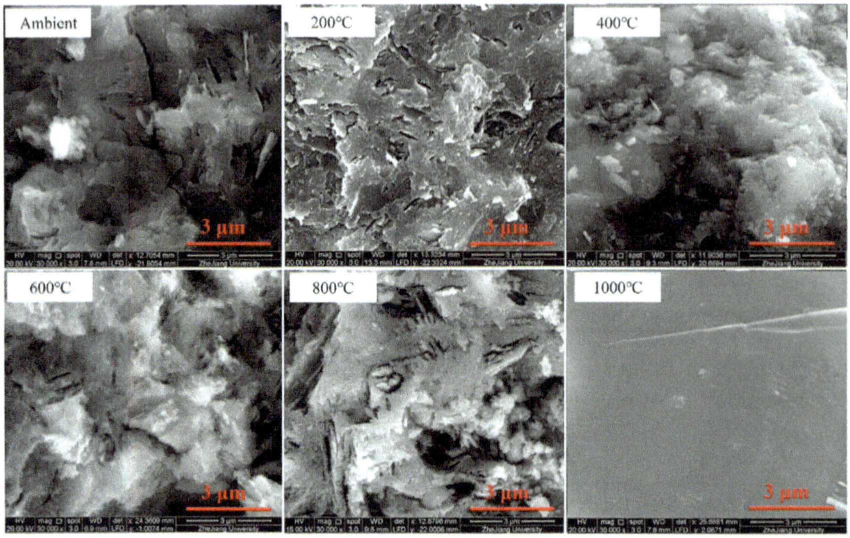

Fig. 6.9 SEM micrographs of geopolymer mortars of GM-NaK at various elevated temperatures. Reproduced from [Effects of Activator Types on Degradation Mechanisms of Metakaolin Geopolymer Mortars Exposed to High Temperature] by [Yajun Zhang] with permission from [Journal of Materials in Civil Engineering]

800 °C, higher than that of the GM-NaNa specimens. Both the GM-NaNa and GM-NaK specimens completely melted and coagulated above 1000°C, a phenomenon similar to that reported by Kljajević et al. (2017).

Because K^+ encouraged SiO and AlO chains to crosslink, strengthening AlO bonds, the GM-NaK group's SEM pictures showed more resistance to sintering than the GM-NaNa group did (Cioffi et al. 2003; Duxson et al. 2007c; Yachao 2014). At 1000 °C, the microstructure of both the GM-NaK and GM-NaNa groups underwent significant changes. The matrix exhibited higher homogeneity and smoother texture, likely as a result of sintering at this temperature, which led to particle packing becoming tighter, microcracks recovering (Duxson et al. 2006b; Lahoti et al. 2018), and an increase in the compressive strength of the specimens (Lin et al. 2009).

6.4.8 TGA

The TG/DSC curves of geopolymer pastes with (Na-Na)-based activators (GP-NaNa) and (Na–K)-based activators (GP-NaK) are shown in Fig. 6.10. The mass losses of the two types of geopolymer pastes were noticeably different. The TG curves demonstrate that up to 1000 °C, the mass losses of the GP-NaNa and GP-NaK groups were 36.42% and 33.03%, respectively, due to water loss from dihydroxylation and free water evaporation. Based on the TG results, the mass loss of the GP-NaNa group was 4.09% greater than that of the GP-NaK group up to 200 °C. This additional mass loss occurred mainly due to loss of free water during the heating process from room temperature to 200 °C. This suggests that higher porosity of the GM-NaNa group may have reduced its ability to retain water (Assaedi et al. 2016). If the mass loss was significant, the drying shrinkage of the GM-NaNa group became severe, leading to cracks forming and progressing. This finding was consistent with the μCT results. In addition, the mass losses of both GP-NaNa and GP-NaK groups remained stable above 500 °C.

The geopolymers underwent dehydroxylation between 200 and 500 °C, which resulted in some mass losses in both the GP-NaNa and GP-NaK groups. The DSC curves of the examined specimens show that there was no noticeable peak between 200 and 800 °C. However, SEM scans revealed that in this temperature range, the GM-NaNa and GM-NaK groups' microstructures experienced modifications. The GM-NaNa group partially sintered at 600 °C, while the microstructure of the GM-NaK group changed at 800 °C. This process, referred to as sintering, resulted in adjacent gel particles bonding together and agglomerating. An evident endothermic peak above 850–1050 °C was observed, possibly due to the crystallization process of GP-NaNa group phases. However, its endothermic peak was slightly less prominent than that

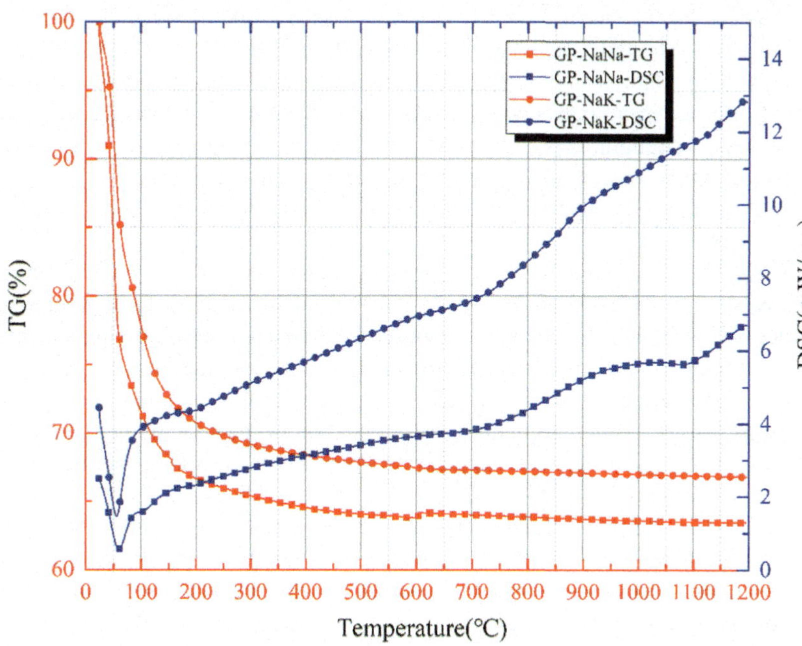

Fig. 6.10 TG and DSC results of geopolymer pastes containing different activators at various elevated temperatures. Reproduced from [Effects of Activator Types on Degradation Mechanisms of Metakaolin Geopolymer Mortars Exposed to High Temperature] by [Yajun Zhang] with permission from [Journal of Materials in Civil Engineering]

of the GP-NaK group. Therefore, mutually promoted effects between Na^+ and K^+ led to more efficient geopolymerization in the GP-NaK group, and an insufficient amount of free alkalis lowered the crystallization temperature of nepheline in the GP-NaK group.

6.4.9 Classification of Degradation Mechanisms

Based on the physical and chemical changes observed from both macrostructural and microstructural perspectives, four distinct stages were identified as the temperature increased to 1000 °C. The phases observed during each stage can be attributed to the following reasons:

(1) Stage I—Strength Adjustment (Temperature range: Room temperature–200 °C): Due to significant drying shrinkage and the appearance and growth of fractures, (Na-Na)-based geopolymers exhibit larger mass losses than (Na–K)-based geopolymers. Because compressive and flexural strengths in (Na-Na)-based geopolymers grow in distinct ways, it is possible that flexural strength is more susceptible to microstructural changes in geopolymer mortars at this stage.
(2) Stage II—Strength Degradation (Temperature range: 200–400 °C): Due to high temperatures and different thermal expansion coefficients, this stage saw the exterior extension of cracks, which resulted in more cracks and decreased strength, especially in (Na-Na)-based geopolymers. (Na-Na)-based geopolymers were more susceptible to extreme heat at this time.
(3) Additionally, (Na–K)-based geopolymers demonstrated better resistance to high temperatures.
(4) Stage III—Rapid Degradation (Temperature range: 400–800 °C): The compressive strength of the geopolymers significantly decreased due to high temperatures, with similar rates of decrease observed in both groups. Sintering and dihydroxylation further induced cracking on the surface and modified the number of pores. However, the strength of the geopolymer matrix including the (Na–K)-based activator was marginally greater than that of the geopolymer containing the (Na-Na)-based activator because the sintering time was later and the volume of the cracks was lower.
(5) Stage IV—Strength Enhancement (Temperature range: 800–1000 °C): Geopolymer mortars melted and solidified due to high temperatures and recovered microcracks. High-temperature sintering improved both the uniformity and density of specimens, resulting in a slight enhancement of the average compressive strength. The chemical stability of the newly formed crystalline phases, however, was decreased, making the (Na-Na)-based geopolymers less robust than their (Na–K)-based counterparts. The modest increase in strength shows that, rather than changes in pore structure and cracking below 800°C, sintering and crystallization are to blame for strength loss in geopolymers over 800°C.

6.5 Conclusion

In this chapter, the effects of different types of activator cations on the properties of geopolymers under high temperatures were investigated. The following conclusions can be made:

(1) With an increase in exposure temperature, the porosity of geopolymers was observed to increase. The degradation of strength was consistently attributed to material properties and crack development patterns. However, the mechanisms of strength deterioration below 800 °C (such as pore structure and fractures) and beyond 800°C (such as sintering and crystallization) are quite different from one another.

(2) Geopolymer containing (Na–K)-based activator exhibited better heat resistance at temperatures above 200 °C compared to their (Na-Na)-based counterparts. This was evidenced by higher compressive strength, lower porosity, less mass loss, and a reduced tendency to develop cracks.

(3) The mutually promoted effects between Na^+ and K^+ delayed sintering time after high temperatures and prevented the formation of new crystalline phases. In comparison to their Na-Na counterparts, geopolymer mortars using (Na–K)-based activator showed superior chemical resilience to high temperatures. Additionally, the microstructures of the specimens heated to 1000 °C were denser and more homogeneous.

It is significant to emphasize that further research should be done on this subject, and additional data should be gathered to confirm the results provided in this chapter, due to the complexity of chemical and physical changes in geopolymer mortar at high temperatures.

References

Abdulkareem OA, Mustafa Al Bakri AM, Kamarudin H et al (2014) Effects of elevated temperatures on the thermal behavior and mechanical performance of fly ash geopolymer paste, mortar and lightweight concrete. Constr Build Mater 50:377–387. https://doi.org/10.1016/j.conbuildmat.2013.09.047

Ali FA, O'Connor D, Abu-Tair A (2001) Explosive spalling of high-strength concrete columns in fire. Mag Concr Res 53:197–204. https://doi.org/10.1680/macr.2001.53.3.197

Assaedi H, Shaikh FUA, Low IM (2016) Effect of nano-clay on mechanical and thermal properties of geopolymer. J Asian Ceram Soc 4:19–28. https://doi.org/10.1016/j.jascer.2015.10.004

Barbosa VFF, MacKenzie KJD (2003a) Synthesis and thermal behaviour of potassium sialate geopolymers. Mater Lett 57:1477–1482. https://doi.org/10.1016/S0167-577X(02)01009-1

Barbosa VFF, MacKenzie KJD (2003b) Thermal behaviour of inorganic geopolymers and composites derived from sodium polysialate. Mater Res Bull 38:319–331. https://doi.org/10.1016/S0025-5408(02)01022-X

Benito P, Leonelli C, Medri V, Vaccari A (2013) Geopolymers: a new and smart way for a sustainable development. Appl Clay Sci 73:1. https://doi.org/10.1016/j.clay.2013.03.008

Bernal SA, Rodríguez ED, Mejía De Gutiérrez R et al (2011) Mechanical and thermal characterisation of geopolymers based on silicate-activated metakaolin/slag blends. J Mater Sci 46:5477–5486. https://doi.org/10.1007/s10853-011-5490-z

Cioffi R, Maffucci L, Santoro L (2003) Optimization of geopolymer synthesis by calcination and polycondensation of a kaolinitic residue. Resour Conserv Recycl 40:27–38. https://doi.org/10.1016/S0921-3449(03)00023-5

Davidovits J (1993) Geopolymer cements to minimise carbon-dioxide greenhouse-warming

Duxson P, Lukey GC, van Deventer JSJ (2006a) Thermal evolution of metakaolin geopolymers: part 1—physical evolution. J Non Cryst Solids 352:5541–5555. https://doi.org/10.1016/j.jnoncrysol.2006.09.019

Duxson P, Lukey GC, van Deventer JSJ (2006b) Thermal conductivity of metakaolin geopolymers used as a first approximation for determining gel interconnectivity. Ind Eng Chem Res 45:7781–7788. https://doi.org/10.1021/ie060187o

Duxson P, Lukey GC, van Deventer JSJ (2007a) Physical evolution of Na-geopolymer derived from metakaolin up to 1000 °C. J Mater Sci 42:3044–3054. https://doi.org/10.1007/s10853-006-0535-4

Duxson P, Mallicoat SW, Lukey GC et al (2007b) The effect of alkali and Si/Al ratio on the development of mechanical properties of metakaolin-based geopolymers. Colloids Surfaces A Physicochem Eng Asp 292:8–20. https://doi.org/10.1016/j.colsurfa.2006.05.044

Duxson P, Provis JL, Lukey GC, van Deventer JSJ (2007c) The role of inorganic polymer technology in the development of "green concrete." Cem Concr Res 37:1590–1597. https://doi.org/10.1016/j.cemconres.2007.08.018

EAPFP (1987) European Association for Passive Fire Protection. EAPFP, Bordon, UK

Huiskes DMA, Keulen A, Yu QL, Brouwers HJH (2016) Design and performance evaluation of ultra-lightweight geopolymer concrete. Mater Des 89:516–526. https://doi.org/10.1016/j.matdes.2015.09.167

Davidovits J (1991) Geopolymers: inorganic polymeric new materials. J Therm Anal 37:1633–1656

Kamseu E, Rizzuti A, Leonelli C, Perera D (2010) Enhanced thermal stability in K_2O-metakaolin-based geopolymer concretes by Al_2O_3 and SiO_2 fillers addition. J Mater Sci 45:1715–1724. https://doi.org/10.1007/s10853-009-4108-1

Kastiukas G, Ruan S, Liang S, Zhou X (2020) Development of precast geopolymer concrete via oven and microwave radiation curing with an environmental assessment. J Clean Prod 255. https://doi.org/10.1016/j.jclepro.2020.120290

Kljajević LM, Nenadović SS, Nenadović MT et al (2017) Structural and chemical properties of thermally treated geopolymer samples. Ceram Int 43:6700–6708. https://doi.org/10.1016/j.ceramint.2017.02.066

Komnitsas K, Zaharaki D (2007) Geopolymerisation: a review and prospects for the minerals industry. Miner Eng 20:1261–1277. https://doi.org/10.1016/j.mineng.2007.07.011

Kong DLY, Sanjayan JG (2008) Damage behavior of geopolymer composites exposed to elevated temperatures. Cem Concr Compos 30:986–991. https://doi.org/10.1016/j.cemconcomp.2008.08.001

Kong DLY, Sanjayan JG, Sagoe-Crentsil K (2007) Comparative performance of geopolymers made with metakaolin and fly ash after exposure to elevated temperatures. Cem Concr Res 37:1583–1589. https://doi.org/10.1016/j.cemconres.2007.08.021

Krivenko PV, Kovalchuk GY (2007) Directed synthesis of alkaline aluminosilicate minerals in a geocement matrix. J Mater Sci 42:2944–2952. https://doi.org/10.1007/s10853-006-0528-3

Kuenzel C, Grover LM, Vandeperre L et al (2013) Production of nepheline/quartz ceramics from geopolymer mortars. J Eur Ceram Soc 33:251–258. https://doi.org/10.1016/j.jeurceramsoc.2012.08.022

Kuenzel C, Vandeperre LJ, Donatello S et al (2012) Ambient temperature drying shrinkage and cracking in metakaolin-based geopolymers. J Am Ceram Soc 95:3270–3277. https://doi.org/10.1111/j.1551-2916.2012.05380.x

Lahoti M, Wong KK, Tan KH, Yang EH (2018) Effect of alkali cation type on strength endurance of fly ash geopolymers subject to high temperature exposure. Mater Des 154:8–19. https://doi.org/10.1016/j.matdes.2018.05.023

Lin TS, Jia DC, He PG, Wang MR (2009) Thermo-mechanical and microstructural characterization of geopolymers with α-Al_2O_3 particle filler. Int J Thermophys 30:1568–1577. https://doi.org/10.1007/s10765-009-0636-9

Mohd Salahuddin MB, Norkhairunnisa M, Mustapha F (2015) A review on thermophysical evaluation of alkali-activated geopolymers. Ceram Int 41:4273–4281. https://doi.org/10.1016/j.ceramint.2014.11.119

Pan Z, Sanjayan JG, Collins F (2014) Effect of transient creep on compressive strength of geopolymer concrete for elevated temperature exposure. Cem Concr Res 56:182–189. https://doi.org/10.1016/j.cemconres.2013.11.014

Panda B, Ruan S, Unluer C, Tan MJ (2020) Investigation of the properties of alkali-activated slag mixes involving the use of nanoclay and nucleation seeds for 3D printing. Compos Part B Eng 186. https://doi.org/10.1016/j.compositesb.2020.107826

Panda B, Ruan S, Unluer C, Tan MJ (2019) Improving the 3D printability of high volume fly ash mixtures via the use of nano attapulgite clay. Compos Part B Eng 165:75–83. https://doi.org/10.1016/j.compositesb.2018.11.109

Papakonstantinou CG, Balaguru P, Lyon RE (2001) Comparative study of high temperature composites. Compos Part B Engineering 32:637–649. https://doi.org/10.1016/S1359-8368(01)00042-7

Phan LT, Carino NJ (2004) Fire performance of high strength concrete: research needs. Struct Congr 2000 Adv Technol Struct Eng 103. https://doi.org/10.1061/40492(2000)181

Rickard WDA, Temuujin J, Van Riessen A (2012) Thermal analysis of geopolymer pastes synthesised from five fly ashes of variable composition. J Non Cryst Solids 358:1830–1839. https://doi.org/10.1016/j.jnoncrysol.2012.05.032

Sindhunata, Van Deventer JSJ, Lukey GC, Xu H (2006) Effect of curing temperature and silicate concentration on fly-ash-based geopolymerization. Ind Eng Chem Res 45:3559–3568. https://doi.org/10.1021/ie051251p

Thomas RJ, Ye H, Radlinska A, Peethamparan S (2016) Alkali-activated slag cement concrete. Concr Int 33–38

Xu H, Deventer JV (2002) The geopolymerisation of alumino-silicate minerals. Int J Miner Process 59:247–466

Yachao W (2014) Investigations on reinforcing, toughening and durability of alkali-activated fly ash-based geopolymer

Ye H, Radlińska A (2016) Fly ash-slag interaction during alkaline activation: influence of activators on phase assemblage and microstructure formation. Constr Build Mater 122:594–606. https://doi.org/10.1016/j.conbuildmat.2016.06.099

Zhang HY, Kodur V, Qi SL et al (2014) Development of metakaolin-fly ash based geopolymers for fire resistance applications. Constr Build Mater 55:38–45. https://doi.org/10.1016/j.conbuildmat.2014.01.040

Zhang HY, Kodur V, Wu B et al (2016) Thermal behavior and mechanical properties of geopolymer mortar after exposure to elevated temperatures. Constr Build Mater 109:17–24. https://doi.org/10.1016/j.conbuildmat.2016.01.043

Chapter 7
Freezing–Thawing Resistance of MKG

Abstract Understanding the composition-dependent features of geopolymer materials as well as their freezing–thawing (F-T) resistance is essential for their effective application in cold climates. In this work, the Si/Al ratio and the Na/Al ratio were controlled to create geopolymer mortars based on metakaolin (MKG). The pore structure was evaluated through mercury intrusion porosimetry, while compression tests were conducted to determine strength; both parameters showed obvious correlations with material composition. Mass loss, strength loss, visual rate, and microscopic observation were utilized to evaluate changes in material characteristics and microstructure brought on by F-T stresses. The strength-porosity connection appeared to essentially follow a linear pattern, according to the results. Capillary pore volume rose with increases in the Si/Al ratio, whereas gel pore volume and F-T resistance dropped. Increases in the Na/Al ratio reduced gel pore volume, but generally improved F-T resistance. MKG mortar with a Na/Al ratio of 1.26 exhibited the lowest total pore volume and the best F-T resistance. Our experimental results suggest that air voids connected by capillary pores facilitate relaxation of hydraulic pressures induced by pore liquid freezing. This study help clarify the compositional dependence of pore structure, strength, and F-T resistance of MKG materials, providing fundamental information for their use in engineering applications in cold regions.

7.1 Introduction

The production of ordinary Portland cement (OPC) emits a significant amount of CO_2, around 0.8 kg CO_2 per kg OPC (Monteiro et al. 2017), creating strong incentives to replace it with green cementitious materials that have low CO_2 emissions. Geopolymer is an inorganic alumino-silicate polymer synthesized primarily from silicon (Si) and aluminum (Al) materials found in geologically active minerals, which has promising potential as a binding material for concrete production. Notably, one ton of geopolymer emits only 20% of the CO_2 emissions produced by one ton of OPC (Duxson et al. 2007a). When synthesized with solid waste, geopolymer

D. Yan et al., *Metakaolin-Based Geopolymers*,
https://doi.org/10.1007/978-981-97-0652-5_7

offers additional environmental and economic benefits with high-value added applications (Kastiukas et al. 2020). The geopolymerization process can quickly form a three-dimensional polymeric structure with complex Al–O–Si bonds using alkaline activators, generating continuous material skeletons that offer high strengths (Provis 2014). Geopolymer-based materials exhibit excellent performance, including high mechanical strength, great fire resistance, enhanced leaching resistance, and improved contaminant stabilization, making them suitable for various engineering applications with great promise (Rowles et al. 2003; Roviello et al. 2019; Cui et al. 2020; Peng et al. 2020; Kohout et al. 2021).

The engineering performance of geopolymer materials is intricately linked to their material composition, synthesizing method, and curing scheme (Rowles et al. 2003; White et al. 2013; Provis 2014; Yan et al. 2016; Lahoti et al. 2017; Kubba et al. 2018). The most important component among these factors is material composition since it significantly affects material structures, mechanical characteristics, and durability performances. For instance, the mechanical characteristics of geopolymer materials are frequently developed to maximize the Si/Al ratio and the Na(K)/Al ratio. Researchers have found that metakaolin-based geopolymers (MKG) with Si/Al ratios of 2–2.5 and Na/Al ratios of 1–1.5 show optimized strengths (Rowles and O'Connor 2009), while the Si/Al ratios between 1.5 and 2.2 and Na/Al ratios between 0.8 and 1.5 enable the production of high-strength MKG mortars over 60 MPa (Lahoti et al. 2017). Material composition designs not only impact the material strength but also alter the pore structure of MKG materials. For example, Duxon et al. (2005) observed that changes in Si/Al ratio affected the pore size distribution (PSD) of MKG materials, with a rise in Si/Al ratio leading to narrower pore sizes at the nanoscale. Nevertheless, alterations in pore structure may cause different responses of geopolymer materials to freezing–thawing (F-T) cycles, an area which has yet to be fully investigated in previous studies.

The durability of geopolymer materials against freezing–thawing (F-T) loads presents great challenges, particularly in cold regions where geopolymer-based structures are used in hydraulic engineering applications like dams, aqueducts, and docks. One significant challenge is that the pore structure of geopolymer materials varies with material composition (Duxson et al. 2005), making standard mechanical tests on their F-T performance inadequate for predicting real-world degradation (Sun and Wu 2013; Topçu et al. 2014; Zhao et al. 2019).

Freezing–thawing (F-T) resistance is an essential durability index that has been widely studied in common structural materials such as concrete, mortar, and masonry (Bocca and Grazzini 2013; Carpinteri et al. 2014; Sáez del Bosque et al. 2020). The F-T damaging processes are the same for all porous materials. Typically, pore pressures built up from the transition of pore water to ice exceed material strength, resulting in F-T damages (Zeng et al. 2011, 2013, 2014, 2015; Huang et al. 2020). Freezing pore pressures typically include hydraulic pressures caused by differences in water–ice density and crystallization pressures induced by differences in free energy (Zeng and Li 2019). Pore confinement significantly affects the water freezing process (Engemann et al. 2004; Jähnert et al. 2008; Zeng et al. 2014), which, in turn, affects the frost damage of porous materials. Therefore, evaluating the F-T resistance of porous materials with different pore structures is challenging.

The goal of this work is to look into the connection between MKG materials' freezing–thawing (F-T) resistance and pore structure. By modifying the Si/Al ratio and Na/Al ratio, we created MKG mortars with various pore architectures. Mercury intrusion porosimetry (MIP) was used to measure the pore structure of the MKG mortars, and its correlation with compressive strength was evaluated. The F-T performance of the MKG mortars was assessed, and the results were analyzed in light of the pore structure. This study provides valuable insights into the dependency of pore structure, strength, and F-T resistance on the composition of MKG materials.

7.2 Experiments

7.2.1 Materials and Specimen Preparation

We used commercial metakaolin powder (Metamax, Basf Co. LTD., Shanghai, China) as the aluminosilicate source for geopolymer materials. Laser particle size analysis (LS-230, Coulter) showed that the mean particle size was 5.91 μm, and the 90%-passed particle size was 13.59 μm. The particle size distribution of the metakaolin powder is presented in Fig. 7.1a. The chemical composition of the metakaolin powder was determined using X-ray fluorescence (XRF) spectrometer analysis (SHIMADU XRF-1800, Shimadzu Global Laboratory Consumables Co., Ltd. Shanghai, China), and the packing density of the metakaolin powder was found to be 0.422 g/mL. The results showed that SiO_2 and Al_2O_3 accounted for 57.47 wt% and 39.81 wt% of total oxides, respectively (Table 7.1). The XRD patterns of the metakaolin are illustrated in Fig. 7.1b, displaying only amorphous halo and anatase peaks, indicating that the metakaolin contained highly reactive amorphous SiO_2 and Al_2O_3.

The alkaline activator utilized in this investigation was formulated via the use of liquid sodium silicate (comprising $SiO_2 = 27.35\%$, $Na_2O = 8.42\%$, and $H_2O = 64.23\%$) and pellet sodium hydroxide (comprising $Na_2O = 77.4\%$, $H_2O = 22.5\%$, and impurity = 0.1%). Fine sands characterized by a fineness modulus of 1.75 and a density of 2510 kg/m^3 were employed as the fine aggregates for the synthesis of MKG mortars. To discern the impact of material composition, specifically the Si/Al ratio and Na/Al ratio, on the pore structure and mechanical properties of MKG materials, two groups of mortars were designed. In the first group (denoted as MKG-1 to MKG-3), the Si/Al ratios ranged from 2.01 to 2.62, while the Na/Al ratio remained fixed at 1.01. The second group (termed MKG-3 to MKG-5) maintained a constant Si/Al ratio of 2.62, with an increase in the Na/Al ratio from 1.01 to 1.36. Detailed mix proportions and corresponding nomenclature are presented in Table 7.2. To reduce the impact of w/b and s/b on the outcomes, all mixes used a w/b ratio of 0.62 and a sand-to-binder (s/b) ratio of 3. Prior to being used in the synthesis of geopolymers,

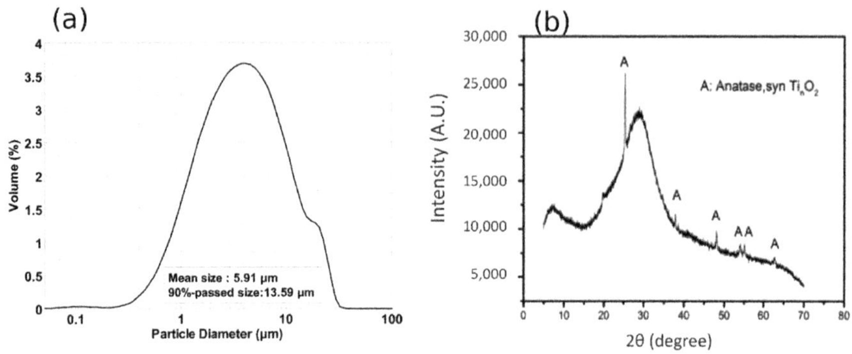

Fig. 7.1 Particle size distribution **a** and XRD patterns **b** of the metakaolin powder. Reproduced from [Compositional Dependence of Pore Structure, Strengthand Freezing–Thawing Resistance of Metakaolin-Based Geopolymers] by [Dongming Yan] with permission from [Alkali-Activated Materials for Sustainable Construction]

Table 7.1 Chemical components of the metakaolin powder

Composition	SiO_2	Al_2O_3	TiO_2	Fe_2O_3	Na_2O	K_2O	CaO
Mass content (%)	57.47	39.81	1.79	0.43	0.27	0.21	0.04

Reproduced from [Compositional Dependence of Pore Structure, Strengthand Freezing–Thawing Resistance of Metakaolin-Based Geopolymers] by [Dongming Yan] with permission from [Alkali-Activated Materials for Sustainable Construction]

the alkaline activators were prepared in accordance with the stoichiometric balances determined by the various Si/Al and Na/Al ratios.

We started by adding a settled alkali activator solution into a Hobart mixer bowl, then we added the precisely weighed metakaolin powder and water to make the fresh MKG mortars. For three minutes, low-speed stirring was done to create the MKG paste slurries. Subsequently, sands were added to the paste slurries and homogenized with another round of 3-min stirring to achieve a uniform MKG mortar. The resulting

Table 7.2 Mix proportions of the metakaolin-based geopolymer (MKG) concretes

Mix ID	Metakaolin	Water Glass	NaOH	Water	Si/Al	Na/Al
MKG-1	1016	640	193	410	2.01	1.01
MKG-2	936	993	138	197	2.32	1.01
MKG-3	868	1290	84	17	2.62	1.01
MKG-4	842	1251	136	31	2.62	1.26
MKG-5	832	1233	156	38	2.62	1.36

Reproduced from [Compositional Dependence of Pore Structure, Strengthand Freezing–Thawing Resistance of Metakaolin-Based Geopolymers] by [Dongming Yan] with permission from [Alkali-Activated Materials for Sustainable Construction]

mortar slurries were cast into cuboid molds with dimensions of $40 \times 40 \times 160\ mm^3$. High-frequency vibrations were applied to the fully filled molds to eliminate any entrapped air bubbles in the mortars. To prevent any loss of water that may cause microstructural alterations, we enveloped the specimens together with the molds in plastic film. Standard curing was carried out at 20 ± 1 °C and 90% relative humidity for a period of 2 days to enhance the strength. Following demolding, all the mortar specimens were stored in a curing chamber for a duration of 28 days.

7.2.2 Testing Methods

7.2.2.1 Strength

A TYE-300D (Jianyi experiment instrument Co. LTD., WuXi, Jiangsu, China) automatic mechanical testing equipment with a loading speed of 2.4 kN/s was used to measure the compressive strengths. To ensure data reliability and productivity, six specimens were tested for each mix and the results were averaged.

7.2.2.2 Freezing–Thawing

In hydraulic engineering structures located in cold regions, the F-T (freeze–thaw) durability of MKG materials is a crucial consideration. F-T tests were performed according to the Test Code for Hydraulic Concrete (DL/T5150-2017 2017) using a CDR-5 rapid freeze–thaw machine. Prior to the F-T tests, the specimens were immersed in tap water for a period of 4 days to enhance the water saturation degrees. Weighed by a highly accurate balance, the mass of the water-filled specimens was referred to as the initial mass m_0. Subsequently, the specimens were loaded into vessels containing water with the water surface positioned 20 mm higher than the top surfaces of the specimens. The vessels were then transferred to the F-T testing machine. The temperature was controlled within the range of − 17 to 8 °C, with each freezing/thawing cycle taking approximately 1.5–2.5 h (1–1.5 h) to decrease (raise) the temperature, and one complete F-T course lasting for 2.5–4 h.

Upon completion of 50 F-T cycles, excess surface water was removed from the specimens using absorbent papers. The mass of the surface-saturated specimens was measured and denoted as m_{FT}. We then determined the relative mass loss (Δm) using the following formula:

$$Am = \frac{m_0 - m_{FT}}{m_0} \quad (7.1)$$

In instances where the F-T loads resulted in complete damage to the specimens, neither their strength nor mass was measured.

7.2.2.3 Pore Structure

To conduct MIP (mercury intrusion porosimetry) tests, the central part of each MKG mortar was crushed into small particles with a diameter of approximately 10 mm. Due to the reproducibility and broad pore ranges of MIP, only one test was conducted for each mortar. We utilized an Autopore IV 9510 (Micromeritics Instrument Corp., Norcross, GA, USA) to perform the MIP tests. The applied intrusion pressures varied from 1.4 kPa to 207 MPa, with each pressure step's equilibrium duration being 10 s. If the forces exerted are sufficient to overcome the surface forces on the pore curvatures, mercury fronts will invade pores or cavities. The pore structure of a porous material may be evaluated by precisely measuring the mercury intrusion (or extrusion) volume (or mass) at each pressurization phase. Generally, the Washburn equation (Washburn 1921) is used to link the applied pressures (P) to the pore sizes D, i.e., $D = -4\gamma \cos\theta/P$, where y is the surface tension of mercury and θ is the contact angle between mercury and the pore wall. Keep in mind that the Washburn equation relies on a linked, progressively growing cylindrical pore structure. We calculated the minimum and maximum accessible pore sizes, according to the Washburn equation, as 6 nm and 360 m, respectively, corresponding to the maximum and minimum applied pressures, based on the commonly used physical parameters of mercury, namely the contact angle of 130° and the surface tension of 485 N/m (Zeng et al. 2019). It should be noted that if the specimens were completely damaged during the F-T loads, neither strength nor mass was measured.

7.2.2.4 Micro Morphology

To examine the microstructural alterations induced by F-T cycles in MKG mortars, we utilized scanning electron microscopy (SEM) in back-scattered electron (BSE) mode using a Quanta FEG 650 equipment (Thermo Fisher Scientific, Beijing, China). BSE images have different electron back-scattering coefficients and thus can more clearly distinguish the pore phase (including cracks) from the solid skeletons. The quick-hardening epoxy glue was infused into the MKG mortar particles after they had been oven-dried at 40 °C for 24 h. After that, they were ground and polished with different sandpaper grades and diamond suspensions. The flat and smooth samples were put onto a sample platform for SEM/BSE testing after a short carbon treatment. To acquire BSE pictures, we employed a voltage of 30 kV with a spot size of 5.0 nm.

7.3 Materials Properties and Pore Structure

7.3.1 Pore Structure

Table 7.3 provides a summary of some characteristic pore parameters of the MKG mortars obtained from MIP tests, including the total porosity Φ_T, average pore size ($D_a = 4V/A$), specific surface area (A), and threshold pore size (D_t). The characteristic pore parameters varied with different mix proportions. A comparison of the total porosity with the specific surface area and that of the average pore size with the threshold pore size revealed similar trends, which generally reflected the pore structure features of a porous material. A more porous structure is often indicated by a higher total porosity or specific surface area, but a bigger average pore size denotes a coarser pore structure. Threshold pore sizes measure the connected throats formed from the interparticle continuum (Katz and Thompson 1986), which is generally an indicator of permeability. More detailed discussions about the pore data, along with material compositions, are presented below.

Figure 7.2 illustrates the MIP pore data obtained from the MKG mortars, including the pore size distributions (PSDs) in accumulative (APSD) and differential forms (DPSD). Despite exhibiting varying PSD shapes, similar physical interactions between mercury fronts and pores were observed in the MKG mortars. Surface conformance effect, which is when mercury first invades open fractures, gaps, holes, and irregularities on the sample surfaces, first occurs during the MIP test (Zeng et al. 2019). Given that sample pretreatments like cutting and drying often increase surface roughness, this impact could be unavoidable. The rapid filling of mercury at low pressures contributed to the rise in APSDs (Fig. 7.2a) and peaks in DPSDs at 100 μm (Fig. 7.2b). Analysis revealed that the mercury volumes required to cover surface roughness were approximately 0.01 mL/g (Fig. 7.2c), accounting for 12–17% of the total intrusion volumes (Fig. 7.2d). These results suggest that the MKG samples had comparable surface roughness.

Table 7.3 Characteristic pore parameters of the MKG mortars form mercury intrusion porosimetry (MIP) tests

Sample	Total porosity (%)	Average pore size (nm)	Specific surface area (m^2/g)	Threshold pore size (nm)
MKG-1	17.59	32.0	11.83	350.1
MKG-2	15.81	27.1	12.43	350.1
MKG-3	18.08	25.2	15.46	553.7
MKG-4	13.01	84.5	3.25	675.9
MKG-5	13.15	89.9	3.14	1049.4

Reproduced from [Compositional Dependence of Pore Structure, Strengthand Freezing–Thawing Resistance of Metakaolin-Based Geopolymers] by [Dongming Yan] with permission from [Alkali-Activated Materials for Sustainable Construction]

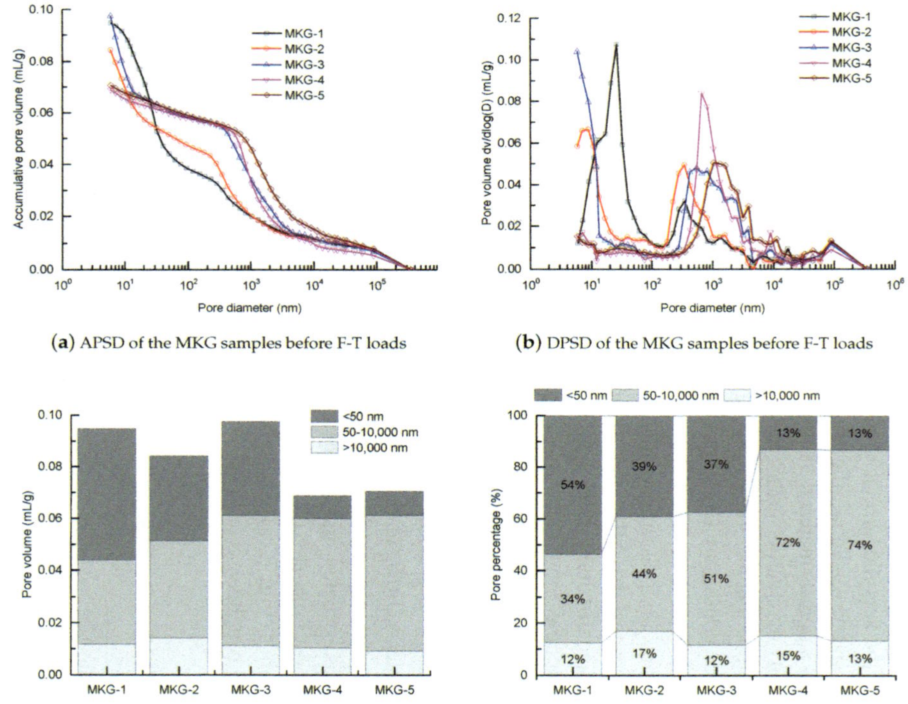

(**a**) APSD of the MKG samples before F-T loads

(**b**) DPSD of the MKG samples before F-T loads

(**c**) Pore volumes in different sizes

(**d**) Pore ratios in different sizes

Fig. 7.2 Pore structure of the MKG mortars before freezing–thawing (F-T) loads: **a** accumulative pore size distribution (APSD) spectra; **b** differential PSD (DPSD) spectra; **c** pore volumes in different sizes; and **d** pore ratios in different sizes. Reproduced from [Compositional Dependence of Pore Structure, Strengthand Freezing–Thawing Resistance of Metakaolin-Based Geopolymers] by [Dongming Yan] with permission from [Alkali-Activated Materials for Sustainable Construction]

The mercury fronts entered smaller holes with higher applied pressure. Between 50 and 10,000 nm, there were sharp mercury increases that may be categorized as capillary holes (Fig. 7.2a). These are inter-particle gaps, which link metakaolin particles, hydration products, and fine aggregates. Upon mercury's entry into these inter-particle gaps, intrusion peaks between 100 and 2000 nm were produced (Fig. 7.2b and Table 7.3). It should be noted that MIP PSD tends to underestimate pore sizes due to the "ink-bottle" effect (Diamond 2000). Nonetheless, it measures pore throats that establish pore connections (Katz and Thompson 1986). The analysis of pores indicated that capillary pores ranging from 50 to 10,000 nm occupied varying ratios and volumes in different MKG mortars, as illustrated in Fig. 7.2c, d. More than 70% of the pores in the MKG-4 and MKG-5 samples were capillary pores (Fig. 7.2d).

Under higher pressures, mercury invaded pores below 50 nm (gel pores). Significant differences in PSDs among the MKG mortars were observed without uniform trends (Fig. 7.2a, b). In Fig. 7.2c, d, it is evident that the MKG-1 sample exhibited higher gel pore volumes and coarser PSDs than MKG-2 and MKG-3, whereas the MKG-4 and MKG-5 samples demonstrated limited distributions and volumes of the gel pores.

7.3.2 Strength

Figure 7.3 presents the statistical compressive strengths of the MKG mortars. It is evident that all mortars exhibited relatively high strengths (> 57 MPa), indicating that the utilized mixes can manufacture geopolymers with good mechanical properties. Similar strength data have been reported in other studies (Rowles and O'Connor 2009; Lahoti et al. 2017).

The majority of granular-compacted materials, such as concrete and geopolymer materials, gain strength by augmenting the solidity of solids or reducing pore volume. Hence, total porosity can serve as an indicator to anticipate material strength. Figure 7.4 presents compressive strength plots against total porosity. Generally, a linear decrease in compressive strength was observed with increasing total porosity, which can be mathematically expressed as:

$$\sigma = \sigma_0(1 - \alpha\Phi) \quad (7.2)$$

In Eq. (7.2), where σ and σ_0 represent the compressive strengths at current porosity Φ and zero porosity, respectively, and α is the coefficient. This equation is commonly referred to as the Hasselman equation for the strength-porosity relationship of porous materials (Hasselman and Fulrath 1964).

The fitting of compressive strength data using Eq. (7.2) resulted in $\sigma_0 = 99.74$ MPa and $\alpha = 0.022$. The obtained value of σ_0 was substantially greater than that of alkali-activated slag mortars (Shi 1996) but similar to that of geopolymer mortars made from slag, fly ash, and palm oil fuel ash (Kubba et al. 2018). The coefficient α was significantly lower than those reported in previous studies (Shi 1996; Kubba et al.

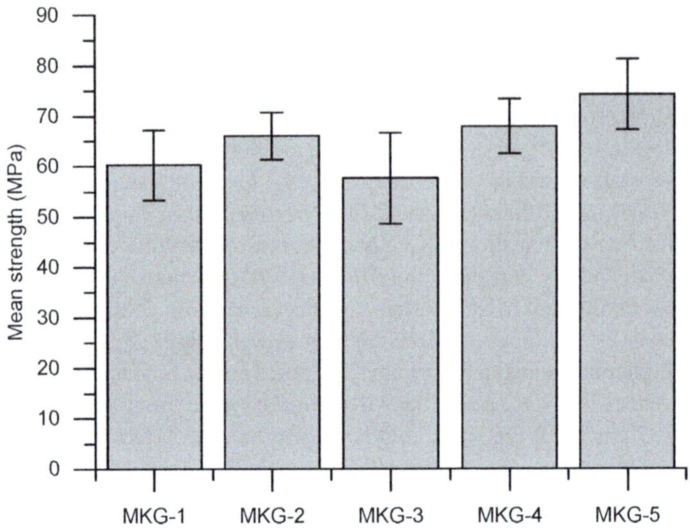

Fig. 7.3 Statistical strengths of the MKG mortars. Reproduced from [Compositional Dependence of Pore Structure, Strengthand Freezing–Thawing Resistance of Metakaolin-Based Geopolymers] by [Dongming Yan] with permission from [Alkali-Activated Materials for Sustainable Construction]

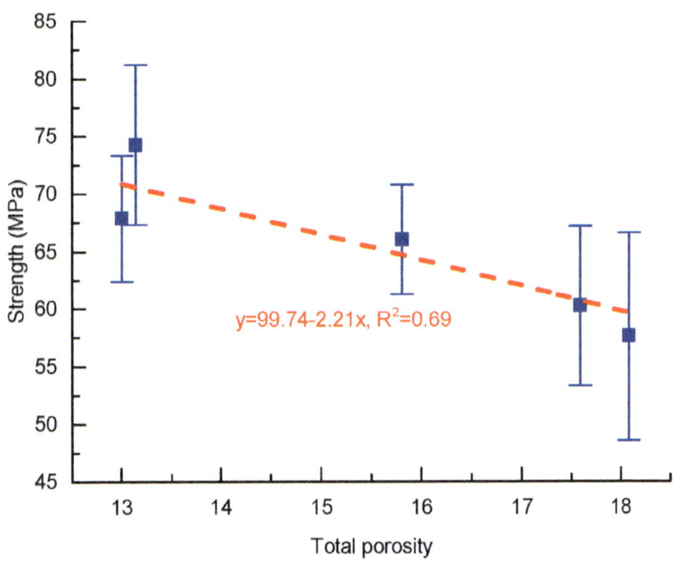

Fig. 7.4 Relationships between strength and porosity. Reproduced from [Compositional Dependence of Pore Structure, Strengthand Freezing–Thawing Resistance of Metakaolin-Based Geopolymers] by [Dongming Yan] with permission from [Alkali-Activated Materials for Sustainable Construction]

2018). These outcomes suggest the high strength of the solid matrix and the limited influence of total porosity.

7.3.3 Roles of the Material Composition

The strength and pore structure of the MKG mortars were thoroughly analyzed in order to better understand the impacts of the Si/Al ratio and the Na/Al ratio. The compressive strengths and pore volumes for the Si/Al and Na/Al ratios are shown in Fig. 7.5. While the other Si/Al ratios produced slightly lower compressive strengths, the mixture with a Si/Al ratio of 2.32 showed the maximum compressive strength (= 66.04 MPa) (Fig. 7.5a). Additionally, a rise in the Si/Al ratio increased the capillary pore volume but decreased the gel pore volume (Fig. 7.5c).

Variations in the Na/Al ratio led to monotonic changes in compressive strengths and pore volumes when the Si/Al ratio was set at 2.62. For instance, compressive strength rose from 57.63 to 74.28 MPa by 29% when the Na/Al ratio climbed from 1.01 to 1.36, whereas gel pore volume dropped from 0.0365 to 0.0093 mL/g by 75% (Fig. 7.5b, d). The absolute capillary volume showed only minor changes as the Na/

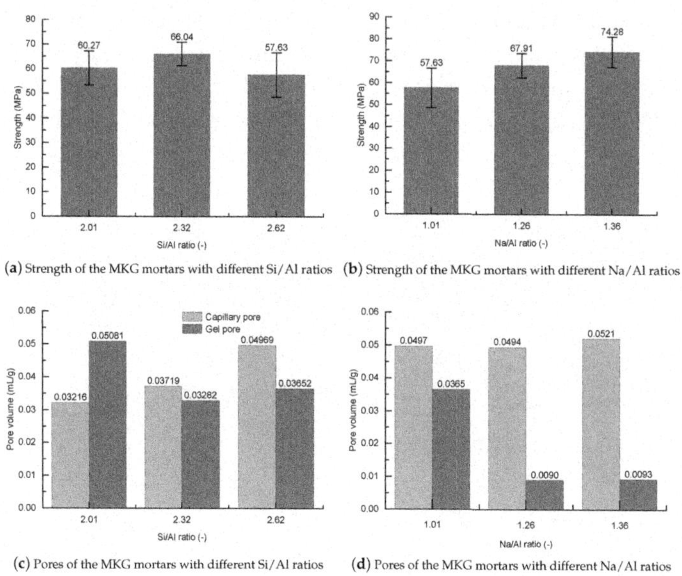

(a) Strength of the MKG mortars with different Si/Al ratios (b) Strength of the MKG mortars with different Na/Al ratios

(c) Pores of the MKG mortars with different Si/Al ratios (d) Pores of the MKG mortars with different Na/Al ratios

Fig. 7.5 Compositional dependence of the strength and pore structure of the MKG mortars: strength of the MKG mortars versus **a** the Si/Al ratio; **b** Na/Al ratio, pores of the MKG mortars versus the **c** Si/Al ratio and **d** Na/Al ratio. Reproduced from [Compositional Dependence of Pore Structure, Strengthand Freezing–Thawing Resistance of Metakaolin-Based Geopolymers] by [Dongming Yan] with permission from [Alkali-Activated Materials for Sustainable Construction]

Al ratio varied (Fig. 7.5d), although the relative ratio of capillary pores increased considerably (Fig. 7.2d).

The alterations in the strength and pore structure of the material system were a consequence of the physico-chemical interactions between the metakaolin particles and alkaline activator. As both the Si/Al ratio and Na/Al ratio were optimally selected based on previous literature (Lahoti et al. 2017) and our previous work (Yan et al. 2016), all MKG mortars exhibited high compressive strengths. Our results contradicted the observation made by Akono et al. (2019) that materials with higher capillary pore volumes exhibit lower strengths. This disagreement might be explained by the improved geopolymer polymerization extents at higher Na/Al ratios (or Na/Si ratios), which would increase the nanostructure's compactness and hence inhibit nanopores and encourage capillary pores (Duxson et al. 2005).

7.4 Freezing–Thawing Damages

7.4.1 Morphology, Mass Loss, and Strength Loss

Images of the MKG mortars after 50 freeze–thaw (F-T) cycles are shown in Fig. 7.6. The MKG-1 specimens showed severe surface spalling (Fig. 7.6a), which led to a mass loss of 21.04% (Fig. 7.7a). Figure 7.6b, c shows that the MKG-2 and MKG-3 specimens were entirely broken or fragmented into tiny mortar pieces, suggesting a 100% mass loss (Fig. 7.7a). Conversely, the MKG-4 and MKG-5 specimens displayed minimal or negligible visual alterations (Fig. 7.6d, e) and mass loss (Fig. 7.7a).

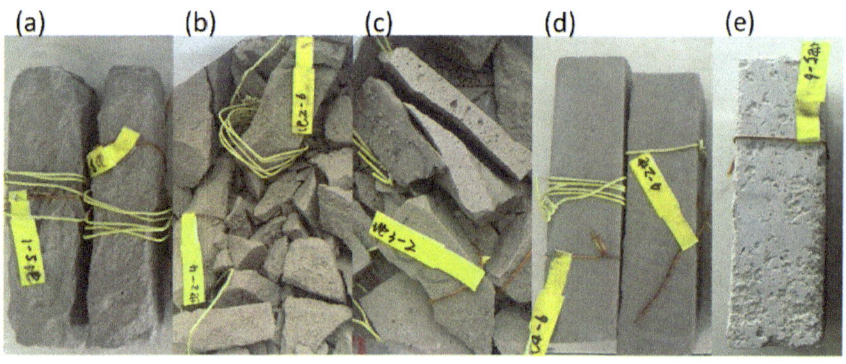

Fig. 7.6 Pictures of the MKG mortars after 50 F-T cycles: **a** MKG-1; **b** MKG-2; **c** MKG-3; **d** MKG-4; and **e** MKG-5. Reproduced from [Compositional Dependence of Pore Structure, Strengthand Freezing–Thawing Resistance of Metakaolin-Based Geopolymers] by [Dongming Yan] with permission from [Alkali-Activated Materials for Sustainable Construction]

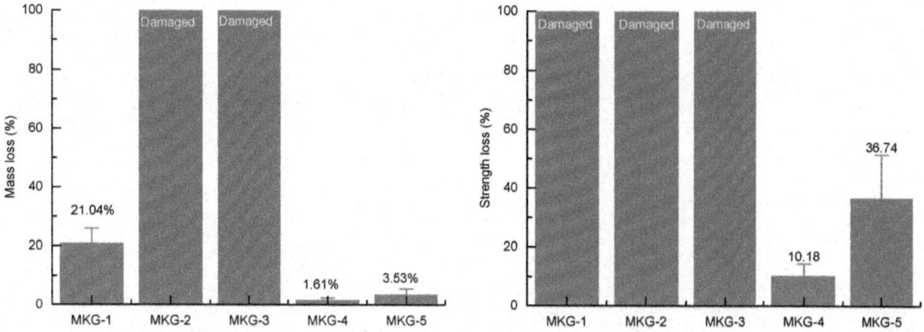

(**a**) Mass loss (**b**) Strength loss

Fig. 7.7 Mass loss **a** and strength loss **b** of the MKG mortars after 50 F-T cycles. Reproduced from [Compositional Dependence of Pore Structure, Strengthand Freezing–Thawing Resistance of Metakaolin-Based Geopolymers] by [Dongming Yan] with permission from [Alkali-Activated Materials for Sustainable Construction]

Due to the severe spalling and pulverization, it was unsuitable to conduct strength testing for the MKG-1, MKG-2, and MKG-3 specimens. Consequently, those mortar mixes were deemed to have a 100% strength loss following F-T loads (Fig. 7.7b). Meanwhile, although the MKG-4 and MKG-5 mortars exhibited minor surface spalling, they experienced significant strength losses, namely 10.18% for MKG-4 and 36.74% for MKG-5 (Fig. 7.7b). This indicated that the materials were severely damaged by F-T cycles. Based on the observations made in Figs. 7.6 and 7.7, it can be inferred that the F-T resistance of the MKG mortars could be classified in the following order: MKG-4 > MKG-5 > MKG-1 > MKG-2 = MKG-3. The data on strength and porosity reported in Sect. 7.3 showed that materials with higher strength (or lower porosity) had superior F-T resistance, and this ranking was broadly consistent with those results.

The MKG-4 and MKG-5 specimens with full appearance underwent SEM/BSE investigations to look into the F-T damages. Selective BSE images of the MKG-4 and MKG-5 mortars are shown in Fig. 7.8, which reveals many tortuous microcracks in the materials connecting some air spaces and/or huge holes. These cracks differed from drying cracks, which should be distributed within the geopolymer matrix with smeared cracking meshes (Balczár et al. 2015). The cracks significantly reduced compressive strength but maintained specimen mass before maturing into larger cracks that caused material spalling. Although the MKG-5 specimen displayed fewer cracks than the MKG-4 specimen, cracks in the former material intersected, forming percolated cracks. This explains why the MKG-5 mortar exhibited more severe mass and strength losses (Fig. 7.7).

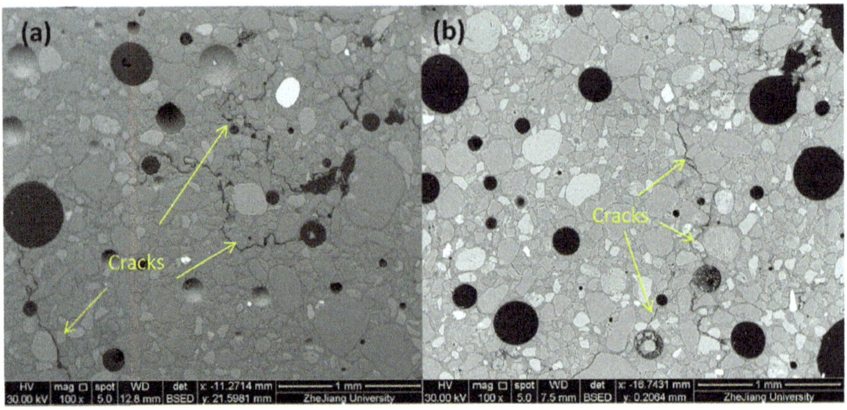

Fig. 7.8 Typical SEM/BSE pictures of **a** MKG-4 and **b** MKG-5 specimens after F-T loads. Reproduced from [Compositional Dependence of Pore Structure, Strengthand Freezing–Thawing Resistance of Metakaolin-Based Geopolymers] by [Dongming Yan] with permission from [Alkali-Activated Materials for Sustainable Construction]

7.4.2 Pore Structure Alterations

The changes in the pore structure of MKG mortars subjected to freeze–thaw (F-T) loads were analyzed using mercury intrusion porosimetry (MIP) tests. The PSD spectra and pore segmentation of the MKG mortars following F-T loads are presented in Fig. 7.9. Initially, both APSD and DPSD spectra showed curves similar to those prior to F-T loads, which is reasonable as the spatial compactness of the material system would remain unaffected by F-T loads. However, a slight increase in the first mercury rise in APSD and the peak in DPSD was observed for the MKG-5 mortar after F-T loads (Fig. 7.9e). This can be attributed to F-T cracking, which enhances surface roughness and subsequently increases the surface-conformance effect (Fig. 7.9f).

The impact of F-T loads on the pore structure of MKG mortars was thoroughly investigated by analyzing two phases of DPSD spectra. Phase A and B referred to gel pores below 50 nm and capillary pores ranging from 100 to 10,000 nm, respectively (Fig. 7.9b). Comparative plots of the two phases before and after F-T loads for the mortars are presented in Fig. 7.9c and d. Surprisingly, all mortar mixes demonstrated an increase in gel pores (Fig. 7.9c), while capillary pores became narrower with decreased pore volumes (Fig. 7.9e). These observations contradicted previous reports of pore structure degradation (Zhao et al. 2019) and suggested continual polymerization of materials immersed in water induced pore refinement rather than material degradation caused by F-T loads. A literature survey revealed that the initial few F-T cycles might enhance the geopolymerization process, refining the pore structure, and even increasing strength (Zhao et al. 2019; Aygörmez et al. 2020a, b). Figure 7.9f illustrates how the geopolymerization products reduced the capillary pore fractions in MKG-4 and MKG-5 by filling coarse holes. Notably, the severe F-T damages

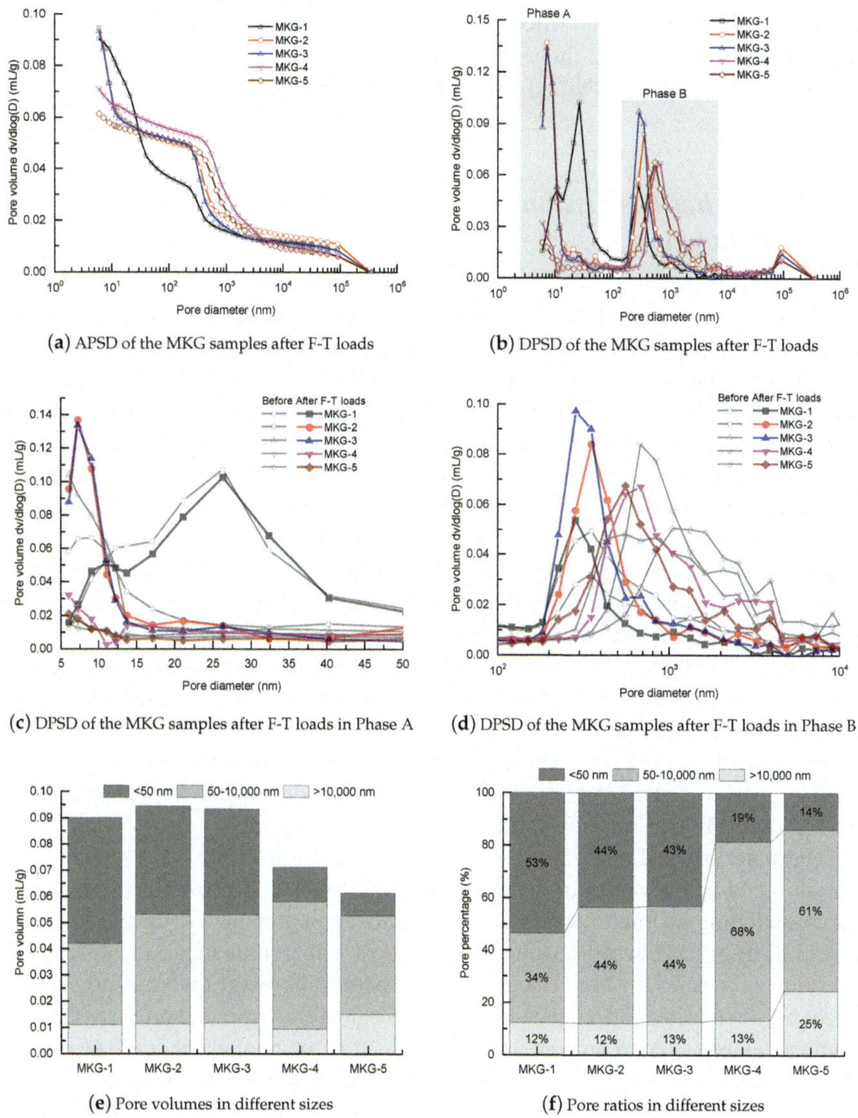

(a) APSD of the MKG samples after F-T loads

(b) DPSD of the MKG samples after F-T loads

(c) DPSD of the MKG samples after F-T loads in Phase A

(d) DPSD of the MKG samples after F-T loads in Phase B

(e) Pore volumes in different sizes

(f) Pore ratios in different sizes

Fig. 7.9 Pore structure of the MKG mortars after F-T loads: **a** APSD spectra; **b** DPSD spectra; comparison of the pores in Phase A **c** and Phase B **d**; **e** pore volumes in different sizes; and **f** pore ratios in different sizes. Reproduced from [Compositional Dependence of Pore Structure, Strengthand Freezing–Thawing Resistance of Metakaolin-Based Geopolymers] by [Dongming Yan] with permission from [Alkali-Activated Materials for Sustainable Construction] observed in the MKG-2 and MKG-3 mixes did not correspond to a decrease in their pore structures but instead showed increases in gel pores, with ratios rising from 39 and 37% (Fig. 7.2d) to 44% and 43% (Fig. 7.9f), respectively.

7.4.3 Further Discussion: Permeability Associated Pressure Relaxation

The sensitivity of strength and pore structure to composition has been demonstrated by experimental tests on MKG mortars. With a linear connection between compressive strength and total porosity, increasing the Na/Al ratio tends to increase material strengths (Fig. 7.4). With the exception of volumes induced by the surface-conformance effect showing similar results, both capillary pores and gel pores depended heavily on the Si/Al ratio and the Na/Al ratio (Fig. 7.5). While our experimental data demonstrated that geopolymer materials with higher strength also exhibited better F-T resistance, we observed anomalies in pore structure alterations following F-T loads (Fig. 7.9). These data raise a question: Why does an MKG material with a coarser pore structure exhibit better F-T resistance? To answer this question, we must understand the roles of different pores during freezing.

Figure 7.10 provides schematic snapshots of a thin pore system and a coarse pore system during freezing. The thin pore (or coarse pore) system comprises a large pore chamber connected by thin channels (or coarse channels), with all pores filled with water prior to freezing. The thin pore system may reflect the pore structures of MKG-1, MKG-2, and MKG-3, while the coarse pore system may represent those of MKG-4 and MKG-5, as shown in Fig. 7.2.

The water trapped in the huge pore chamber began to freeze when the temperature dropped below zero, but the pore channels remained unfrozen. Freezing pressures, including hydraulic pressure and crystallization pressure (Zeng and Li 2019), resulting from the phase transition of water to ice in confinement, increased. The crystallization pressure would locally exert on pore walls owing to free-energy differences between water and ice (Scherer 1999). Meanwhile, hydraulic pressure that was homogeneously exerted on pore walls through unfrozen pore water could be relaxed through viscous flow of pore water (Fig. 7.10). Concrete scientists have utilized this feature by homogeneously entraining air voids in the material matrix to shorten water flow distance and reduce hydraulic pressures (Sun and Scherer 2010).

In general, hydraulic pressure relaxation is inversely proportional to the permeability of pore channels for water flow, which can be expressed as (Zhang et al. 1998):

$$\Delta P_h \propto L_t / K_h \tag{7.3}$$

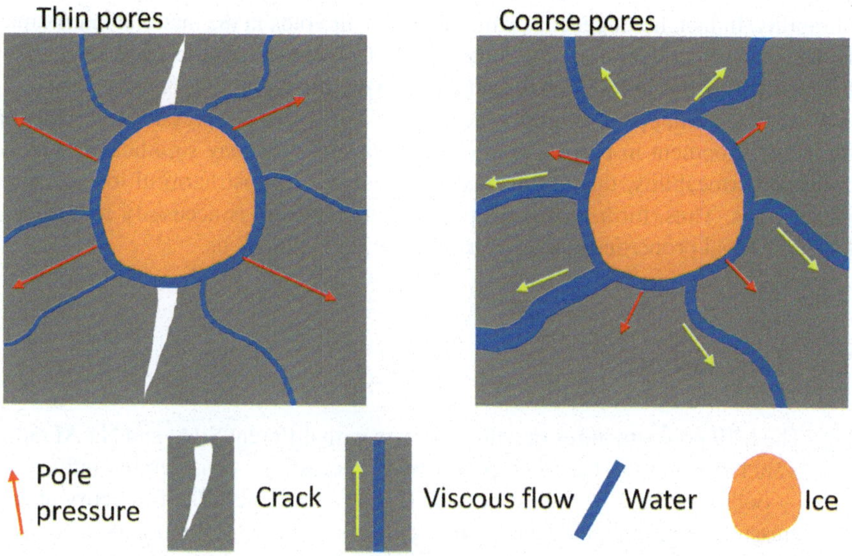

Fig. 7.10 Schematic illustration of the freezing resistance differences between a material with thin pores and that with coarse pores. Reproduced from [Compositional Dependence of Pore Structure, Strengthand Freezing–Thawing Resistance of Metakaolin-Based Geopolymers] by [Dongming Yan] with permission from [Alkali-Activated Materials for Sustainable Construction]

where ΔP_h represents the changes in hydraulic pressure due to relaxation, L denotes pore length, t refers to relaxation time, and K_h represents permeability. The permeability of a porous material can be predicted by the threshold pore size, $K_h \propto D_T^2$, according to Katz and Thompson (1986). Equation (7.3) can be expressed as:

$$\Delta P_h \propto D_T^2 L_t \tag{7.4}$$

The pore pressure relaxation time t can be presented as:

$$t \propto \Delta P_h / L D_T^2 \tag{7.5}$$

According to Eqs. (7.4) and (7.5), a thin-pore system may have less pore pressure relaxation with the same relaxation time, or it may take longer time to relax the same pore pressures. For instance, assuming ice formation in the big pores accumulated hydraulic pressures of 10 MPa while pore length L was constant, MKG-2 mortar would require approximately 9 times longer than MKG-5 to relax hydraulic pressure at 10 MPa, according to Eq. (7.5), using pore structure data of MKG-2 and MKG-5 as examples. This observation might explain why MKG materials with coarser pore structures exhibit better F-T resistance.

However, it is essential to note that experimental data presented in this study do not imply that a material with a coarser pore structure would necessarily have better durability. In fact, homogeneous entrainment of air voids in the material matrix may be the most effective routine for enhancing F-T resistance. Figure 7.8 demonstrates that the MKG mortars (MKG-4 and MKG-5) with strong F-T resistance had air spaces to accommodate ejected water and lessen hydraulic pressures brought on by ice development in pores. Additionally, coarser pores may significantly reduce material permeability, thereby decreasing resistance against harmful ion transport in materials. Thus, further rigorous pore structure design is necessary to optimize MKG material properties before their engineering applications.

7.5 Conclusions

(1) The MIP pore structures of MKG mortars with different Si/Al and Na/Al ratios were found to vary. Capillary pores and gel pores were considerably influenced by both Si/Al and Na/Al ratios, which showed surface fills of mercury at low pressures with insignificant volume changes in the initial mercury increases. Increases in both ratios resulted in decreased gel pores but promoted capillary pores.
(2) All MKG mortars exhibited relatively high strengths. However, strength decreased as total porosity increased, roughly following a linear plot.
(3) F-T resistance varied among the different MKG mortars, with MKG-4 exhibiting better F-T resistance than MKG-5, MKG-1, MKG-2, and MKG-3. Mortars with Na/Al ratios less than 1.26 (i.e., MKG-1, MKG-2, and MKG-3) showed severe F-T damage. Increasing the Na/Al ratio improved F-T resistance, although F-T loads also caused notable cracking of MKG-4 and MKG-5.
(4) MIP studies showed that the pore architectures were improved during F-T stresses, most likely as a result of continuous material curing. Due to the slower pore pressure relaxation rates, materials with smaller pore shapes showed lesser F-T resistance.

References

Akono AT, Koric S, Kriven WM (2019) Influence of pore structure on the strength behavior of particle- and fiber-reinforced metakaolin-based geopolymer composites. Cem Concr Compos 104:103361. https://doi.org/10.1016/j.cemconcomp.2019.103361

Aygörmez Y, Canpolat O, Al-mashhadani MM (2020a) Assessment of geopolymer composites durability at one year age. J Build Eng 32. https://doi.org/10.1016/j.jobe.2020.101453

Aygörmez Y, Canpolat O, Al-mashhadani MM, Uysal M (2020b) Elevated temperature, freezing-thawing and wetting-drying effects on polypropylene fiber reinforced metakaolin based geopolymer composites. Constr Build Mater 235. https://doi.org/10.1016/j.conbuildmat.2019.117502

Balczár I, Korim T, Dobrádi A (2015) Correlation of strength to apparent porosity of geopolymers—understanding through variations of setting time. Constr Build Mater 93:983–988. https://doi.org/10.1016/j.conbuildmat.2015.05.059

Bocca P, Grazzini A (2013) Mechanical properties and freeze-thaw durability of strengthening mortars. J Mater Civ Eng 25:274–280. https://doi.org/10.1061/(asce)mt.1943-5533.0000597

Carpinteri A, Grazzini A, Lacidogna G, Manuello A (2014) Durability evaluation of reinforced masonry by fatigue tests and acoustic emission technique. Struct Control Heal Monit 21:950–961. https://doi.org/10.1002/stc.1623

Cui Y, Gao K, Zhang P (2020) Experimental and statistical study on mechanical characteristics of geopolymer concrete. Materials (Basel) 13. https://doi.org/10.3390/ma13071651

Diamond S (2000) Mercury porosimetry. An inappropriate method for the measurement of pore size distributions in cement-based materials. Cem Concr Res 30:1517–1525. https://doi.org/10.1016/S0008-8846(00)00370-7

DL/T5150-2017 (2017) Test code for hydraulic concrete; DL/T 5150-2017; The Electric Power Ministry of the People's Republic of China, Beijing, China, 2017

Duxson P, Provis JL, Lukey GC et al (2005) Understanding the relationship between geopolymer composition, microstructure and mechanical properties. Colloids Surfaces A Physicochem Eng Asp 269:47–58. https://doi.org/10.1016/j.colsurfa.2005.06.060

Duxson P, Fernández-Jiménez A, Provis JL et al (2007a) Geopolymer technology: the current state of the art. J Mater Sci 42:2917–2933. https://doi.org/10.1007/s10853-006-0637-z

Engemann S, Reichert H, Dosch H et al (2004) Interfacial melting of ice in contact With SiO_2. Phys Rev Lett 92. https://doi.org/10.1103/PhysRevLett.92.205701

Hasselman DPH, Fulrath RM (1964) Effect of small fraction of spherical porosity on elastic moduli of glass. J Am Ceram Soc 47:52–53. https://doi.org/10.1111/j.1151-2916.1964.tb14644.x

Huang S, Ye Y, Cui X et al (2020) Theoretical and experimental study of the frost heaving characteristics of the saturated sandstone under low temperature. Cold Reg Sci Technol 174. https://doi.org/10.1016/j.coldregions.2020.103036

Jähnert S, Vaca Chávez F, Schaumann GE et al (2008) Melting and freezing of water in cylindrical silica nanopores. Phys Chem Chem Phys 10:6039–6051. https://doi.org/10.1039/b809438c

Kastiukas G, Ruan S, Liang S, Zhou X (2020) Development of precast geopolymer concrete via oven and microwave radiation curing with an environmental assessment. J Clean Prod 255. https://doi.org/10.1016/j.jclepro.2020.120290

Katz AJ, Thompson AH (1986) Quantitative prediction of permeability in porous rock. Phys Rev B 34:8179–8181. https://doi.org/10.1103/PhysRevB.34.8179

Kohout J, Koutník P, Hájková P et al (2021) Effect of k/al molar ratio on the thermo-mechanical properties of metakaolinite-based geopolymer composites. Polymers (Basel) 13. https://doi.org/10.3390/polym13213754

Kubba Z, Fahim Huseien G, Sam ARM et al (2018) Impact of curing temperatures and alkaline activators on compressive strength and porosity of ternary blended geopolymer mortars. Case Stud Constr Mater 9. https://doi.org/10.1016/j.cscm.2018.e00205

Lahoti M, Narang P, Tan KH, Yang EH (2017) Mix design factors and strength prediction of metakaolin-based geopolymer. Ceram Int 43:11433–11441. https://doi.org/10.1016/j.ceramint.2017.06.006

Monteiro PJM, Miller SA, Horvath A (2017) Towards sustainable concrete. Nat Mater 16:698–699. https://doi.org/10.1038/nmat4930

Peng X, Shuai Q, Li H et al (2020) Fabrication and fireproofing performance of the coal fly ash-metakaolin-based geopolymer foams. Materials (Basel) 13. https://doi.org/10.3390/ma13071750

Provis JL (2014) Geopolymers and other alkali activated materials: why, how, and what? Mater Struct Constr 47:11–25. https://doi.org/10.1617/s11527-013-0211-5

Roviello G, Chianese E, Ferone C et al (2019) Hybrid geopolymeric foams for the removal of metallic ions from aqueous waste solutions. Materials (Basel) 12:4091. https://doi.org/10.3390/ma12244091

Rowles M, O'Connor B, Matthew R et al (2003) Chemical optimisation of the compressive strength of aluminosilicate geopolymers synthesised by sodium silicate activation of metakaolinite. J Mater Chem 13:1161–1165. https://doi.org/10.1039/b212629j
Rowles MR, O'Connor BH (2009) Chemical and structural microanalysis of aluminosilicate geopolymers synthesized by sodium silicate activation of metakaolinite. J Am Ceram Soc 92:2354–2361. https://doi.org/10.1111/j.1551-2916.2009.03191.x
Sáez del Bosque IF, Van den Heede P, De Belie N et al (2020) Freeze-thaw resistance of concrete containing mixed aggregate and construction and demolition waste-additioned cement in water and de-icing salts. Constr Build Mater 259. https://doi.org/10.1016/j.conbuildmat.2020.119772
Scherer GW (1999) Crystallization in pores. Cem Concr Res 29:1347–1358. https://doi.org/10.1016/S0008-8846(99)00002-2
Shi C (1996) Strength, pore structure and permeability of alkali-activated slag mortars. Cem Concr Res 26:1789–1799. https://doi.org/10.1016/S0008-8846(96)00174-3
Sun P, Wu HC (2013) Chemical and freeze-thaw resistance of fly ash-based inorganic mortars. Fuel 111:740–745. https://doi.org/10.1016/j.fuel.2013.04.070
Sun Z, Scherer GW (2010) Effect of air voids on salt scaling and internal freezing. Cem Concr Res 40:260–270. https://doi.org/10.1016/j.cemconres.2009.09.027
Topçu IB, Toprak MU, Uygunoğlu T (2014) Durability and microstructure characteristics of alkali activated coal bottom ash geopolymer cement. J Clean Prod 81:211–217. https://doi.org/10.1016/j.jclepro.2014.06.037
Washburn EW (1921) Note on a method of determining the distribution of pore sizes in a porous material. Proc Natl Acad Sci 7:115–116. https://doi.org/10.1073/pnas.7.4.115
White CE, Page K, Henson NJ, Provis JL (2013) In situ synchrotron X-ray pair distribution function analysis of the early stages of gel formation in metakaolin-based geopolymers. Appl Clay Sci 73:17–25. https://doi.org/10.1016/j.clay.2012.09.009
Yan D, Chen S, Zeng Q et al (2016) Correlating the elastic properties of metakaolin-based geopolymer with its composition. Mater Des 95:306–318. https://doi.org/10.1016/j.matdes.2016.01.107
Zeng Q, Li K (2019) Quasi-liquid layer on ice and its effect on the confined freezing of porous materials. Crystals 9. https://doi.org/10.3390/cryst9050250
Zeng Q, Fen-Chong T, Dangla P, Li K (2011) A study of freezing behavior of cementitious materials by poromechanical approach. Int J Solids Struct 48:3267–3273. https://doi.org/10.1016/j.ijsolstr.2011.07.018
Zeng Q, Fen-Chong T, Li KF (2013) Elastic behavior of saturated porous materials under undrained freezing. Acta Mech Sin Xuebao 29:827–835. https://doi.org/10.1007/s10409-013-0082-6
Zeng Q, Fen-Chong T, Li K (2014) Freezing behavior of cement pastes saturated with NaCl solution. Constr Build Mater 59:99–110. https://doi.org/10.1016/j.conbuildmat.2014.02.042
Zeng Q, Li K, Fen-Chong T (2015) Heterogeneous nucleation of ice from supercooled NaCl solution confined in porous cement paste. J Cryst Growth 409:1–9. https://doi.org/10.1016/j.jcrysgro.2014.09.045
Zeng Q, Wang X, Yang P et al (2019) Tracing mercury entrapment in porous cement paste after mercury intrusion test by X-ray computed tomography and implications for pore structure characterization. Mater Charact 151:203–215. https://doi.org/10.1016/j.matchar.2019.02.014
Zhang D, Weinbaum S, Cowin SC (1998) Estimates of the peak pressures in bone pore water. J Biomech Eng 120:697–703. https://doi.org/10.1115/1.2834881
Zhao R, Yuan Y, Cheng Z et al (2019) Freeze-thaw resistance of Class F fly ash-based geopolymer concrete. Constr Build Mater 222:474–483. https://doi.org/10.1016/j.conbuildmat.2019.06.166

Chapter 8
Aggregate Influence on MKG Concrete

Abstract In order to incorporate geopolymer in concrete, comprehending the compatibility between the coarse aggregate and the geopolymer binder is imperative. Experimental investigations were carried out to examine the influence of coarse aggregate size on both the mechanical properties and microstructure of MKG concrete. Additionally, the impact of coarse aggregate size on the bond between Carbon Fiber-Reinforced Polymer (CFRP) and MKG concrete was investigated.

8.1 Introduction

Concrete, being a quasi-brittle material, requires a comprehensive study of its mechanical parameters. These parameters encompass essential aspects such as compressive strength, tensile strength, modulus of elasticity, and fracture energy (Sucharda et al. 2020). Concrete is a composite material that strongly depends on the interaction of its elements and their chemical and physical characteristics to operate. Cement paste, coarse or fine aggregate, and the interfacial transition zone (ITZ) between the coarse aggregate and matrix make up the three phases of concrete, which is a heterogeneous material (Xu et al. 2015). The coarse aggregate plays a crucial role in determining the mechanical behavior of concrete, accounting for approximately 70% of the concrete volume (Chen and Liu 2004). Mahmoud and Ahmad (2015) emphasized that the coarse aggregate, cement paste, and the ITZ between them significantly influence the mechanical properties of concrete. The effect of the coarse aggregate on the characteristics of regular Portland cement (OPC) concrete has been the subject of much investigation. Previous research has shown that a key element influencing the growth of the ITZ area and the subsequent creation and spread of microcracks is the size of the coarse aggregate (Akçaoğlu 2017; Yuan et al. 2019a). Furthermore, the size of the coarse aggregate can significantly impact the mechanical properties and durability of OPC concrete (Ćosić et al. 2015; Zhong and Wille 2016). Ćosić et al. (2015) discovered that smaller-sized coarse aggregates can lead to higher flexural strength in OPC concrete. However, Yu et al. (2019) reported

D. Yan et al., *Metakaolin-Based Geopolymers*,
https://doi.org/10.1007/978-981-97-0652-5_8

that contrary to the findings of Zhong and Wille (2016), the compressive strength of OPC concrete increases with an increase in the size of the coarse aggregate.

However, the initial investigations on geopolymer mainly concentrated on mortar and paste. Numerous research have been done to determine how various variables, such as the ratio of the alkali precursor (such as fly ash), the Si/Al ratio, the alkali concentration, and the characteristics of the coarse aggregate, affect geopolymer concrete (GPC) (Lahoti et al. 2017; Farhan et al. 2020). By Nikoloutsopoulos et al. (2021), the physical and mechanical characteristics of fly ash-based geopolymer concrete were contrasted with those of Portland cement concrete. Their findings indicated that the binder-to-aggregate ratio significantly affects geopolymer concrete properties. Moreover, the influence of coarse aggregate content in GPC exhibits considerable variation (Joseph and Mathew 2012; Zhang et al. 2020). Although some studies have explored the influence of various factors on geopolymer, the size effect of coarse aggregate on GPC remains uncertain, despite its acknowledged significance and impact on the mechanical properties and durability of OPC concrete. Additionally, a robust and reliable mix design procedure specifically tailored for GPC is yet to be established. Therefore, this study aims to investigate the influence of coarse aggregate size on metakaolin-based geopolymer (MKG) concrete.

Despite the extensive research conducted in various areas, further research, regulation, and systemization are still necessary for the use of MKG concrete in the construction industry. Previous research has demonstrated that the coarse aggregate size directly influences the microstructure of ordinary Portland cement (OPC) concrete, which in turn impacts the material's mechanical characteristics. However, existing research on geopolymer concrete (GPC) has primarily focused on the content and properties of coarse aggregates, rather than their size. Considering the significant influence of these factors on the performance and durability of OPC concrete, it is essential to expand the literature by investigating the effect of coarse aggregate size on the strength and microstructure of MKG concrete. To the best of our knowledge, this aspect has not been studied yet, and ongoing debate and further work are needed to explore the impact of coarse aggregate size on the mechanical properties and microstructure of MKG concrete. This study offers data acquired under ambient temperature, which mimics real-world application settings, whereas prior investigations on the mechanical behavior of geopolymer concrete were generally done under increased temperature conditions.

Crack formation and the subsequent impact on concrete performance are significant concerns in terms of concrete durability (Junaid et al. 2020). However, when exposed to harsh conditions, intensive fracture control methods might not be enough to stop steel bar corrosion in concrete (Junaid et al. 2020). Consequently, the long-term deterioration of steel-reinforced concrete (RC) structures, along with more stringent safety requirements and design capacity, has led to a growing demand for retrofitting or repairing existing structural members to enhance their durability, load resistance, and serviceability (Godat et al. 2007; Fazli et al. 2018b). Geopolymer RC constructions are susceptible to early degradation, especially those that are exposed to harsh conditions and deicing agents. Therefore, the use of sustainable and durable reinforcement materials, such as fiber-reinforced polymer (FRP), can be employed to improve and enhance their performance. In theory, the combination of metakaolin-based geopolymer (MKG) concrete with FRP offers a more durable system (Godat et al. 2007; Fazli et al. 2018b).

FRP sheets, specifically carbon fiber-reinforced polymer (CFRP), possess several advantageous properties such as corrosion and chemical resistance, lightweight, high strength-to-weight ratio, low thermal conductivity, and ease of application (Chen et al. 2017). These characteristics make FRPs a reliable alternative to conventional materials like steel and concrete for repairing buildings and infrastructure (Karbhari and Ghosh 2009; Shen et al. 2015a). In the realm of structural improvement under loading, FRPs have been widely utilized to enhance the structural capacity of reinforced concrete (RC) members (Pham and Hao 2016; Chen et al. 2018). Wrapping concrete columns with FRP can significantly increase their ductility compared to the conventional method of confinement with steel reinforcement (Lokuge et al. 2011). Moreover, the application of FRP is relatively straightforward when done in situ. To strengthen or repair concrete structures and infrastructure, various conventional techniques are employed including external post-tensioning, externally bonded steel plates, steel or concrete jacketing, and externally bonded composites (Fazli et al. 2018b). Building structures and infrastructures often face exposure to harsh acidic environmental conditions caused by industrial effluents and acidic rain (Akcil and Koldas 2006). Therefore, it is important for such structures to exhibit resistance to acid attack (Bakharev 2005). Applying CFRP to MKG concrete structures can help enhance their resistance to acid attack and improve the durability of deteriorated buildings subjected to harsh environmental conditions.

Externally bonded fiber-reinforced polymer (FRP) is a widely recognized technique for strengthening and repairing structures in the construction industry. This method offers numerous advantages, such as easy installation, cost reduction, minimal clearance losses, and improved resistance against chloride-ion intrusion into concrete. In a research conducted by Junaid et al. (2020), it was observed that the use of carbon fiber-reinforced polymer (CFRP) effectively restores the original moment capacities of damaged glass fiber-reinforced polymer cement (GPC) beams when externally strengthened. Additionally, geopolymer beams reinforced with CFRP exhibited significantly reduced ultimate deflection (around 50% reduction) and fewer cracks compared to reference beams (Junaid et al. 2020). Another study by Alzeebaree et al. (2019) investigated the mechanical properties and durability of GPC externally bonded with FRP in the presence of sulfuric acid. The results indicated that specimens bonded with CFRP exhibited superior resistance to sulfuric acid attacks compared to GPC samples, thanks to their enhanced resilience to acid solutions.

Previous research has extensively demonstrated that debonding from the concrete substrate is the predominant failure mode for externally bonded fiber-reinforced polymer (FRP). This failure occurs suddenly and in a brittle manner due to high stress concentrations, resulting in shear flexural or localized flexural cracks within the substrate (Franco and Royer-Carfagni 2014; Rathinam et al. 2016; Junaid et al. 2020). This limitation arises as the debonding occurs at strains below the ultimate strain of FRP (Mostofinejad and Shameli 2013), emphasizing the crucial role played by the FRP and its interfacial bond with the substrate in determining the failure mode of FRP-bonded concrete structures (Nakaba et al. 2001; Dai et al. 2005). It is essential to note that any stress placed on the substrate is transferred via the interfacial connection to the FRP (Fazli et al. 2018a). An epoxy glue is used to transfer the applied load to a concrete element that is FRP-bonded across a short distance close to the applied load, mostly in the form of shear stress. Increasing the bond length over a particular point does not further enhance the bond strength since the majority of the bond shear stress is transmitted in this short length (Ben Ouezdou et al. 2009; Diab and Farghal 2014). The "effective bond length" is the name given to this brief measurement. It is crucial for figuring out the maximum bond capacity between the FRP and concrete substrate interface as well as for calculating the failure load (debonding load) of a bonded specimen. As a result, models of bond strength that rely their predictions on the effective bond length have been suggested, underscoring their significance (Neubauer and Rostasy 1997; Shen et al. 2015d; Fazli et al. 2018b). Although several analytical models have been proposed to determine the effective bond length (Neubauer and Rostasy 1997; Khalifa et al. 1998; Chen and Teng 2001; Yang et al. 2001; Lu 2004; ACI 440.2R-08 2008; Wu et al. 2009; Canadian Standard Association 2012; Shen et al. 2015b; Fazli et al. 2018b) between OPC concrete and CFRP, there has been little experimental work on the effective bond length between CFRP and GPC. The length of the bond, the stiffness of the FRP, and the mechanical characteristics of the adhesive and substrate all have a role in determining the interfacial bond capacity (Wu et al. 2009; Diab and Farghal 2014). Because the designed bond capacity is affected by the properties of the substrate (Neubauer and Rostasy 1997; Wu et al. 2009; Bilotta et al. 2011; Diab and Farghal 2014), the use of MK instead of cement in concrete can affect interfacial bond behavior due to its different microstructural and mechanical properties, as reported by Singh et al. (2015).

The interfacial bond in concrete is primarily influenced by the tensile behavior of the material, while the size of the coarse aggregate plays a crucial role in determining its tensile strength. Akçaoğlu et al. (2002) highlighted that an increase in the size of the coarse aggregate leads to a reduction in splitting tensile strength due to the proliferation of microcracks around it. Furthermore, previous research has shown that concrete's fracture energy rises with the size of the coarse aggregate since it largely depends on the material's mechanical characteristics (Wolinski et al. 1987; Mihashi et al. 1991; Tasdemir et al. 1996; Chen and Liu 2004). As the strain placed on concrete increases, fractures spread along the stronger interfacial zone or along the matrix's big pores. A piece of coarse aggregate can either cause a spreading fracture to deflect or cause it to continue through it. Typically, cracks tend to follow a path around the aggregate-mortar interface, deflecting away from the coarse aggregate. This occurs because the toughness of the interface with the coarse aggregate is lower than that of the concrete paste. Consequently, a tortuous cracking path is formed, requiring higher fracture energy to overcome the interfacial bond (Chen and Liu 2004). Notably, a previous study indicated that the mechanical properties and performance of GPC are significantly influenced by the size of the coarse aggregate (Guades 2017). Specifically, if the size of the coarse aggregate exceeds 18.8 mm, the compressive strength of GPC tends to decrease. Because it influences the development and spread of microcracks in both the substrate and the interfacial transition zone, the size of the coarse aggregate has an impact on the mechanical characteristics of the substrate (Perry and Gillott 1977; Tasong et al. 1999; Akçaoğlu et al. 2002). While the impact of coarse aggregate size on the bond behavior between CFRP and OPC concrete has been extensively investigated (Pan and Leung 2007; Yuan et al. 2019a), to the best of the authors' knowledge, the influence of coarse aggregate size on the interfacial bond behavior of CFRP-GPC remains unexplored.

The use of metakaolin (MKG) concrete as a substitute for ordinary Portland cement (OPC) concrete offers potential solutions for the porosity and permeability weaknesses of OPC in aggressive environments. When carbon fiber-reinforced polymer (CFRP) is combined with MKG reinforced concrete (RC) members, they not only enhance shear and flexural capacities but also improve durability and corrosion resistance. However, to effectively implement CFRP in MKG concrete, extensive studies on the bond performance between CFRP and MKG concrete are necessary, including a comparison with OPC counterparts. The bond strength plays a critical role in the performance and durability of CFRP applications. A strong bond between concrete and CFRP leads to better application performance and increased durability in aggressive environments. The type of binder and the size of coarse aggregate have a significant impact on the thickness and porosity of the interfacial transition zone (ITZ). As a result, the characteristics of the ITZ can affect the mechanical and transport qualities of concrete. Additionally, the formation of the epoxy adhesive bond between CFRP and concrete relies on a combination of mechanical interlocking and chemical bonding. Therefore, different types of concrete are expected to have varying effects on the interfacial bond between CFRP and the substrate due to their distinct physical and microstructural properties. Furthermore, when utilizing composite structural systems, ensuring compatibility among different materials is crucial. Thus, investigating the stress transfer mechanisms between CFRP and MKG concrete is essential for design purposes and contributes to understanding the engineering potential of MKG concrete.

An experimental study was conducted to examine the influence of coarse aggregate size on the mechanical properties of MKG concrete. The strength performance of MKG concrete was assessed using a variety of techniques, such as failure mode analysis, compressive strength testing, splitting tensile strength testing, scanning electron microscopy/energy-dispersive X-ray spectroscopy (SEM/EDS) analysis, and pore structure characterization. The results were compared with those of ordinary Portland cement (OPC) concrete to provide insights for potential engineering applications. These findings contribute to a better understanding of how coarse aggregate size affects the properties of geopolymer concrete (GPC). Additionally, the experimental study aimed to enhance the understanding of the effects of coarse aggregate size on the bond between carbon fiber-reinforced polymer (CFRP) and MKG concrete. By considering the impact of coarse aggregate size, models were created to forecast the effective bond length and bond-slip behavior of CFRP-MKG concrete. The bond between CFRP and MKG concrete was examined using the single-lap shear testing method. Strain gauges and linear variable displacement transducers (LVDTs) were used to measure strain distribution during specimen loading. Analyzing the experimental results provided bond parameters, including effective bond length, interfacial bond-slip relationship, failure load, and local slip at the maximum interfacial shear bond stress.

8.2 Aggregate Influence on Mechanical Properties of MKG Concrete

8.2.1 Experimental Procedures

8.2.1.1 Materials

Typical OPC concrete specimens were prepared using the cement of CEM I 42.5 N (British Standard Institution 2011). Metakaolin (MK), of the industrial grade, was purchased from BASF MetaMax. The chemical compositions of OPC and MK as determined by X-ray fluorescence (XRF) studies are shown in Table 8.1. The XRF spectrum was recorded using a Malvern PANalytical Epsilon 1 spectrometer (Malvern PANalytical, Almelo, Netherlands). XRF spectroscopy is a non-destructive and accurate technique used to identify material composition and sample preparation is mostly not required. The chemical analysis revealed that the sum of SiO_2, Al_2O_3, and Fe_2O_3 content was 95.46%. This is above the minimum value of 70% as required by ASTM C618 (ASTM International 2019). The MgO concentration was also under 5%. The alkaline activator solution was made using a pellet of sodium hydrate and sodium silicate (water glass, WG). Table 8.2 lists the chemical make-up of the water glass. The sodium hydrate pellet had an analytical reagent purity of 96%. The dry river sand and limestone coarse aggregate were used for all the specimens in this study. Figure 8.1a shows the grain size distribution of the sand and coarse aggregate. The fineness modulus of the sand and coarse aggregate was 2.8 and 4.34, respectively. The water absorption capacity of the coarse aggregate was 0.9% and the density was 2620 kg/m^3. The three groups of coarse aggregate were prepared using the usual sieves, as indicated in Fig. 8.1b. Figure 8.11c shows the distribution curves for their particle sizes. The specimens were made using coarse aggregate with three size ranges (5–10 mm, 10–16 mm, and 16–20 mm) and a fineness modulus of 2.73.

8.2.1.2 Test Specimen Preparation

According to Table 8.3, concrete mixes with a grade of 40 MPa compressive strength after 28 days were created using 5–10 mm coarse aggregate size. The coarse aggregates were submerged in water for 24 h before to casting in order to limit the impact of water absorption on the test findings, and their surface water was then dried by laying them over a big screen for two hours and allowing water to evaporation

Table 8.1 Chemical composition of OPC and MK (wt%)

Chemicals	SiO_2	Al_2O_3	Fe_2O_3	CaO	MgO	SO_3	K_2O	TiO_2	Na_2O	LOI *
OPC	23.81	10.79	3.36	50.58	5.31	2.75	0.92	0.73	0.61	1.14
MK	53.29	41.64	0.53	1.09	0.28	-	0.14	1.13	0.07	1.83

Reproduced from [Effect of Size of Coarse Aggregate on Mechanical Properties of Metakaolin-Based Geopolymer Concrete and Ordinary Concrete] by [Hamed Fazli] with permission from [Materials]

* *LOI* loss on ignition

Table 8.2 Oxide composition of sodium silicate solution (WG)

Oxide	SiO_2	Na_2O	H_2O
Mass content (%)	26.00	8.20	65.80

Reproduced from [Effect of Size of Coarse Aggregate on Mechanical Properties of Metakaolin-Based Geopolymer Concrete and Ordinary Concrete] by [Hamed Fazli] with permission from [Materials]

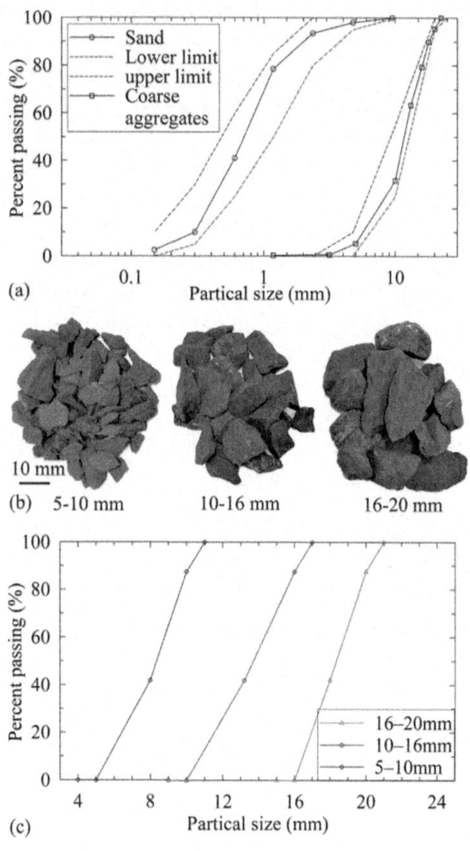

Fig. 8.1 Coarse aggregate and sand used in this study: **a** particle size distribution curve for the coarse aggregate and sand; **b** three range of the sizes of the coarse aggregates; **c** the grain size distribution curve for the three size ranges of the coarse aggregates. Reproduced from [Effect of Size of Coarse Aggregate on Mechanical Properties of Metakaolin-Based Geopolymer Concrete and Ordinary Concrete] by [Hamed Fazli] with permission from [Materials]

(Gonilho Pereira et al. 2009). The coarse aggregate and sand were first dry-mixed for two minutes. Then, for casting the MKG concrete, the alkaline activator solution and water were added gradually, and mixing continued for five minutes. The alkaline activator solution was made by combining a liquid sodium silicate (water glass, WG) and a pellet sodium hydrate in accordance with the mixture's design at least 24 h prior to the synthesis of the geopolymer. After the coarse aggregate, sand, and cement had been well mixed dry for five minutes, water was progressively added for the OPC concrete casting. The fresh MKG and OPC concrete were cast into respective molds in three layers. Each layer was compacted 25 times with a standard steel rod according to the procedure reported in BS EN 12390-2 (British Standard Institution 2019).

Three concrete cubes with the dimensions of 150 mm × 150 mm × 150 mm and 100 mm × 100 mm × 100 mm were made for each group to measure the splitting tensile strength (f_t) and compressive strength (f_{cu}), according to BS EN 12390-6 (British Standard Institution 2009) and BS EN 12390-3 (British Standard Institution 2019), respectively. All OPC and MKG concrete samples were covered in plastic film to avoid water loss, demolded after 48 h, and then kept in a curing room at 20–25 °C and 90–95% relative humidity until the testing day. The results of the mechanical tests and their standard deviations (SDs) are presented in Table 8.4. A specimen label was assigned to each cube as X-Y-Z. The first term (X) refers to the material type, which is either the OPC or the MKG. The second term (Y) denoted as 1, 2, and 3 consists of the coarse aggregate size range of 5–10 mm, 10–16 mm, and 16–20 mm, respectively. The last term (Z) refers to the number of specimens.

In the study, a total of 18 specimens were tested for compressive strength and splitting tensile strength after being cured for 28 days. The loading rate applied during the tests was 0.3 MPa/s, which indicates the rate at which the load was increased. Figure 8.2 illustrates the dimensions of the specimens and the loading configuration used in the testing process.

Table 8.3 Chemical composition of OPC and MK (wt%)

Group	MK	WG		Cement	Coarse aggregate	Sand	Water
		SiO_2	Na_2O				
OPC	–	–	–	487	1115	560	208
MKG	268	92.82	106.77	–	1115	560	255

Reproduced from [Effect of Size of Coarse Aggregate on Mechanical Properties of Metakaolin-Based Geopolymer Concrete and Ordinary Concrete] by [Hamed Fazli] with permission from [Materials]

Table 8.4 Chemical composition of OPC and MK (wt%)

Series		Specimens	Coarse Aggregate Size (mm)	f_{cu} (MPa)			f_t (MPa)		
				Indiv. *	Ave. *	SD	Indiv. *	Ave. *	SD
OPC	OPC1	OPC-1-1	5–10	31.80	31.59	0.22	4.82	4.77	0.05
		OPC-1-2		31.36			4.72		
		OPC-1-3		31.62			4.77		
	OPC2	OPC-2-1	10–16	34.53	34.96	0.48	4.08	4.14	0.08
		OPC-2-2		34.87			4.11		
		OPC-2-3		35.47			4.23		
	OPC3	OPC-3-1	16–20	36.29	36.40	0.25	3.82	3.74	0.09
		OPC-3-2		36.22			3.65		
		OPC-3-3		36.68			3.75		
MKG	MKG1	MKG-1-1	5–10	36.11	36.39	0.29	3.00	3.13	0.15
		MKG-1-2		36.68			3.29		
		MKG-1-3		36.38			3.10		
	MKG2	MKG-2-1	10–16	33.27	33.05	0.27	2.69	2.85	0.15
		MKG-2-2		32.75			2.88		
		MKG-2-3		33.12			2.98		
	MKG3	MKG-3-1	16–20	26.95	26.81	0.15	1.44	1.46	0.03
		MKG-3-2		26.65			1.45		
		MKG-3-3		26.82			1.49		

Reproduced from [Effect of Size of Coarse Aggregate on Mechanical Properties of Metakaolin-Based Geopolymer Concrete and Ordinary Concrete] by [Hamed Fazli] with permission from [Materials]

* Indiv: individual; Ave.: average

8.2.1.3 Microstructural Analysis

On the broken surfaces of the concrete specimens made of OPC and MKG, microstructural studies were done. The microanalytical technique SEM/EDS analysis was used to examine the interfacial morphology between the paste and coarse aggregate. The Quanta FEG650 (FEI, Hillsboro, OR, USA) SEM/EDS apparutus was employed at the 5 or 8 or 10-kV accelerating volt (FEI) model. hrough microstructural examinations of the ITZ area in specimens of the OPC and MKG concrete taking into account the size of the coarse aggregate, the interfacial areas between the paste and the coarse aggregate were analyzed. Additionally, the mercury intrusion porosimeter (MIP; Micromeritics AutoPoreIV9500, Norcross, GA, USA) test was employed to determine the porosity and pore size distribution of the tested specimens.

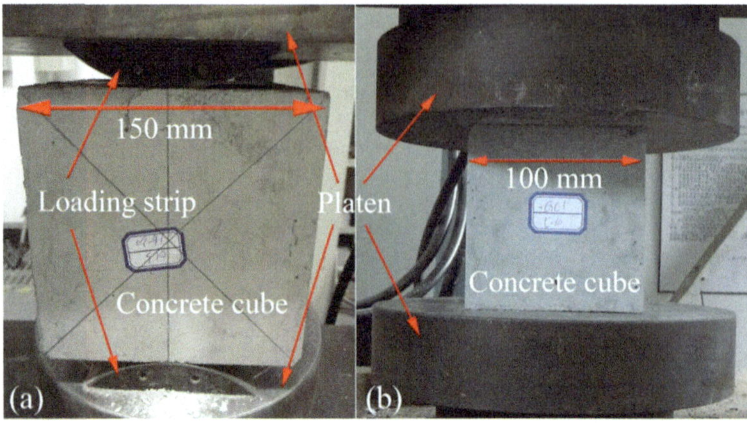

Fig. 8.2 Test set-up for **a** splitting tensile test; **b** compression test. Reproduced from [Effect of Size of Coarse Aggregate on Mechanical Properties of Metakaolin-Based Geopolymer Concrete and Ordinary Concrete] by [Hamed Fazli] with permission from [Materials]

8.2.2 Results and Discussion

8.2.2.1 Failure Behavior

Figure 8.3 depicts the failure types observed in the specimens after the compression test. In OPC concrete specimens, an increase in the size of the coarse aggregate resulted in a relatively lower occurrence of spalled material, which can be attributed to the higher compressive strength of these specimens. Conversely, MKG concrete specimens exhibited a different behavior, with a greater extent of crushing observed as the size of the coarse aggregate increased, indicating a decrease in compressive strength. Although MKG specimens demonstrated more fracture and brittle behavior compared to OPC specimens during the compression test, they generally displayed similar cracking and failure patterns to OPC concrete specimens (Guades 2016). The examined specimens showed cracking during the compression test as a result of splitting along the height of the cubes brought on by the development of tensile stresses. The compression test results showed that the coarse aggregate quality was good since cracks appeared in the cement paste rather than the coarse aggregate itself.

8.2.2.2 Compressive Strength

The average compressive strength test results of OPC concrete and MKG concrete are presented in Table 8.4 and depicted in Fig. 8.4. The OPC concrete's compressive strength improved as the size of the coarse aggregate increased. This trend varied between 31.59 MPa for 5–10 mm coarse aggregate size and 36.40 MPa for 16–20 mm coarse aggregate size. The findings are in line with the research of

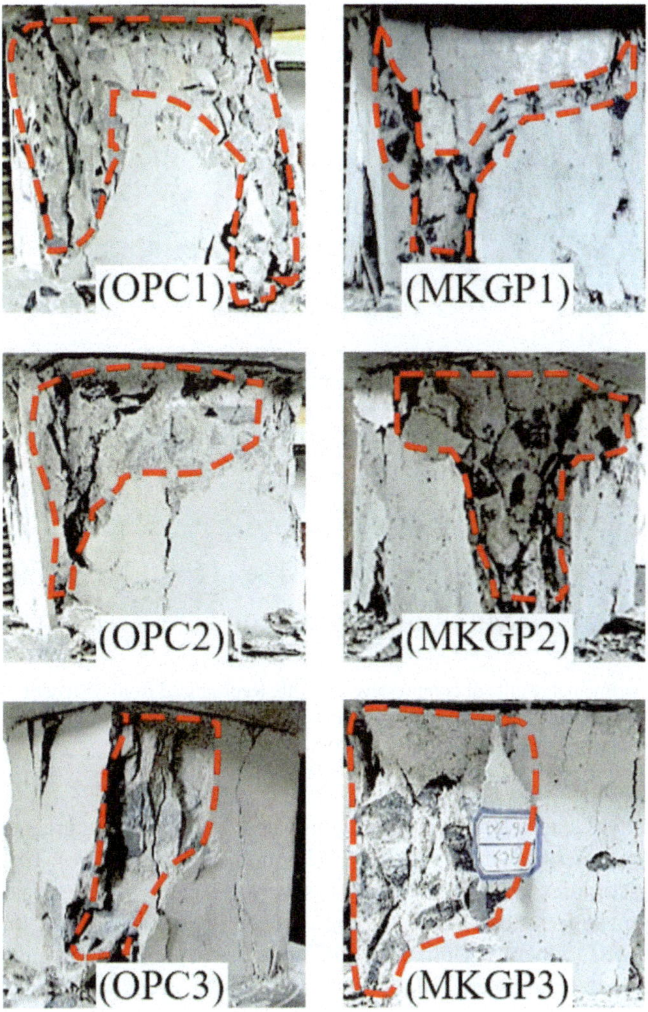

Fig. 8.3 OPC (left) and MKG (right) concrete cube specimens after compression test. Reproduced from [Effect of Size of Coarse Aggregate on Mechanical Properties of Metakaolin-Based Geopolymer Concrete and Ordinary Concrete] by [Hamed Fazli] with permission from [Materials]

Yuan et al. (2019b) in which the compressive strength increases when the aggregate size increases. The bigger coarse aggregate results in a less-specific surface area with the same weight of coarse aggregate in OPC specimens, therefore it is surrounded by a thicker OPC paste. Consequently, the paste between the larger coarse aggregate would have better quality and fewer microcracks (Torres et al. 2015), which yielded a higher compressive strength.

According to the MKG concrete results, compressive strength values decreased as the size of the coarse aggregate increased. The findings support the findings of

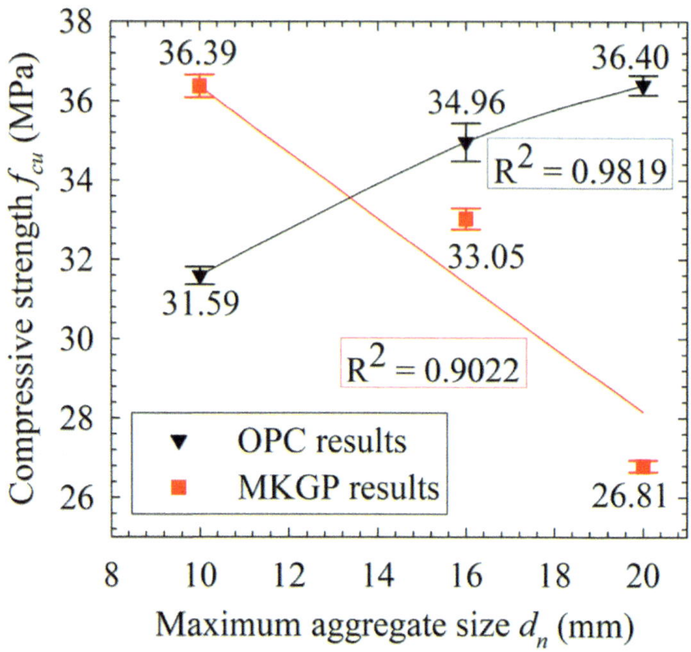

Fig. 8.4 Experimental results of compressive strength. Reproduced from [Effect of Size of Coarse Aggregate on Mechanical Properties of Metakaolin-Based Geopolymer Concrete and Ordinary Concrete] by [Hamed Fazli] with permission from [Materials]

Guades' study (Guades 2017, which demonstrated the impact of aggregate size variation on the compressive strength of the GPC at ambient temperature). The substantial reduction in compressive strength shown in specimens made with aggregates between 16 and 20 mm suggests that the coarse aggregate size may have a major impact on the strength of MKG concrete. Unlike the OPC specimens, the large volume of the paste around the coarse aggregate reduces the compressive strength. It could contribute to the higher shrinkage of the MKG paste compared to the OPC concrete, resulting in a highly porous paste of the MKG concrete, which will be discussed further in Sect. 8.4.

8.2.2.3 Splitting Tensile Strength Test Results

Table 8.4 and Fig. 8.5 present the splitting tensile strength test results of the OPC and MKG concrete. These findings show that the splitting tensile strength reduced despite the aggregate size increasing, which is consistent with other research (Akçaoğlu et al. 2002). Additionally, it implies that concrete's ability to split tensile strength cannot be increased by using a greater coarse aggregate.

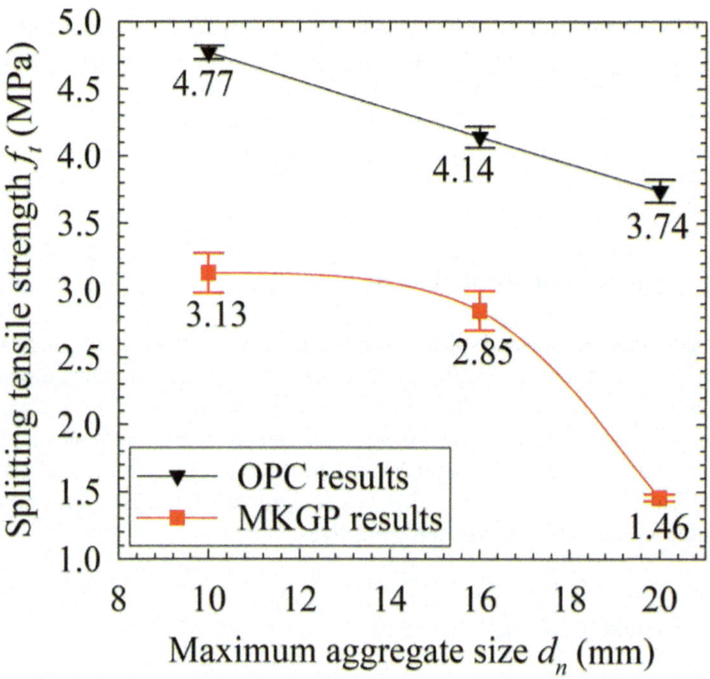

Fig. 8.5 Experimental results of compressive strength. Reproduced from [Effect of Size of Coarse Aggregate on Mechanical Properties of Metakaolin-Based Geopolymer Concrete and Ordinary Concrete] by [Hamed Fazli] with permission from [Materials]

Due to the higher coarse aggregate volume in relation to the specimen volume, the greater volume of the paste between the bigger coarse aggregate causes a more obvious difference between the elastic modulus of the coarse aggregate and paste (Akçaoğlu et al. 2004). Consequently, it increases the stress concentration and results in more microcracks near the coarse aggregate. It, therefore, yields a higher reduction in the splitting tensile strength of concrete with a lower w/c ratio of 0.4, as reported by Akçaoğlu et al. (2004). They revealed that the interfacial bond is a critical factor for tensile strength compared to its role in compressive strength.

Physically, the tiny surface area may prevent the gel bond from developing as much, which would increase the amount of shrinkage fractures near the coarse aggregate. Moreover, the possible internal bleeding underside of a larger coarse aggregate could contribute to decreasing the splitting tensile strength by reducing the interfacial bond strength inside the concrete.

The results demonstrate that the rate of f_t value reduction in MKG surpasses that observed in OPC concrete. This discrepancy can be ascribed to the greater occurrence of shrinkage cracks in the MKG paste, in comparison to those observed in OPC. Moreover, this disparity might be attributed to the inferior microstructure of the MKG paste relative to the OPC specimens.

The mechanical test results indicate distinct microstructural characteristics between MKG concrete and OPC concrete. Additionally, the interfacial bond development may differ in MKG concrete with varying coarse aggregate sizes compared to OPC concrete. However, compared to MKG, this impact is more pronounced in OPC concrete, most likely because of a better compatibility between the OPC and coarse aggregate phases.

8.2.2.4 Proposed Empirical Models

The compressive strength and splitting tensile strength of concrete are crucial material properties utilized for assessing design requirements and understanding the structural performance of concrete elements. Furthermore, the tensile strength of concrete can be estimated based on its compressive strength. In this investigation, novel empirical equations are introduced to establish a relationship between the compressive and splitting tensile strengths of both MKG concrete and OPC concrete, taking into account the influence of coarse aggregate size.

Effect on the OPC Concrete

A model based on Bazant's law of size effect and the calibrated model by Kim et al. was proposed by Jiang et al. (2017) to forecast the compressive strength of concrete by taking the coarse aggregate size impact into consideration (Kim et al. 1999).

$$f_c = f_c' \cdot \delta(d_{max}, \mathrm{h}, \mathrm{d}_a^m) \quad (8.1)$$

$$\delta(d_{max}, \mathrm{h}, \mathrm{d}_a^m) = a + \frac{B}{\sqrt{1 + \frac{d_{max}}{\lambda_0 d_a^m \left(\frac{h}{d} - \beta\right)}}} \quad (8.2)$$

where, in consideration of the size effect, f_c and f_c' present the actual cylinder compressive strength of the concrete specimen and the strength of the concrete specimen of standard size, respectively (designed compressive strength = 40 MPa). α, B, λ_0 and β are the coefficients that can be determined via experimental results. d_{max} is the maximum coarse aggregate size of concrete. h and d define the specimen size. The regression analysis of Kim et al. (1999) yielded $d_a^m \approx 1$. In light of the coarse aggregate size, the cylindrical compressive strength of concrete may be represented as follows:

$$f_c = \alpha f_c' + \frac{B f_c'}{\sqrt{1 + \frac{d_{max}}{\lambda_0 \left(\frac{h}{d} - \beta\right)}}} \quad (8.3)$$

The designed compressive strength (f_c'), in this study, is 40 MPa, and the h/d ratio is equal to 1, as the tested specimens were concrete cubes. Additionally, using 0.8 f_c', the measured cube compressive strength may be changed to the cylinder compressive strength. Equation (8.3) for the tested specimens can therefore be expressed as follows:

$$f_c = 0.8 \times \left(\alpha f_c' + \frac{B f'_{cu}}{\sqrt{1 + \frac{d_{max}}{\lambda_0 \left(\frac{h}{d} - \beta \right)}}} \right) \tag{8.4}$$

Figure 8.4 illustrates the link between compressive strength and the maximum coarse aggregate size as determined by the experimental cube compressive strength measurements (Table 8.4) and Eq. 8.4. A Levenberg–Marquardt algorithm in Matlab (The MathWorks, 2018), a curve fitting toolbox, was used to establish the fitting parameters. It yields that α, B, λ_0 and β are equal to 1.570, – 1.351, 2.171, and 1.052, respectively. According to ACI 318-14, the compressive strength of concrete may be used to determine its splitting tensile strength (f_t) (American Concrete Institute, 2014) and CEB-FIP (CEB-FIP 1990):

$$f_t = \phi (f_c)^c \tag{8.5}$$

Figure 8.6 shows that the splitting tensile strength findings from the trials are consistent with the results for compressive strength. The acquired fitting parameters led to the results that and c are, respectively, equal to 1334 and – 1.63.

Effect on the MKG Concrete

The impact of coarse aggregate size on the compressive strength of MKG concrete is also depicted in Fig. 8.4. The intriguing aspect is how the bigger coarse aggregate size has the opposite effect on the compressive strength of MKG specimens compared to OPC specimens. Therefore, Eq. (8.4) was employed to describe the relationship between the MKG compressive strength and the maximum coarse aggregate size. After fitting the analysis of the experimental results, the coefficients α, B, λ_0 and β are determined as – 17.43, 18.83, 2.336, and 1.993, respectively. The calibrated model is employed to predict the compressive strength, as given in Table 8.5. Comparing the experimental results (f_{cu}) with the calibrated values (f_c) revealed that the predicted result matches well with the experimental results, as its mean value and the standard deviation (SD) are 0.98 and 0.041, respectively.

It might be challenging to experimentally determine the splitting tensile strength of concrete specimens directly. In order to prevent the labor-intensive and time-consuming direct measurements, the connection between the splitting tensile strength and compressive strength of tested MKG concrete specimens is therefore supplied (Fig. 8.6). Equation (8.5) allows for the following relationship to be created:

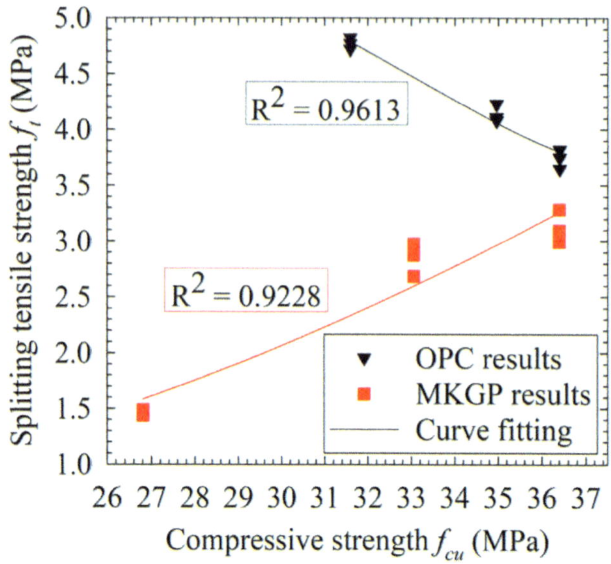

Fig. 8.6 Relationship between Eq. (8.5) and experimental results. Reproduced from [Effect of Size of Coarse Aggregate on Mechanical Properties of Metakaolin-Based Geopolymer Concrete and Ordinary Concrete] by [Hamed Fazli] with permission from [Materials]

Table 8.5 Comparison of theoretical and experimental results of compressive strength

Specimens	Calibrated model (f_c) (mm)	Proposed model (f_{tg}) (mm)	f_{cu}/f_c	f_t/f_{tg}
MKG-1-1	35.97	3.44	1.00	0.87
MKG-1-2	35.97	3.57	0.98	0.92
MKG-1-3	35.97	3.5	0.99	0.89
MKG-2-1	30.85	2.83	0.93	0.95
MKG-2-2	30.85	2.73	0.94	1.05
MKG-2-3	30.85	2.8	0.93	1.06
MKG-3-1	27.52	1.71	1.02	0.84
MKG-3-2	27.52	1.67	1.03	0.87
MKG-3-3	27.52	1.7	1.03	0.88
Average			0.98	0.93
SD			0.041	0.077

Reproduced from [Effect of Size of Coarse Aggregate on Mechanical Properties of Metakaolin-Based Geopolymer Concrete and Ordinary Concrete] by [Hamed Fazli] with permission from [Materials]

$$f_{tg} = 0.0006775(f_{cg})^{2.359} \tag{8.6}$$

where f_{tg} and f_{cg} are the splitting tensile strength and cube compressive strength of the geopolymer, respectively. The proposed model is employed to predict the splitting tensile strength, as given in Table 8.5. Comparing the experimental results (f_{cu}) with the model values (f_c) yielded acceptable prediction values, as its mean value and the standard deviation (SD) are 0.93 and 0.077, respectively.

8.2.2.5 Microstructure Analysis

SEM Observation and EDS Analysis

Investigations using SEM and EDS were done on the specimens' fracture surfaces. Figure 8.7 illustrates the location of the ITZ area, paste zone, and coarse aggregate zone in the SEM images. The line scan analysis of SEM/EDS presents the element intensity of the OPC and MKG concrete specimens, as shown in Fig. 8.8. The samples shown in Fig. 8.8 have a structure that is comparable to that in Fig. 8.7. To identify the properties of the examined specimens' microstructure, data were collected at 10-μm intervals from the paste to the surface of the coarse aggregate.

The microstructure of the interfacial transition zone (ITZ) surrounding various coarse aggregate sizes in both OPC and MKG concrete was examined, as presented in Fig. 8.8. SEM observations indicated that the ITZ region possesses a less compact structure compared to other parts of the paste. This porous zone observed in MKG specimens signifies a weaker section of the sample. The disparity in the formation mechanisms of the ITZ for OPC and MKG may account for this observation. Microstructural analysis revealed no significant impact of the investigated coarse aggregate sizes on the ITZ area. However, it was apparent that the microstructure of the MKG paste exhibited inferior strength compared to that of the OPC specimens.

Overall, the EDS analysis conducted in this study demonstrated that increasing the coarse aggregate size did not have a significant effect, as depicted in Fig. 8.8. Along the perpendicular direction of the ITZ, data was collected at intervals of 127.5 μm (0.5 μm) from the aggregate to the paste, resulting in 255 data points. The results revealed the presence of a C-S-H phase (calcium-silicate-hydrate) due to the hydration of pure OPC. The cementitious OPC matrix consisted of three distinct phases, which are as follows:

- Portlandite, which the brittle $Ca(OH)_2$;
- Hygrated Ca silicate;
- Hydrated Ca aluminate.

The intensity of the Ca atoms reduced as the concrete paste transitioned into the coarse aggregate, demonstrating that the ITZ's porosity is mostly caused by the proportion of CH particles on the paste side (Al-Bayati et al. 2016). Si and Al are the main components of the N-A-S-H gel in the metakaolin geopolymer (Peng et al. 2019), which were observed in the paste part of MKG concrete specimens. The

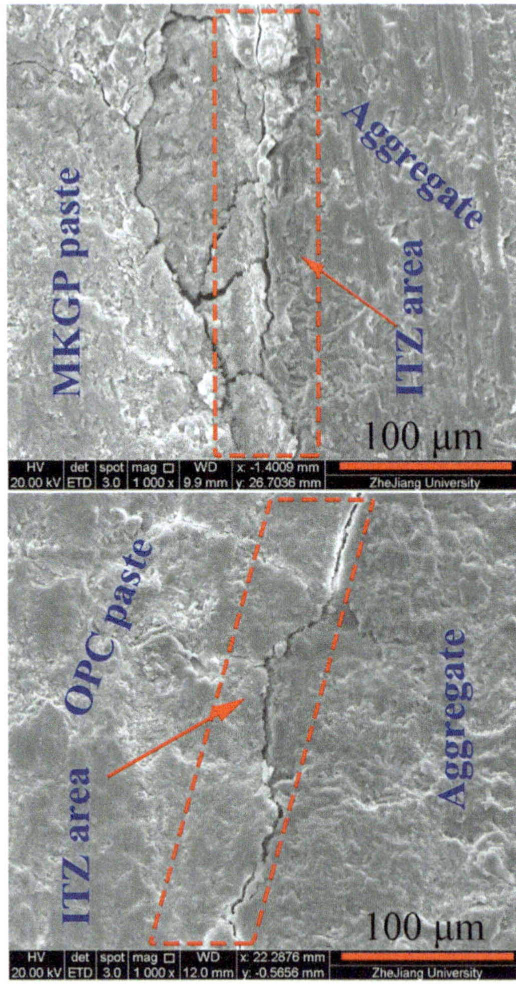

Fig. 8.7 Microstructure of the MKG3 and the OPC3. Reproduced from [Effect of Size of Coarse Aggregate on Mechanical Properties of Metakaolin-Based Geopolymer Concrete and Ordinary Concrete] by [Hamed Fazli] with permission from [Materials]

inclusion of Al and Si causes the production of N-A-S-H, which along with Ca may contribute to a cohesive mix during the mixing of concrete. Visual observations made during specimen preparation confirm this theory. Different Ca/Si and Al/Si ratios are present in the hydration product. Ca and Si atoms were shown to decrease with the size of the coarse aggregate, according to an EDS study of OPC samples. The porous ITZ region may be responsible for the declining Ca/Si ratio. Additionally, the Si/Al ratio rose with the size of the coarse aggregate in MKG samples, indicating a less dense structure.

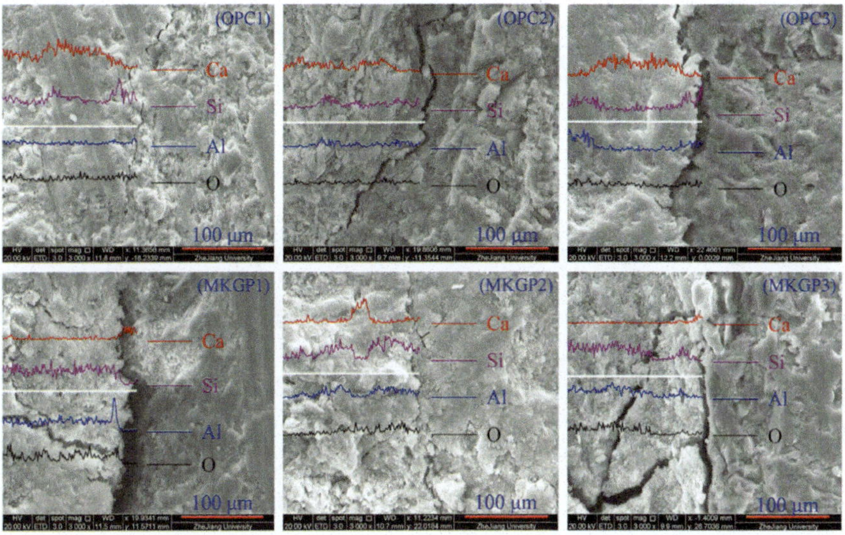

Fig. 8.8 EDS Line scan analysis and SEM images of the OPC concrete and the MKG concrete. Reproduced from [Effect of Size of Coarse Aggregate on Mechanical Properties of Metakaolin-Based Geopolymer Concrete and Ordinary Concrete] by [Hamed Fazli] with permission from [Materials]

Pore Structure Analysis Using MIP

The porosity and pore size distribution of OPC and MKG concrete specimens were determined using MIP (Mercury Intrusion Porosimetry) tests. Figures 8.9 and 8.10 present the pore size distribution of the concrete specimens. The tests were conducted on crushed concrete samples weighing approximately 1.5 g. Figure 8.9 illustrates the cumulative intruded pore volumes plotted against pore size diameters. The results indicate that the intrusion curves for the OPC specimens are nearly identical, as well as the curves for the MKG specimens. However, significant differences in the MKG intrusion curves are observed within the pore diameter range of 0.2–20 μm, indicating variations in pore structure between MKG and OPC specimens. Additionally, the MKG curves demonstrate that an increase in coarse aggregate size slightly affects the pore size distribution, particularly within the pore diameter range of 0.05–6 μm, potentially leading to a decrease in the mechanical properties of the MKG specimens.

In Fig. 8.10, additional insights into the pore structure evolution of the tested specimens are provided. MIP results indicate that MKG concrete exhibits dual pore peaks, with large pores around 10 μm and gel pores around 20 nm. In contrast, OPC concrete displays a single pore peak, corresponding to capillary/gel pores of approximately 100 nm. The majority of pore sizes in OPC specimens are distributed between 0.35 and 32 μm, whereas in MKG specimens, most pore sizes range from 0.06 to 32 μm. Peaks associated with MKG specimens are observed in the pore size range of 1–32 μm, with MKG3 exhibiting the highest peak among the specimens.

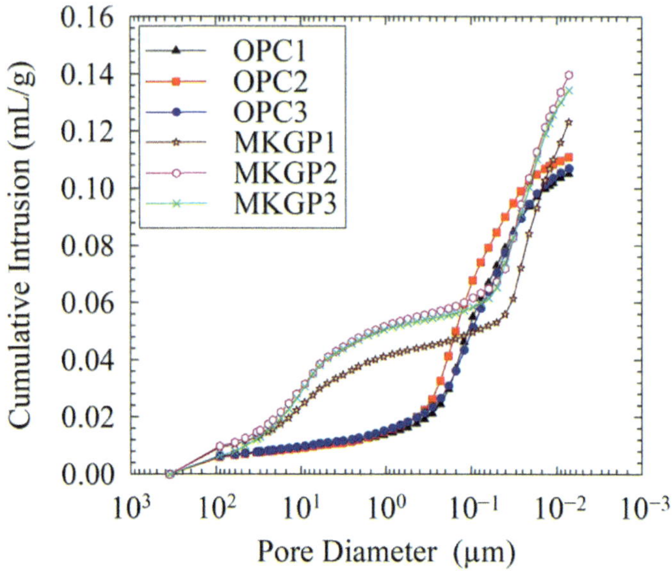

Fig. 8.9 Cumulative intruded pore volume versus pore diameter for OPC and MKG specimens. Reproduced from [Effect of Size of Coarse Aggregate on Mechanical Properties of Metakaolin-Based Geopolymer Concrete and Ordinary Concrete] by [Hamed Fazli] with permission from [Materials]

The results reveal that MKG3 shows a dominant pore size distribution of less than 7 µm. Furthermore, a decrease in the coarse aggregate size leads to a less porous structure in MKG1.

The MIP testing conducted in this study revealed the presence of distinct size ranges for pores in the tested specimens, namely micropores (less than 1 µm) and macropores (larger than 1 µm). Figure 8.11 illustrates the influence of pore structure on the compressive strength of OPC concrete and MKG concrete. It provides information on total porosity, macropore porosity, total pore area, and average (median) pore diameter obtained from the MIP test. The average pore size can be calculated as four times the ratio of pore volume to pore area (4V/A).

The presented data indicates noticeable changes in the pore structure. While the average pore size does not show significant differences, the MKG specimens exhibit higher values for total porosity, macropore porosity, and total pore area compared to the OPC specimens. This demonstrates that the pore structure has a substantial impact on the mechanical properties of the MKG specimens.

The mercury intrusion porosities for OPC1, OPC2, OPC3, MKG1, MKG2, and MKG3 are 20.49%, 21.05%, 20.70%, 22.78%, 24.77%, and 23.11% respectively. Although the effect of coarse aggregate size on total porosity is insignificant, the MKG concrete specimens exhibit higher porosities compared to OPC concrete specimens. In the case of MKG concrete specimens, an increase in coarse aggregate size

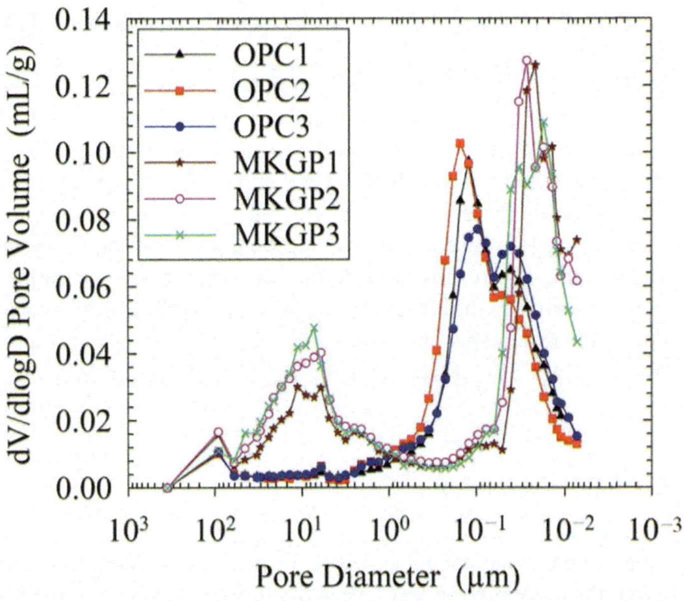

Fig. 8.10 Differential pore size distribution curves of the OPC and MKG specimens. Reproduced from [Effect of Size of Coarse Aggregate on Mechanical Properties of Metakaolin-Based Geopolymer Concrete and Ordinary Concrete] by [Hamed Fazli] with permission from [Materials]

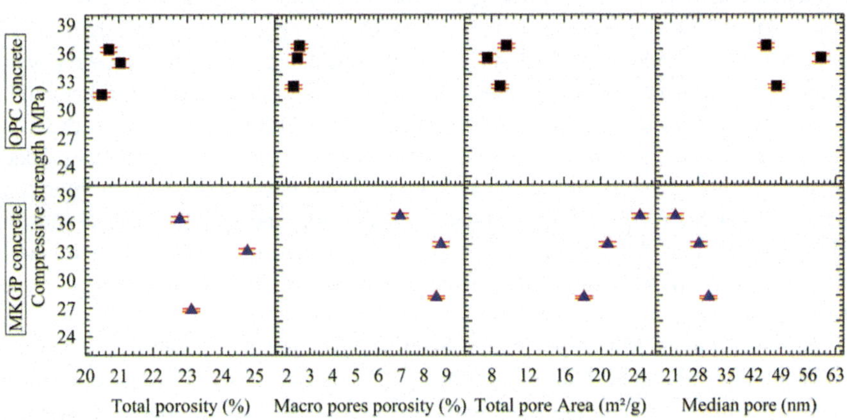

Fig. 8.11 Pore structure characterization of the OPC and MKG specimens. Reproduced from [Effect of Size of Coarse Aggregate on Mechanical Properties of Metakaolin-Based Geopolymer Concrete and Ordinary Concrete] by [Hamed Fazli] with permission from [Materials]

results in larger macropore porosity compared to OPC concrete, potentially leading to a decrease in compressive strength.

Figure 8.11 demonstrates that the variation in total pore area in OPC specimens is not significant, while average pore size tends to fluctuate with changes in coarse aggregate size. Despite OPC2 having a smaller total pore area than OPC3, its average pore size is larger, resulting in a slightly lower compressive strength. Conversely, for MKG specimens, an increase in coarse aggregate size leads to a decrease in total incoming mercury and an increase in average pore size, subsequently decreasing compressive strength. The results indicate that the compressive strengths of MKG specimens tend to decline as the average pore size increases. The order of average pore diameter among the tested specimens, based on coarse aggregate size and concrete type, is as follows: OPC2 > OPC1 > OPC3 > MKG3 > MKG2 > MKG1.

8.2.3 Further Discussion

In comparison to OPC concrete cubes, visual inspection of MKG concrete surfaces of cubes before the compression test revealed a greater number of visible and finer cracks, as depicted in Fig. 8.12. Early surface drying was linked to this incident. In MKG concrete examples, it was shown that larger coarse aggregate size resulted in more surface fractures after hardening. Additionally, it is known that the water content in the MKG concrete mixture includes both water in the activators and added water. Therefore, the increased surface cracking in specimens with larger coarse aggregate sizes could be attributed to concrete shrinkage caused by excessive water evaporation from the MKG specimens. As the applied force increased during uniaxial compression loading, tiny fractures emerged and generally propagated in the same direction as the applied stress. According to van Mier, in such circumstances, a lack of strong bonding between the coarse aggregate and paste may result in interfacial fissures (van Mier 1998). These cracks may develop and generate lateral tension. However, findings from the compression test indicated that cracks predominantly occurred in the MKG paste, which can be attributed to its higher shrinkage compared to OPC concrete.

In contrast to the MKG specimens, the OPC concrete cubes' central core was also substantially unharmed. Due to confinement loads caused by friction between the testing machine's platens and the concrete cubes as the size of the coarse aggregate in OPC concrete grows, continuous crack propagation may be retarded (Di Maio et al. 1996). It can therefore be concluded that in the MKG concrete specimens with a coarse aggregate of larger size, the confinement could not restrain the crack growth, as it yields lower strength compared to the smaller size of the coarse aggregate. It may be because of the high shrinkage of the MKG concrete due to the large surface area of metakaolin (Mastali et al. 2018).

Given that it makes approximately 70–80% of the concrete's volume, the coarse aggregate may have a significant impact on the properties of the material (Hansen and Nielsen 1965). According to reports, the volume of the coarse aggregate will increase

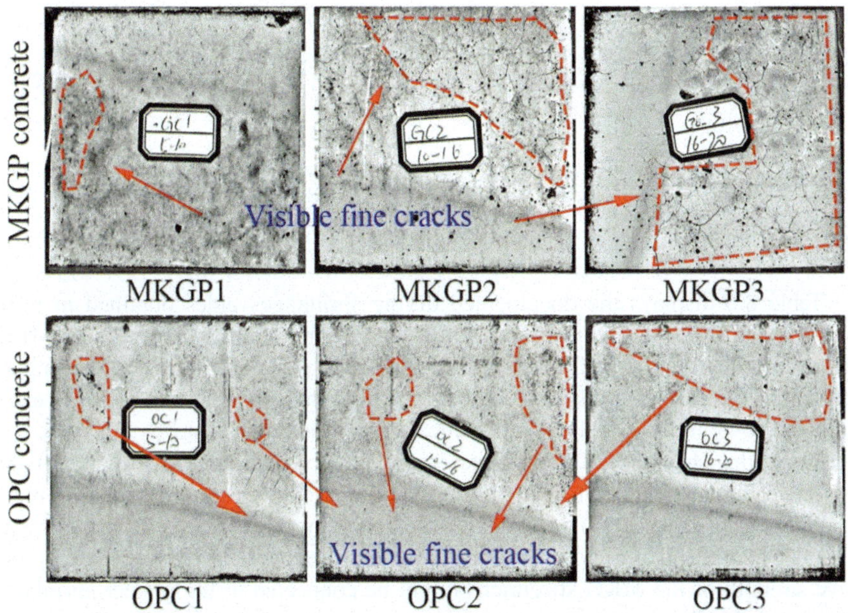

Fig. 8.12 Surface cracks of the specimens before the compression test. Reproduced from [Effect of Size of Coarse Aggregate on Mechanical Properties of Metakaolin-Based Geopolymer Concrete and Ordinary Concrete] by [Hamed Fazli] with permission from [Materials]

in OPC concrete with big coarse aggregate, which will reduce the shrinkage of the concrete (Hansen and Nielsen 1965). It is reported that the larger coarse aggregate restrains the inner strains and prevents the transition of microcracks into macrocracks (Karaguler and Yatagan 2018) and hence decreases shrinkage. Consequently, the compressive strength of concrete increases with the increase of the coarse aggregate size.

The MKG paste's non-uniform microstructure as compared to the microstructure of the OPC concrete was confirmed by SEM studies. It could be because water is not mixed directly with the gel product while making geopolymers; as a result, some water is left in the gel as interstitial water (Perera et al. 2007). The MKG paste's non-uniform microstructure as compared to the microstructure of the OPC concrete was confirmed by SEM studies. It could be because water is not mixed directly with the gel product while making geopolymers; as a result, some water is left in the gel as interstitial water (Zuhua et al. 2009). Besides, the water absorption increases as a result of increasing the Si/Al ratio and decrease in the rate of the geopolymerization process, which, in turn, affects the MKG concrete and leads to a porous and less dense microstructure and ITZ area. Therefore, the results of the previous studies, including the findings from Mastali et al. (2018), confirm that the MKG paste tended toward a high shrinkage. It follows that the MKG specimens show reduced strength as a result of increased shrinking.

Table 8.6 Values of drying shrinkage were reported in previous studies

OPC paste drying shrinkage ($\times 10^{-6}$)		MKG paste drying shrinkage ($\times 10^{-6}$)	
Bakharev et al. (2000)	600	Yang, et al. (2017)	5976
Neupane (2016)	550	Xiang, et al. (2019)	2505

Reproduced from [Effect of Size of Coarse Aggregate on Mechanical Properties of Metakaolin-Based Geopolymer Concrete and Ordinary Concrete] by [Hamed Fazli] with permission from [Materials]

Table 8.6 displays the documented drying shrinkage values obtained in prior studies. Notably, the findings indicate that the drying shrinkage results for MKG paste surpass those of OPC paste. This discrepancy can be attributed to the impact of metakaolin's extensive surface area and particle shape, which subsequently exert influence over the physical and mechanical properties of the paste.

The higher shrinkage of MKG concrete, which produced a porous structure in the MKG specimens, may be responsible for the observed pore structure measurement changes. Yang et al. (2017) investigation of geopolymers revealed a stronger correlation between shrinkage and micropore structure. Moreover, the decreased compressive strength of the MKG specimens might be connected to the greater shrinkage (Part et al. 2017).

The impact of shrinking on the MKG specimens is seen in Fig. 8.13. As the coarse aggregate size grows, a higher MKG paste volume is seen surrounding the coarse aggregate. Because of its porous microstructure, the increased MKG paste volume results in higher shrinkage. As a result, greater shrinkage leads to more macropores (>1 μm), as seen in Fig. 8.13b. It therefore causes a reduction in compressive strength.

8.3 Aggregate Influnce on Bonding Between CFRP Sheets and MKG Concrete

8.3.1 Notations in This Section

The following symbols are used in this section:

c coefficient;
C2 coefficient;
d coarse aggregate size;
dn maximum coarse aggregate size;
E_c modulus of elasticity of concrete;
E_f modulus of elasticity of FRP;
f_c' cylinder compressive strength;
f_{cu}' cube compressive strength;
f_{ctm} mean splitting tensile strength;

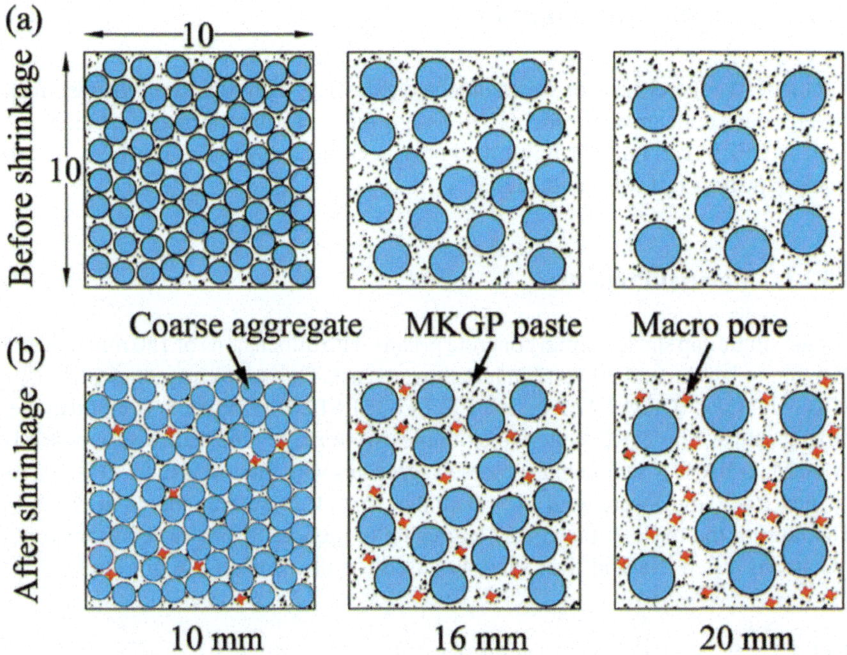

Fig. 8.13 Schematic of MKG specimens. **a** Before shrinkage; **b** after shrinkage. Reproduced from [Effect of Size of Coarse Aggregate on Mechanical Properties of Metakaolin-Based Geopolymer Concrete and Ordinary Concrete] by [Hamed Fazli] with permission from [Materials]

f_t splitting tensile strength;
L_e effective bond length;
$L_{e,exp}$ experimental effective bond length;
$L_{e,theo}$ theoretical effective bond length;
$L_{e,pre}$ predictive effective bond length;
n coefficient;
s local slip;
s_0 local slip at maximum interfacial shear bond stress;
t_f thickness of CFRP sheet;
x_i Point i at local axis;
α coefficient;
β coefficient;
ε_i measured strains at Point i;
τ interfacial shear bond stress;
τ_i interfacial shear bond stress at point i; and.
τ_{max} maximum interfacial shear bond stress.

8.3.2 Experimental Program

A total of 12 concrete prisms were subjected to testing, with each series of specimens undergoing duplicate experiments. To assess the performance of the specimens, a layer of CFRP (carbon fiber-reinforced polymer) sheet was bonded to them, and the single-lap shear test was conducted.

8.3.2.1 Material Properties

For the single-lap shear testing, concrete prisms with dimensions of 150 mm in width, 150 mm in height, and 300 mm in length were cast. The samples were created using MKG and OPC (CEM I 42.5 N; BS EN 197-1, BSI 2011). Table 8.7 provides their mixing ratios. The MKG concrete examples were made with industrial-grade MK (MetaMax, BASF). Sedimentary coarse aggregates with three size ranges (5–10, 10–16, and 16–20 mm) were used to prepare the concrete prisms, as shown in Fig. 8.14. The ratio of the mass of the coarse aggregate to the total mass was constant for all specimens tested. Natural river sand was used for all specimens in this study.

Table 8.7 OPC and MKG concrete mix design (kg/m^3)

Series	MK	WG	NaOH	Cement	Coarse	Sand	Water
OPC	–	–	–	487	1115	560	208
MKG	268	357	50	–	1115	560	20

Reproduced from [Influence of Coarse Aggregate Size on the Bonding between CFRP Sheets and Metakaolin-Based Geopolymer Concrete and Ordinary Concrete] by [Hamed Fazli] with permission from [Journal of Composites for Construction]

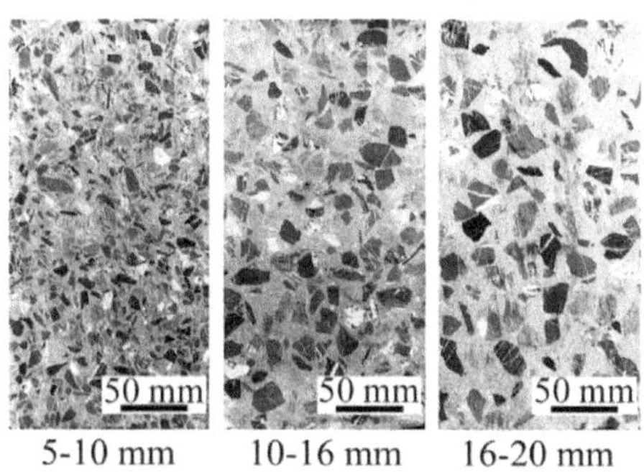

Fig. 8.14 Concrete prisms with different coarse aggregate sizes. Reproduced from [Influence of Coarse Aggregate Size on the Bonding between CFRP Sheets and Metakaolin-Based Geopolymer Concrete and Ordinary Concrete] by [Hamed Fazli] with permission from [Journal of Composites for Construction]

The specimens were demolded 48 h after casting and subsequently cured for 28 days at 90–95% relative humidity and a temperature of 20–25 °C. The average cubic compressive strength (f_{cu}') and average splitting tensile strength (f_t) of the concrete specimens were measured experimentally using three standard cubic specimens of 100 × 100 × 100 mm and 150 × 150 × 150 mm according to BS EN 12390-3 (BSI 2019) and BS EN 12390-6 (BSI 2009). The mechanical test results after 28 days are shown in Table 8.8 along with their standard deviations (SDs). When the single-lap shear tests were carried out at the age of 29 days, the concrete had the mechanical characteristics indicated in Table 8.8.

Unidirectional carbon-fiber sheets (300 g/m^2), 50 mm wide, and 0.135 mm thick, were externally bonded onto the specimens along the axial direction using epoxy resin. To ascertain the material characteristics of the CFRP sheets, CFRP coupon tensile tests based on (ASTM D3039, ASTM 2017) were carried out. The CFRP sheets' rupture strain, Young's modulus, and tensile strength were 2547 MPa, 236 GPa, and 1.1%, respectively. The epoxy resin utilized in this investigation was HM-180C3P. Two-part epoxy-based impregnating resins are present. It has two

Table 8.8 Mechanical properties of cubic concrete specimens

Specimens	Coarse aggregate size (mm)	f_{cu} (MPa)			f_t (MPa)		
		Indiv.*	Ave.*	SD	Indiv.	Ave.	SD
OPC	OPC-1-1	5–10	31.8	31.6	0.2	4.8	0.1
	OPC-1-2		31.4				
	OPC-1-3		31.6				
	OPC-2-1	10–16	34.5	35	0.5	4.1	0.1
	OPC-2-2		34.9				
	OPC-2-3		35.5				
	OPC-3-1	16–20	36.3	36.4	0.3	3.7	0.1
	OPC-3-2		36.2				
	OPC-3-3		36.7				
MKG	GP1-1	5–10	36.1	36.4	0.3	3.1	0.2
	GP1-1		36.7				
	GP1-1		36.4				
	GP2-1	10–16	33.3	33.1	0.3	2.9	0.2
	GP2-2		32.8				
	GP2-3		33.1				
	GP3-1	16–20	27	26.8	0.2	2.1	0.1
	GP3-2		26.7				
	GP3-3		26.8				

components, part A (bisphenol A) and part B (cycloaliphatic amine), mixed in a weight ratio of 2:1. The epoxy resin tested had an ultimate tensile strength of 45 MPa, an elastic modulus of 3.1 Pa, and a rupture tensile strain of 3.4% (ASTM D638, ASTM 2014).

8.3.2.2 Specifications of Test Specimens

The test specimens were categorized based on the size of the coarse aggregate and concrete type, as outlined in Table 8.9. In the identifier for each concrete specimen, the first term indicates the type of concrete (OPC or MKG). The letter "C" denotes the presence of CFRP (carbon fiber-reinforced polymer) sheet. The third term (Numbers 1, 2, or 3) signifies the size range of the coarse aggregates: 5–10 mm, 10–16 mm, and 16–20 mm, respectively. The final number is used to distinguish nominally identical specimens. The research investigated the effects of different coarse aggregate sizes, comparing the results between MKG concrete specimens and OPC specimens.

Table 8.9 Details of the specimens

Specimen	Failure load (kN)		Maximum strain ($\mu\varepsilon$)		τ_{max} (MPa)		L_e (mm)	
	Indiv.	Ave.	Indiv.	Ave.	Indiv.	Ave.	Indiv.	Ave.
MKG-C-1-1	11.8	11.2	8395	8180	6	5.8	80.1	80.7
MKG-C-1-2	10.5		7965		5.7		81.2	
MKG-C-2-1	11	10.6	7582	7234	5.8	5.5	93.7	96.8
MKG-C-2-2	10.2		6885		5.1		99.98	
MKG-C-3-1	10.1	9.6	5641	6064	4.8	4.4	123.9	118.5
MKG-C-3-2	9.2		6487		3.9		113.1	
OPC-C-1-1	11.8	11.6	8650	8405	7.7	6.4	49.2	52.6
OPC-C-1-2	11.4		8159		5.1		56	
OPC-C-2-1	10.7	10.9	7984	7338	6.3	5.8	51.5	58.3
OPC-C-2-2	11		6692		5.2		65.1	
OPC-C-3-1	10.5	10.3	6249	5587	5.9	5	58.1	65.5
OPC-C-3-2	10.1		4925		4.1		72.8	

Reproduced from [Influence of Coarse Aggregate Size on the Bonding between CFRP Sheets and Metakaolin-Based Geopolymer Concrete and Ordinary Concrete] by [Hamed Fazli] with permission from [Journal of Composites for Construction]
Note Indiv. = individual; Ave. = average

8.3.2.3 Externally Bonded FRP Application

The efficiency of a bond depends on the surface roughness of the concrete substrate (Júlio et al. 2005; Elarbi 2011). The top surface of the substrate with regard to the casting direction was first prepared to reveal the coarse aggregates before the FRP was externally attached to the concrete substrate. This phase involved removing the thin concrete paste coating that was present on the top surface. The surface of each specimen was prepared manually using a grinder to remove all loose particles and reduce the unevenness of the surface, after which sandpaper was used to smooth the surface. Before using the HM-180C3P epoxy glue to connect CFRP sheets to the surface of a specimen, the specimen's surface was cleaned with alcohol. 50 mm of the FRP were not attached to the concrete prisms in order to prevent the concentration of stress at the loaded end (Mazzotti et al. 2008). Prior to testing, all concrete prisms were adhered to a single layer of a 200-mm-long CFRP sheet and allowed to cure for at least two weeks at room temperature. The bond length of 200 mm was longer than the estimated effective bond length obtained from the model proposed by Lu (2004), who measured the full development of the bond capacity. A schematic diagram of a specimen is shown in Fig. 8.15.

To ensure uniform tension distribution and prevent premature fiber rupture, the remaining free length of the FRP (fiber-reinforced polymer) sheet was impregnated with epoxy resin. Figure 8.15 depicts the implementation of anchorage plates, custom-made for each specimen, to prevent early slippage of the FRP sheets from the top wedge of the universal testing machine. The inner surface of each anchorage plate was firmly bonded to the CFRP sheet using epoxy resin.

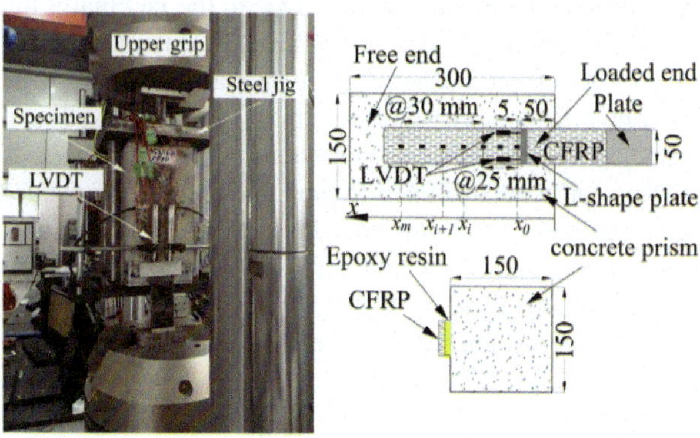

(a) (b)

Fig. 8.15 Concrete specimens: **a** universal testing machine setup; and **b** schematic diagram (dimensions in mm). Reproduced from [Influence of Coarse Aggregate Size on the Bonding between CFRP Sheets and Metakaolin-Based Geopolymer Concrete and Ordinary Concrete] by [Hamed Fazli] with permission from [Journal of Composites for Construction]

8.3.2.4 Loading Procedure and Instruments

Tests were conducted using a universal testing machine with a capacity of 250 kN, as depicted in Fig. 8.15a. The specimens were subjected to displacement-controlled loading at a rate of 0.005 mm/s until failure. Measurements of the crosshead displacement were made utilizing a data recorder and the actuator's inbuilt LVDT. Seven 5-mm-long strain gauges were used to measure the axial strain in the CFRP along the reference axis x. The origin of this measurement (i.e., $x_0 = 0$) was located on the front side of the concrete prism. Displacement during loading was measured using two LVDTs with a $\pm$ 10 mm range. The resolution and precision of the LVDTs were determined to be 1.00×10^{-12} m and 0.05%, respectively. A sketch of the tested concrete prisms can be found in Fig. 8.15b. The strain gauge locations along the FRP sheet are indicated by the letters xi. A data collection system was used to gather information from the load cell, strain gauges, and LVDTs. Displacement, strain, and force measurements were recorded at a frequency of 10 Hz.

8.3.3 Test Results and Discussion

8.3.3.1 Mechanical Properties of Concrete with Various Sizes of Coarse Aggregate

Concrete's mechanical properties must be established while taking the effects of the size of the coarse aggregate into consideration in order to investigate the interfacial bond behavior between CFRP and concrete. Due to the fact that it makes up about 70% of the concrete's volume, the coarse aggregate is a crucial factor in determining how the material will behave mechanically (Chen and Liu 2004). The findings in Table 8.8 present that the compressive strength of OPC increased on increasing the size of the coarse aggregate, which is consistent with the previous study by Yuan et al. (2019b). This could be attributed to the larger coarse aggregate having a less specific surface area. Hence, it is surrounded by a thicker layer of OPC paste. There would be greater quality and fewer microcracks in the paste separating the bigger coarse aggregate as a result (Torres et al. 2015), increasing compressive strength. The MKG specimens, however, behaved differently. As the size of the coarse aggregate was increased, their compressive strength decreased. These results are consistent with the studies conducted by Karakoç et al. (2014) and Guades (2017). Isabella et al. (2003) found that larger pieces of coarse aggregate might not provide sufficient interfacial area for a geopolymer paste to bind sufficiently. Therefore, larger coarse aggregate results in lower strength. Moreover, the difference in aggregate size can affect the ingress of water into the ITZ.

As presented in Table 8.8, the splitting tensile strength of OPC and MKG specimens decreased despite an increase in the size of the coarse aggregate, which is consistent with previous studies (Akçaoğlu et al. 2002; Yuan et al. 2019b). This could be due to the fact that growing the coarse aggregate causes the ITZ to grow and creates additional microcracks close to the coarse aggregate. According to Akçaoğlu et al. (2004), this therefore results in a greater fall in the splitting tensile strength of concrete with a lower water to cement ratio of 0.4 (Akçaoğlu et al. 2004). Moreover, previous studies reported that the larger coarse aggregate might result in a weak bond zone inside the concrete due to internal bleeding (Akçaoğlu et al. 2002).

8.3.3.2 Failure Mode

The evaluation of the bond between CFRP and the concrete substrate was performed to assess performance and efficiency, primarily by examining the failure mode. A successful bond between CFRP and concrete is indicated when the concrete substrate layer fails (Fazli et al. 2018b). In this study, all tested specimens displayed failure modes resulting from CFRP debonding from the concrete substrate. Figure 8.16a illustrates the typical debonding failure mode observed in MKG specimens. Following debonding, a thin layer of concrete detached from the substrate and adhered to the CFRP, consistent with previous research findings (Yuan et al. 2004, 2019b). Debonding is a significant factor influencing the quality and efficiency of FRP applications. The study results indicate that aggregate size had an insignificant effect on the failure mode of the specimens. Figure 8.16b presents images obtained using the image thresholding technique, where the white and black areas represent paste and coarse aggregates, respectively. These images depict the distribution of coarse aggregates and show areas where the paste detached from the concrete but remained attached to the debonded CFRP sheet. Notably, the specimen with a coarse aggregate size of 16–20 mm showed a denser distribution than the specimen with a coarse aggregate size of 5–10 mm. In comparison to the specimen with the smallest coarse aggregate size, the area of detached paste was greater for the specimen with a coarse aggregate size of 16–20 mm. In contrast, after debonding failure, more aggregate particles were affixed to the CFRP sheets in the specimens with the least coarse aggregate size.

8.3.4 Load and Displacement Relationships

The debonding process of CFRP from the concrete substrate typically involves three stages: elasticity, softening, and the debonding plateau, as described by Teng et al. (2003). In the elastic stage, an increase in shear slip causes the initiation of microcracks in the interfacial bond between the concrete surface and adhesive (ASTM D3039, ASTM 2017). Debonding begins at the loaded end before the completion of the softening stage, followed by a gradual increase in displacement, indicating progressive debonding. Yuan et al. (2004) reported that a longer bond length of the CFRP sheet results in a longer debonding plateau, as the length of the plateau is primarily influenced by the bond length of the CFRP.

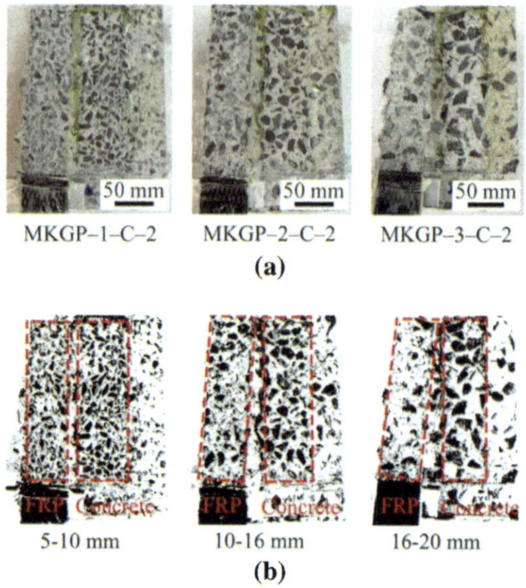

Fig. 8.16 Concrete specimens after single-lap shear test: **a** failure modes of debonding; and **b** aggregate distributions. Reproduced from [Influence of Coarse Aggregate Size on the Bonding between CFRP Sheets and Metakaolin-Based Geopolymer Concrete and Ordinary Concrete] by [Hamed Fazli] with permission from [Journal of Composites for Construction]

Graphs of the load versus global slip (displacement) for the MKG–CFRP and OPC–CFRP interfaces were obtained as the average of measurement from the two LVDTs. Figure 8.17 shows an example of these graphs. Five points (i.e., A-E, F-J, and K-O) were selected on each load–displacement curve in different stages of loading, as shown in Fig. 8.17. A, F, and K coincide with the end of the elastic stage. Points E, J, and O are at the maximum applied load (higher than 96% of the failure load). Points B, G, and N and B, H, and N, respectively, denote the effective bond length for the MKG and OPC specimens. The interfacial shear bond stress and strain distributions along the CFRP were tracked using additional sites. The elongation of the CFRP sheets' unbonded portion and the shear slip of the bonded portion were two additional displacements acquired from the LVDTs (Yao et al. 2005).

The specimens MKG-C-1, MKG-C-2, MKG-C-3, OPC-C-1, OPC-C-2, and OPC-C-3, respectively, had average failure loads of 11.2, 10.6, 9.6, 11.6, 10.9, and 10.3 kN. According to Table 8.9's stated failure loads for the specimens, the failure loads for the MKG and OPC specimens somewhat reduced as the aggregate size grew. This suggests that the concrete-CFRP interfacial debonding load is influenced by the size of the coarse aggregate. Additionally, according to the results, the splitting tensile strength of the tested specimens decreased with failure load, which is consistent with earlier investigations (Neubauer and Rostasy 1997; Lu et al. 2005).

Figure 8.18 is the outcome of an idealization rather than a firsthand observation. A schematic illustration shows how the coarse aggregate's size affects how well it interlocks. Due to their equal mass ratios, aggregate particle spacing increases with particle size. In comparison to the larger coarse aggregates, the distribution of the smallest coarse aggregate was denser and more consistent. The interfacial shear interlocking action was strengthened by this denser dispersion. The larger particle

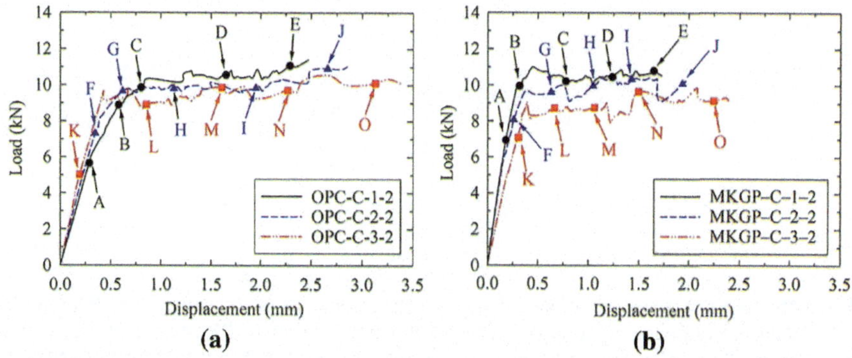

Fig. 8.17 Failure loads versus displacement relationship of the CFRP—concrete interface for **a** OPC concrete specimens; and **b** MKG concrete specimens. Reproduced from [Influence of Coarse Aggregate Size on the Bonding between CFRP Sheets and Metakaolin-Based Geopolymer Concrete and Ordinary Concrete] by [Hamed Fazli] with permission from [Journal of Composites for Construction]

spacing of the larger coarse aggregates, on the other hand, resulted in a relatively weak interlocking action. The interlocking action of a coarse aggregate depends on its size, and, thus, interfacial shear interlocking is among the main factors that determine the debonding failure of CFRP (Pan and Leung 2007). The interfacial bond strength between the concrete substrate and the CFRP was higher when the interlocking action was stronger, owing to the higher fracture energy required for crack propagation (Yuan et al. 2019b). This may be because the splitting tensile strength of concrete is the main factor affecting the interfacial bond strength between the concrete substrate and CFRP, and the larger coarse aggregates had a lower splitting tensile strength (Yuan et al. 2019b). This impact was influenced by the larger surface area of the biggest coarse aggregate, which led to a higher stress concentration and more microcracks near the coarse aggregate (Akçaoğlu et al. 2002, 2005; Yuan et al. 2019b). The splitting tensile strength of the concrete is inversely correlated with the interfacial bond strength between the concrete substrate and CFRP. Therefore, a weaker interfacial connection between the CFRP and the concrete is caused by a lower splitting tensile strength. Neubauer and Rostasy (1997) found findings that were comparable (Neubauer and Rostasy 1997).

8.3.4.1 Strain Distribution Along CFRP Sheet

Figure 8.19 illustrates the strain distribution along the surface of the CFRP sheets in relation to the distance from the loaded end, along with the axial strain distribution along the bonded CFRP at a specific loading stage. The axial strain was measured using seven strain gauges positioned on the CFRP sheet surface. The strain distribution gradually decreased from the loaded end towards the free end (Ko and Sato 2007). Local strain deviations with zigzag curves were observed due

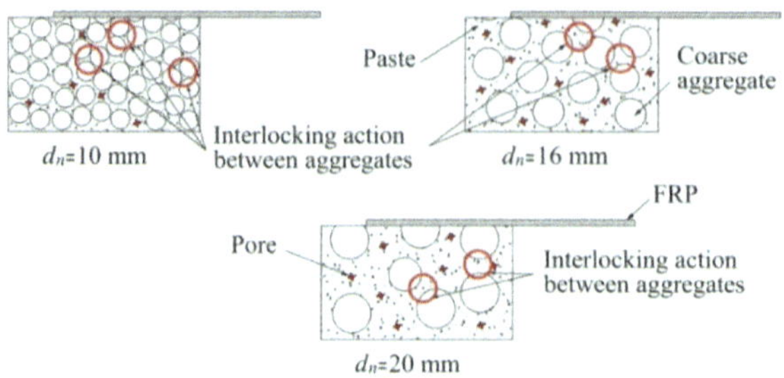

Fig. 8.18 Effect of coarse aggregate size on the coarse aggregate interlocking action. Reproduced from [Influence of Coarse Aggregate Size on the Bonding between CFRP Sheets and Metakaolin-Based Geopolymer Concrete and Ordinary Concrete] by [Hamed Fazli] with permission from [Journal of Composites for Construction]

to stress concentration induced by aggregate particles present between the substrate and the CFRP sheet (Shen et al. 2015d). In addition, nonuniform distributions of epoxy and variations in local material could have caused spatial variations in the measured surface strain (Ali-ahmad et al. 2004, 2006; Ghiassi et al. 2013). As the load increased, the strain also increased until changes in the bond resulted in a gradual shift in the curves from the maximum strain recorded by the first strain gauge (Fazli et al. 2018a; Fawzia et al. 2020). This gradual increase in strain persisted until the applied loading reached the debonding load. Table 8.9 presents the maximum strain on the CFRP surface for all tested specimens. The average maximum bond strains for specimens OPC-C-1, OPC-C-2, OPC-C-3, MKG-C-1, MKG-C-2, and MKG-C-3 were measured as 8405 $\mu\varepsilon$, 7338 $\mu\varepsilon$, 5587 $\mu\varepsilon$, 8180 $\mu\varepsilon$, 7234 $\mu\varepsilon$, and 6064 $\mu\varepsilon$, respectively. The results indicate that as the size of the concrete aggregate increased, the maximum strain decreased. Compared to OPC-C-1, the average maximum strain of OPC-C-2 and OPC-C-3 decreased by 12.7% and 33.5%, respectively. In the MKG series, the average maximum strain of MKG-C-2 and MKG-C-3 decreased by 11.6% and 25.9%, respectively, compared to MKG-C-1.

More microcracks developed in the layer of the link between the epoxy and the substrate after the softening phase. Due to the coarse aggregate's interlocking properties, the CFRP sheets were still able to resist the shear stress. The MKG-C-1 and OPC-C-1 specimens in each series thus had the highest strain, which is consistent with a recent research (Yuan et al. 2019b), since the smaller aggregates had a larger fracture energy due to a stronger interlocking action (Yuan et al. 2019b). The substantial deformation of CFRP sheets with the same stiffness as a result of the greater shear force is responsible for this outcome. The bond strain was affected by the size of the coarse aggregate, which demonstrates the influence of the latter on the strain in a CFRP–concrete bond.

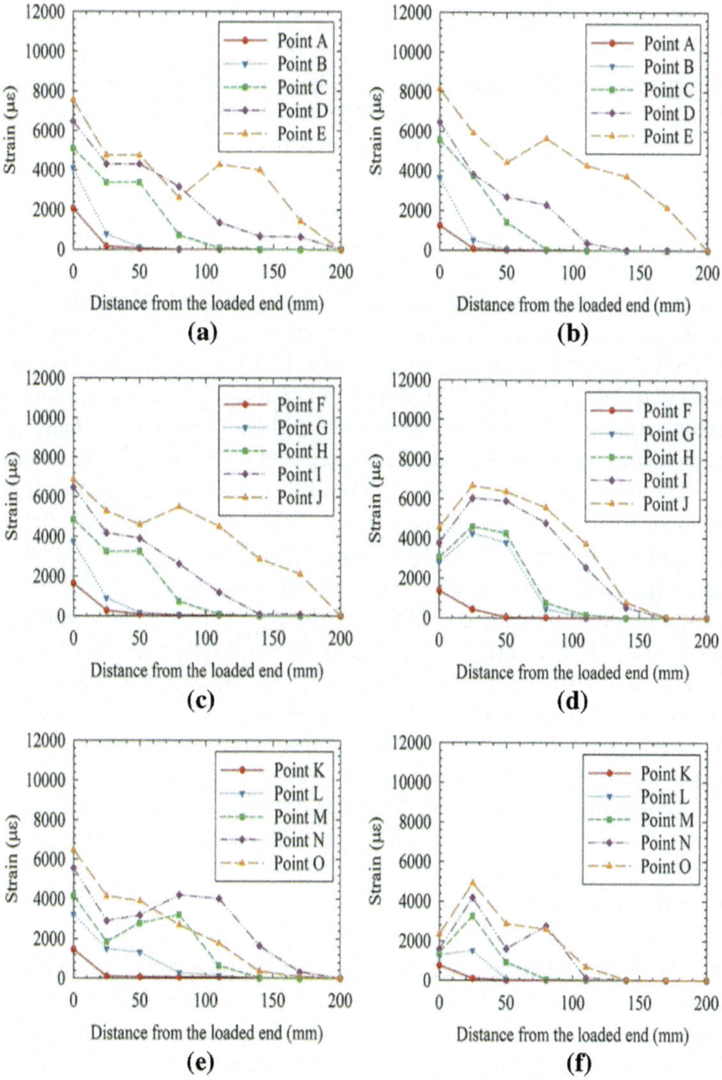

Fig. 8.19 Strain distribution on CFRP: **a** MKG-C-1-2 specimen; **b** MKG-C-2-2 specimen; **c** MKG-C-3-2 specimen; **d** OPC-C-1-2 specimen; **e** OPC-C-2-2 specimen; and **f** OPC-C-3-2 specimen. Reproduced from [Influence of Coarse Aggregate Size on the Bonding between CFRP Sheets and Metakaolin-Based Geopolymer Concrete and Ordinary Concrete] by [Hamed Fazli] with permission from [Journal of Composites for Construction]

8.3.4.2 Bond Strength

The FRP–concrete bond was assessed with the single-lap shear test. The positions of the strain gauges are denoted by x_i in Fig. 8.2. The interfacial shear bond stress between consecutive strain gauges was calculated as

$$\hat{\tau}_{i+1/2} = -\frac{E_f t_f (\varepsilon_{i+1} - \varepsilon_i)}{(x_{i+1} - x_i)} \tag{8.7}$$

where E_f and t_f = elastic modulus and the thickness of the FRP sheet, respectively; and ε_i and ε_{i+1} = consecutively measured strains at x_i and x_{i+1}, respectively.

The connection between the interfacial shear bond stress and the separation from the loaded end is seen in Fig. 8.20. All of the studied specimens' findings display a consistent pattern. The maximum interfacial shear bond stress changed further from the loaded end as the applied load increased, and fractures then spread as a result of debonding. Equation (8.7) was used to determine the highest interfacial shear bond stresses (τ_{max}) for all specimens examined, as shown in Table 8.9. The findings demonstrate that the interfacial shear bond stress decreased as the size of the coarse aggregate increased. This shows that the interfacial shear bond tension is significantly influenced by the size of the coarse aggregate. For the specimens OPC-C-1, OPC-C-2, OPC-C-3, MKG-C-1, MKG-C-2, and MKG-C-3, the average values of the maximum interfacial shear bond stress were 6.4, 5.8, 5, 5.8, 5.5, and 4.4 MPa, respectively. The average maximum interfacial shear bond stresses of MKG-C-2 and MKG-C-3 decreased by 5.2% and 24.1%, respectively, compared with MKG-C-1. In the OPC series, in comparison with OPC-C-1, the average maximum interfacial shear bond stresses of OPC-C-2 and OPC-C-3 fell by 9.4% and 21.9%, respectively.

8.3.4.3 Effective Bond Length

All calculations for the design of FRP strengthening take the effective bond length (L_e) into consideration. The binding behavior of the CFRP sheet-concrete contact was examined in this case using it. The L_e value determines when the FRP bonding system fails and when the failure load cannot be increased by extending the bond length (Ben Ouezdou et al. 2009). The tension created by the applied load on a CFRP-bonded specimen is transmitted to the CFRP sheet as shear stresses through the adhesive bond during a specified short length close to the applied load. It is an active bond zone at this small length. An increase in the applied load causes the active bond zone to shift farther from the loaded end due to the debonding of the CFRP from the concrete surface. The shifting of the active bond zone continues until the failure of the bonding system between the CFRP and the concrete substrate. Thus, only one part of the bond line is effective, a required length at which the strain tended toward zero (Ben Ouezdou et al. 2009; Franco and Royer-Carfagni 2014). At any applied loading level, an active bond zone or the effective bond length was shown. This is consistent with earlier research (Kang et al. 2012; Neto et al. 2016).

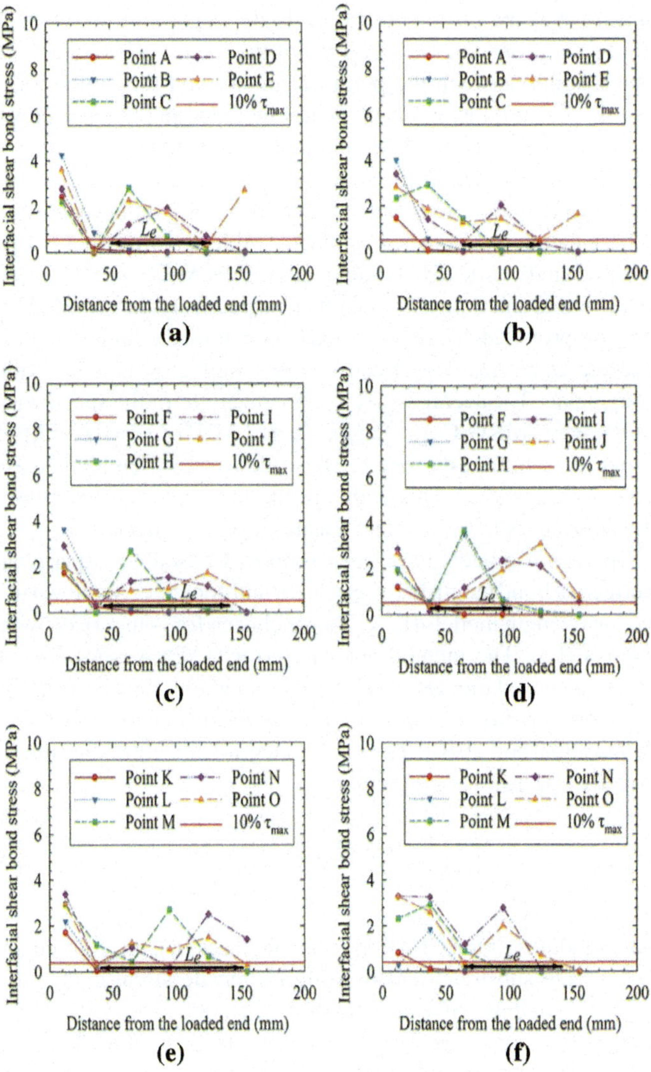

Fig. 8.20 Relationship between interfacial shear bond stress and distance from the loaded end: **a** MKG-C-1-2 specimen; **b** MKG-C-2-2 specimen; **c** MKG-C-3-2 specimen; **d** OPC-C-1-2 specimen; **e** OPC-C-2-2 specimen; and **f** OPC-C-3-2 specimen. Reproduced from [Influence of Coarse Aggregate Size on the Bonding between CFRP Sheets and Metakaolin-Based Geopolymer Concrete and Ordinary Concrete] by [Hamed Fazli] with permission from [Journal of Composites for Construction]

According to Nakaba et al. (2001), L_e is the separation between sites where the interfacial shear bond stresses are 10% of the maximum interfacial shear bond stresses. The lengths between the point of contact and the CFRP location where the strain is minimal and insignificant are not included in the reported values of L_e, which are instead these measured lengths (Athawale and B.E. 2012). As illustrated in Fig. 8.20 the experimentally measured effective bond lengths are given by the intersection of the 10% τ_{max} lines and the curve for τ versus distance from the loaded end lines. These lengths are depicted by black double-arrowed lines in Fig. 8.20. The curves that represent the effective bond lengths of the MKG and OPC specimens are for Points D, I, and O and Points D, H, and O, respectively. In Table 8.9, the effective bond lengths are provided. These outcomes show how the bond length is impacted by the coarse aggregate size. The effective bond length grew together with the size of the coarse aggregate. The average effective bond length for the specimens OPC-C-1, OPC-C-2, OPC-C-3, MKG-C-1, MKG-C-2, and MKG-C-3 were 52.6, 58.3, 65.5, 80.7, 96.8, and 118.5 mm, respectively. The effective bond length increased with the size of the coarse aggregate. This is due to the fact that increasing the size of the coarse aggregate causes the concrete's tensile strength to decrease.

Several theoretical models have been proposed based on experimental data to predict the effective bond length between CFRP and concrete. As shown in Table 8.10, six effective bond length models ($L_{e,theo}$) were chosen for comparison with the experimental findings (L_e). The integral absolute error (IAE) was used to evaluate the accuracy of the proposed models of effective bond length. Since the IAE is sensitive to variations from test findings, it is frequently used for model evaluation (Girgin et al. 2007; Wu and Zhou 2010). The term "integral" indicates the sum. It penalizes all errors equally, regardless of direction:

$$IAE = \sum \frac{\left|L_{e,\,\exp} - L_{e,\,theo}\right|}{\left|L_{e,\,\exp}\right|} \tag{8.8}$$

The elastic modulus of the CFRP (E_f), its thickness (t_f), the compressive strength (f_c') of the concrete substrate, and its splitting tensile strength (f_t and f_{ctm}) were compared for the models. ACI 440.2R-08 (ACI 2008) and CIDAR (2006) were the only models that did not rely on the splitting tensile strength of concrete. The values that the theoretical models projected were often very dispersed. This can be partially ascribed to the models' imprecision (Shen et al. 2015c). Experimental and numerical investigations have both demonstrated that altering the test setting can drastically alter the experimental test findings, leading to erroneous model output (Bilotta et al. 2011). The most accurate model was that put out by Lu (2004), which had the lowest IAE. A bond length of 200 mm between the CFRP and concrete was longer than the effective bond lengths predicted by the proposed models and reflects the full development of the bond capacity (Shen et al. 2015c).

Table 8.10 Performance of proposed effective bond length models

Reference	Equation	Factors	$L_{e,theo}$	IAE (%)
ACI (2008)	$L_e = \frac{23300}{(nE_f t_f)^{0.58}}$	E_f, t_f	MKG-C-1: 56.9	29.4
			MKG-C-2: 56.9	41.2
			MKG-C-3: 56.9	52
			OPC-C-1: 56.9	8.2
			OPC-C-2: 56.9	2.4
			OPC-C-3: 56.9	13.1
FIB B14 (2001)	$L_e = \sqrt{\frac{E_f t_f}{C_2 f_{ctm}}}\ C_2 = 2$	E_f, t_f, f_{ctm}, f_t	MKG-C-1: 71.3	11.6
			MKG-C-2: 74.8	22.8
			MKG-C-3: 86.5	27
Neubauer and Rostasy (1997)	$L_e = \sqrt{\frac{E_f t_f}{2 f_t}}$		OPC-C-1: 57.8	9.9
			OPC-C-2: 62	6.3
			OPC-C-3: 65.3	0.2
BS EN 1998-3 (2005)	$L_e = \sqrt{\frac{E_f t_f}{4 f_{ctm}}}$	E_f, t_f, f_{ctm}	MKG-C-1: 50.4	37.5
			MKG-C-2: 52.9	45.4
			MKG-C-3: 61.2	48.4
			OPC-C-1: 40.9	22.2
			OPC-C-2: 43.9	24.7
			OPC-C-3: 46.1	29.6
CIDAR (2006)	$L_e = \sqrt{\frac{E_f t_f}{\sqrt{f'_c}}}$	E_f, t_f, f'_c	MKG-C-1: 76.8	4.8
			MKG-C-2: 78.7	18.7
			MKG-C-3: 82.9	30
			OPC-C-1: 79.6	51.3
			OPC-C-2: 77.6	33.1
			OPC-C-3: 76.8	17.3
Lu (2004)	$L_e = 1.33\frac{\sqrt{E_f t_f}}{f_t}$	E_f, t_f, f_t	MKG-C-1: 75.8	6
			MKG-C-2: 83.3	14
			MKG-C-3: 111.5	5.9
			OPC-C-1: 49.8	5.3
			OPC-C-2: 57.3	1.7
			OPC-C-3: 63.5	3

Table 8.11 Comparison of calibrated model (Eq. 8.9) and experimental effective bond length (L_e)

Specimens	Maximum aggregate size (mm)	Calibrated model ($L_{e,\ Lu}$) (mm)	$L_e/L_{e,\ Lu}$
MKG-C-1	10	81.5	0.990
MKG-C-2	16	89.5	1.082
MKG-C-3	20	119.7	0.990
OPC-C-1-1	10	53.5	0.984
OPC-C-1-2	16	61.6	0.946
OPC-C-1-3	20	68.2	0.960
Average			0.992
SD			0.047

Lu's (2004) model for the effective bond length is

$$L_e = \alpha \frac{\sqrt{E_f t_f}}{f_t} \tag{8.9}$$

Least-squares fitting gives a value of 1.429 for the constant α ($R^2 = 0.932$). The effective bond length of the study's specimens was then determined using this information. The projected effective bond length from the calibrated model (Eq. 8.9) is shown in Table 8.11, and it agrees well with the experimental findings from this work, resulting in an average value of $L_e/L_{e,Lu}$ of 0.992 and an SD of 0.047. In line with the values obtained from the experimental results, the projected values of the effective bond length rose with the size of the coarse aggregate.

8.3.4.4 Proposed Equations to Determine Effective Bond Length

The experimental effective bond length and the calibrated average effective bond length were used to establish the relationship among the effective bond length, elastic modulus of the CFRP, elastic modulus of the concrete, and the maximum size of the coarse aggregate. The elastic modulus of the CFRP over that of the concrete was determined by findings for the effective bond length normalized with the thickness of the CFRP. According to ACI 318-14 and ACI 2014, the elastic modulus of concrete (E_c, MPa) was calculated

$$E_c = 4730\sqrt{f_c'} \tag{8.10}$$

where f_c' = cylindrical compressive strength (MPa) of concrete ($f_c' = 0.8\ f_{cu}'$). The Levenberg–Marquardt method in MATLAB (MathWorks 2018) was used to determine the fitting parameters based on the nonlinear least-squares algorithm, as

Table 8.12 Fitting parameters of proposed models

Series	Fitting parameters	$L_e = t_f\left(a + b\left(\frac{E_f}{E_c}\right) + cd\right)$	$L_e = t_f\left(a + b\left(\frac{E_f}{\sqrt{f_c'}}\right) + cd\right)$
MKG	a	– 820.7	– 820.7
	b	147.6	0.0312
	d	5.549	5.549
	R^2	0.950	0.950
OPC	a	– 518.2	– 518.2
	b	75.32	0.01592
	d	17.15	17.15
	R^2	0.935	0.935

Reproduced from [Influence of Coarse Aggregate Size on the Bonding between CFRP Sheets and Metakaolin-Based Geopolymer Concrete and Ordinary Concrete] by [Hamed Fazli] with permission from [Journal of Composites for Construction]

reported in Table 8.12, where L_e, t_f, and E_f are the effective bond length (mm), thickness of the CFRP (mm), and its elastic modulus (MPa), respectively; and d is the size of the coarse aggregate. The suggested correlations take into account the elastic modulus of concrete, the size of the coarse aggregate, and the stiffness of the CFRP. In experimental work, the compressive strength of concrete is often measured. In light of this, correlations between the effective bond length, concrete's compressive strength, and the size of the coarse aggregate are suggested.

The equations developed in this study are linear with regard to the thickness of the CFRP and its elastic modulus (CFRP stiffness), in contrast to the equations created for ACI 440.2R-08 (ACI 2008) and CSA S806-02 (CSA 2002). The study's suggested equation demonstrates that Le rises as the size of the coarse aggregate does. The suggested equations' benefits include their simpler use in practice and their consideration of the impact of the coarse aggregate's size.

The proposed equations were verified through a comparison of the predicted values ($L_{e,pre}$) and the experimentally obtained effective bond length (L_e), as presented in Table 8.13. The two sets of a $L_{e,exp}$ for GPC samples was calculated via the Lu (2004) model values agree well. Table 8.14 compares the results for the proposed model ($L_{e,pre}$) against experimental values for the effective bond length ($L_{e,exp}$) obtained from other studies. However, due to the lack of experimental data, the $L_{e,exp}$ for the GPC samples was calculated with the Lu (2004) model by utilizing the material properties obtained from the literature.

8.3.4.5 Local Slip Calculation and Interfacial Bond–Slip Relationship

The following assumptions were made in order to calculate the slip distribution along the CFRP sheet based on measured strain: (1) that the bond between the CFRP sheet and the concrete substrate at x_m was perfect, i.e., $s(x_7) = 0$, (2) that deformation

Table 8.13 Comparison of proposed and experimental effective bond length

Specimens	Maximum aggregate size (mm)	Proposed model ($L_{e,\ pre}$) (mm)	$L_e/L_{e,\ pre}$
MKG-C-1	10	80.9	0.997
MKG-C-2	16	94.4	1.025
MKG-C-3	20	118.9	0.997
OPC-C-1	10	54.1	0.972
OPC-C-2	16	63	0.926
OPC-C-3	20	70.4	0.931
Average			0.975
SD			0.040

Reproduced from [Influence of Coarse Aggregate Size on the Bonding between CFRP Sheets and Metakaolin-Based Geopolymer Concrete and Ordinary Concrete] by [Hamed Fazli] with permission from [Journal of Composites for Construction]

Table 8.14 Comparison of proposed $\left(L_{e,\ pre}\right)$ and experimental $\left(L_{e,\ \exp}\right)$ effective bond length results gathered from literature

Reference	Reference	$L_{e,\ pre}$ (mm)	$L_{e,\ \exp}$ (mm)	IAE%
GPC*	Junaid et al. (2020)	82	79.9	2.6
	Hampshire (2012)	57.4	58.8	2.3
	Ozbakkaloglu and Xie (2016)	77.7	76.9	1.1
	Tang et al. (2020)	48.1	56.1	14.2
	Sarker et al. (2017)	53.7	53	1.3
	Junaid et al. (2019)	63.8	59.7	6.8
	Lokuge and Karunasena (2016)	95.1	85.5	11.2
		52.4	56.3	7.1
Average				5.8
SD				4.9
OPC	Sato et al. (1997)	46.2	45.2	2.2
	Nakaba et al. (2001)	100	99.1	0.9
	Foster and Khomwan (2005)	248.3	270	8
	Boschetto et al. (2006)	85.9	95	9.6
	Fazli et al. (2018a, b)	54.3	58.6	7.4
	Subramaniam et al. (2011)	91.6	81.0	13.1
	Pellegrino et al. (2008)	149.0	130.0	14.6
	Carloni and Subramaniam (2013)	78.9	76.0	3.8
Average				7.5
SD				5

Reproduced from [Influence of Coarse Aggregate Size on the Bonding between CFRP Sheets and Metakaolin-Based Geopolymer Concrete and Ordinary Concrete] by [Hamed Fazli] with permission from [Journal of Composites for Construction]

* $L_{e,\ \exp}$ for GPC samples was calculated via the Lu model

in the concrete specimen far from the external cover was negligible compared to its CFRP counterpart, and (3) that the strain in the CFRP sheet changed linearly between strain gauges. As a result, beginning with the final strain gauge at $x = x_m$, the interfacial bond-slip relationship was calculated using the slip distribution along the x-axis. Finally, the local slip at the reference axis x, where $x_i \leq x \leq x_i + 1$ was determined using (Ferracuti et al. 2007)

$$s(x) = s(x_{i+1}) - \frac{(\varepsilon_{i+1} - \varepsilon_i)}{(x_{i+1} - x_i)} \frac{(x_{i+1} - x)^2}{2} + \varepsilon_i (x_{i+1} - x). \quad (8.11)$$

The average local slip between consecutive strain gauges at x_i and x_{i+1} was calculated as

$$\hat{s}_{i+1/2} = \frac{s(x_{i+1}) + s(x_i)}{2}. \quad (8.12)$$

Table 8.15 summarizes the results for the local slip at the maximum shear stress (s_0). When the coarse aggregate size range expanded from 5–10 to 10–16, and then to 16–20 mm, respectively, for the MKG specimens, s_0 increased from 0.095 to 0.107 mm and then to 0.128 mm. With the size of the coarse aggregate, s_0 for the OPC specimens rose from 0.122 to 0.139 and to 0.145 mm, respectively. The interfacial bond between the CFRP and the concrete substrate was regulated by the splitting tensile strength since the failure mode caused by debonding occurred inside the concrete layers of the tested specimens. According to Akçaoğlu et al. (2002), the splitting tensile strength decreased as the size of the coarse aggregate increased. These findings support the hypothesis that, as the size of the coarse aggregate decreased, the interfacial shear bond tension rose. This is because a large coarse aggregate has a lower splitting tensile strength and, consequently, a lower interfacial shear bond stress.

For studying the behavior of CFRP-bonded concrete structures, the constitutive model of the bond-slip connection between the CFRP and the substrate is crucial. The nonlinear relationship between the local average bond stress and the average slip ($\hat{\tau}_{i+1/2}, \hat{s}_{i+1/2}$) for all specimens are plotted in Figs. 8.8 and 8.9. The results have an ascending branch and a descending branch. They were calculated from the measurements.

The interface behavior between FRP and concrete is known to exhibit significant nonlinearity, which is influenced by various factors such as the structural model adopted and the shearing deformation of both the concrete cover and the epoxy resin. The maximum load capacity is determined by the mode II fracture energy of the interface law. Kinematic variables are employed to describe the interfacial slip between FRP and concrete, as well as the constitutive law applied to the concrete. A nonlinear mode II law for the FRP-concrete interface typically exhibits the following key characteristics: (1) a softening branch corresponding to larger slips, (2) a maximum shear stress and its associated local slip, and (3) a stiffness value for small local slips (Ferracuti et al. 2007). The interface layer, where the constitutive law includes

Table 8.15 Experimental results and fitted results of interfacial shear bond–slip relationship

Specimen	τ_{max} (MPa)		S_0 (mm)		Regression coefficient, n	Correlation coefficient, R^2
	Indiv.	Ave.	Indiv.	Ave.		
MKG-C-1-1	6	5.8	0.087	0.095	2.928	0.964
MKG-C-1-2	5.7		0.102		2.934	0.951
MKG-C-2-1	5.8	5.5	0.110	0.107	3.076	0.952
MKG-C-2-2	5.1		0.104		3.070	0.941
MKG-C-3-1	4.8	4.4	0.129	0.128	3.116	0.964
MKG-C-3-2	3.9		0.126		3.125	0.981
OPC-C-1-1	7.7	6.4	0.115	0.122	2.975	0.957
OPC-C-1-2	5.1		0.128		2.969	0.979
OPC-C-2-1	6.3	5.8	0.143	0.139	2.913	0.983
OPC-C-2-2	5.2		0.135		2.907	0.970
OPC-C-3-1	5.9	5	0.151	0.145	2.783	0.972
OPC-C-3-2	4.1		0.139		2.769	0.959

Reproduced from [Influence of Coarse Aggregate Size on the Bonding between CFRP Sheets and Metakaolin-Based Geopolymer Concrete and Ordinary Concrete] by [Hamed Fazli] with permission from [Journal of Composites for Construction]
Note Indiv. = Individual; Ave. = Average

all compliance and nonlinear contributions from the adhesive and concrete cover, is where the relative displacement between the FRP and concrete is concentrated (Freddi and Sacco 2014).Consequently, Popovics' equation serves as the basis for the constitutive model.

$$\tau = \tau_{\max}\left[\frac{s}{s_0}\frac{n}{(n-1)+(s/s_0)^n}\right] \tag{8.13}$$

where τ and τ_{max} = interfacial shear bond stress (MPa) and the maximum interfacial shear bond stress (MPa), respectively; s and s_0 = local slip (mm) and the local slip at the maximum shear stress (mm), respectively; n depends on the compressive strength of the concrete, being a coefficient that determines the slope of the curve for both the ascending and descending branches (Popovics 1973); and τ_{max}, s, and s_0 were determined directly from the experimental results, with τ_{max} obtained from Eq. (8.1), the local slip values (s and s_0) were calculated using Eqs. (8.11) and (8.12), and n was calculated using the Levenberg–Marquardt algorithm in MATLAB (MathWorks 2018) (Figs. 8.21 and 8.22).

By taking into account stress-dependent factors, Eq. (8.13) takes into account the nonlinear contributions of the concrete cover and adhesive (Popovics 1973; Nakaba et al. 2001; Ferracuti et al. 2007; Shen et al. 2015b). In fact, the compliances for both the concrete cover and the epoxy resin are taken into account to derive the interfacial bond stress and local slip from the measured strain of the FRP-concrete. These are specifically taken into account and are interdependent (Ferracuti et al. 2007).

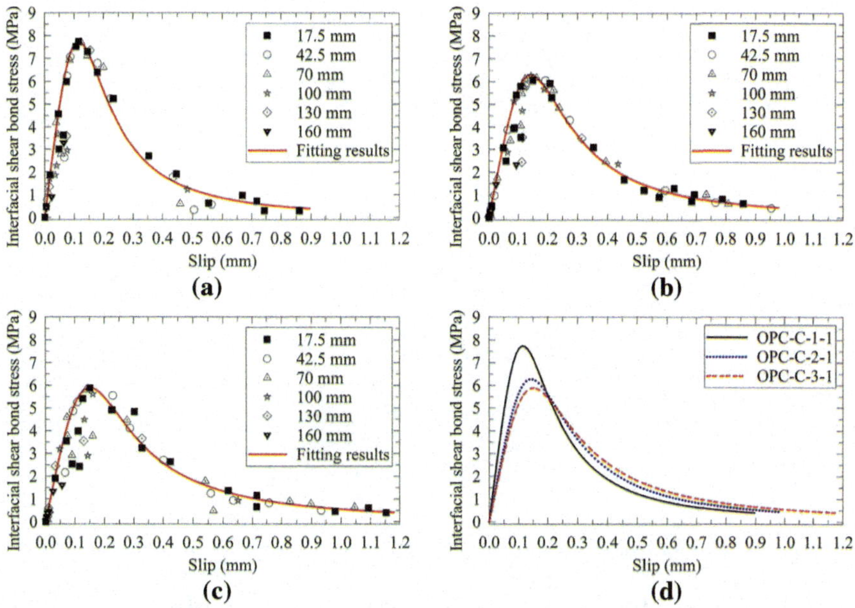

Fig. 8.21 Interfacial shear bond stress versus local slip relationship for **a** OPC-C-1-1 specimen; **b** OPC-C-2-1 specimen; **c** OPC-C-3-1 specimen; and **d** comparison of the predictions. Reproduced from [Influence of Coarse Aggregate Size on the Bonding between CFRP Sheets and Metakaolin-Based Geopolymer Concrete and Ordinary Concrete] by [Hamed Fazli] with permission from [Journal of Composites for Construction]

The mechanical characteristics of the specimens in this study produced a correlation between n and the size of the coarse aggregate, as the latter affected the compressive strengths of both MKG and the OPC specimens, despite some studies (Nakaba et al. 2001; Sato and Vecchio 2003) using a constant value for the coefficient n. Table 8.15 lists the fitting results of the specimens tested. Popovics (1973) also proposed a relationship between n and the maximum size of the coarse aggregate by considering the compressive strength of concrete:

$$n = \alpha + \beta f_c \tag{8.14}$$

where f_c = compressive strength, and α and β are coefficients derived from a nonlinear regression analysis (least-squares approach). Figure 8.23 displays the experimental findings (Table 8.15) that were used to establish the parameters for and. The averages and standard deviations of the calibrated model for the MKG and OPC specimens were 0.993 and 0.013, and 1 and 0.014, respectively, indicating that the findings closely matched the experimentally determined value of n.

Figures 8.8 and 8.9 illustrate the relationship between interfacial shear bond stress and local shear slip for the CFRP-concrete interface. The plots exhibit three distinct phases: linear elastic, nonlinear, and softening. Nonlinear behavior occurs following

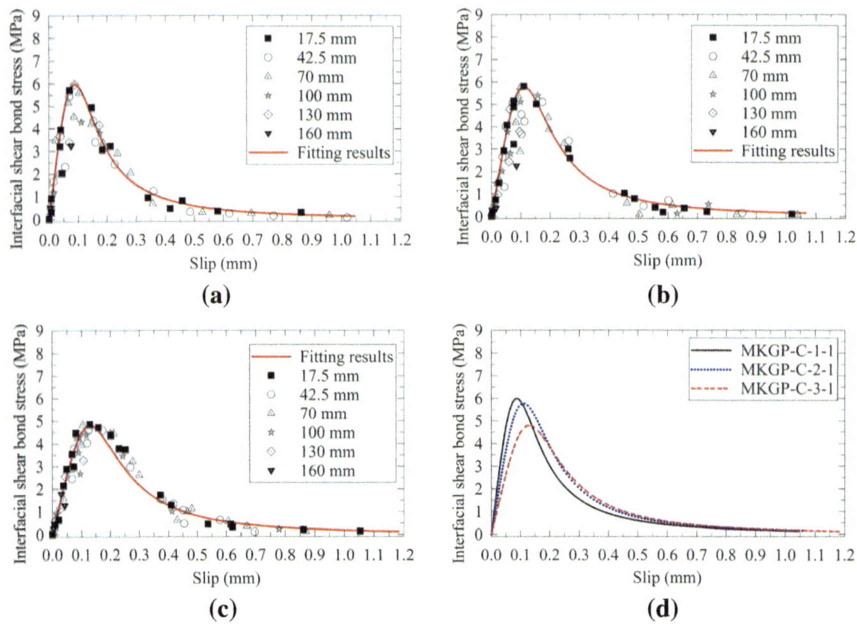

Fig. 8.22 Interfacial shear bond stress versus local slip relationship for **a** MKG-C-1-1 specimen; **b** MKG-C-2-1 specimen; **c** MKG-C-3-1 specimen; and **d** comparison of the predictions. Reproduced from [Influence of Coarse Aggregate Size on the Bonding between CFRP Sheets and Metakaolin-Based Geopolymer Concrete and Ordinary Concrete] by [Hamed Fazli] with permission from [Journal of Composites for Construction]

Fig. 8.23 Fitted results of n. Reproduced from [Influence of Coarse Aggregate Size on the Bonding between CFRP Sheets and Metakaolin-Based Geopolymer Concrete and Ordinary Concrete] by [Hamed Fazli] with permission from [Journal of Composites for Construction]

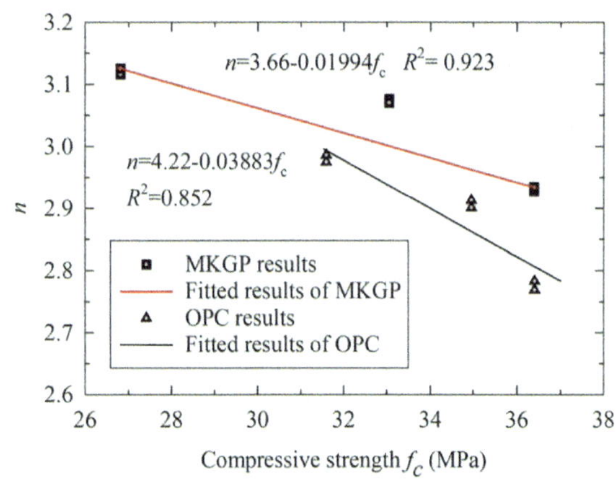

the linear elastic stage, with an increasing shear bond stress up to the peak value as local slip increases. Upon reaching τ_{max}, the descending branch indicates the onset of the softening phase, during which the shear bond stress gradually decreases to zero due to increased local slip. The softening phase begins once the applied slip at the loaded length reaches the elastic limit. Debonding occurs when the local slip reaches its maximum value, resulting in a shear bond stress of zero. The significance of the softening phase lies in its influence on the fracture energy of the interface law and, consequently, the maximum transmissible load (Savoia et al. 2009). Subsequently, more microcracks accumulate, leading to the formation of a destructive crack within the adhesive-concrete layer (Yuan et al. 2019b). The interlocking action of aggregates enables the CFRP sheets to continue resisting shear forces. Specimens with smaller aggregate sizes exhibit higher fracture energy due to stronger interlocking action. Increasing the maximum size of coarse aggregate reduces the maximum interfacial shear bond stress. Additionally, as the aggregate's maximum size increases, the slope of the ascending branches of the curves lowers. This is explained by a decline in the area under the interfacial shear bond-slip curve, also known as the interfacial fracture energy. The specimens MKG-C-3-1, MKG-C-2-1, MKG-C-1-1, OPC-C-3-1, OPC-C-2-1, and OPC-C-1-1, respectively, had interfacial fracture energies of 1.4, 1.5, 1.7, 2.1, 2.3, and 2.4 N/mm. Taking into account the varying sizes of coarse aggregate, Popovics's equation can be employed to predict the relationship between interfacial shear bond stress and local slip for CFRP bonded to MKG concrete, as the predicted results align well with the experimental findings.

8.4 Conclusions

This research examines the influence of coarse aggregate size on the mechanical properties of MKG concrete and OPC concrete. The compressive strength and splitting tensile strength of the specimens were evaluated after a 28-day curing period and subjected to analysis. Based on the findings obtained from the experimental investigation conducted in this study, the following conclusions can be drawn:

(1) As the size of the coarse aggregate grew, the compressive strength of MKG concrete and OPC concrete declined and rose, respectively. This phenomenon can be attributed to the MKG concrete's greater shrinkage.
(2) The highest decrease and increase in compressive strength caused by the expansion of the coarse aggregate in OPC concrete and MKG concrete, respectively, were 13 and 36%.
(3) Because coarse aggregates with bigger diameters produce more microcracks around the coarse aggregate, the splitting tensile strength fell in both groups.
(4) For OPC concrete and MKG concrete, the highest drop in splitting tensile strength caused by an increase in the size of the coarse aggregate was 28% and 114%, respectively.

(5) SEM/EDS analyses showed that the ITZ area in this study is not significantly impacted by the coarse aggregate's size. According to MIP data, the pore diameter grew as the coarse aggregate's size increased. Therefore, when the size of the coarse aggregate rises, the role of paste between the coarse aggregates becomes more prominent.

(6) It can be inferred that the strength development of the tested specimens is significantly influenced by changes in pore structure with coarse aggregate size. A greater shrinkage of MKG concrete and consequently more macropores are produced when the size of the coarse aggregate is increased. Consequently, macropores have a big impact.

(7) According to the tests, the size of the coarse aggregate has little bearing on how the CFRP-concrete interface fails.

(8) As the maximum size of the coarse aggregate increased from 10 to 16 mm and then to 20 mm, the average failure loads of the MKG and OPC specimens decreased by 5.4%, 14.3%, 6%, and 11.2%, respectively.

(9) The MKG and OPC specimens exhibited a considerable decrease in the average maximum interfacial shear bond stresses of up to 24.1% and 21.9%, respectively, when the maximum size of the coarse aggregate was increased from 10 to 20 mm.

(10) The studies reveal that, compared to the specimens with lesser coarse aggregate, the MKG specimens with the highest coarse aggregate saw a considerable increase in the effective bond length of up to 46.8%, whereas this increase was 24.5% for the OPC specimens.

(11) For all studied specimens, there was a noticeable inverse relationship between the size of the coarse aggregate and the local slip (s_0) at the highest interfacial shear bond stress. The average S_0 of MKG-C-3 and OPC-C-3 rose by 34.7% and 18.9%, respectively, in comparison to MKG-C-1 and the OPC-C-1.

(12) By taking into account the impact of the coarse aggregate size, empirical formulae for the effective bond length of CFRP sheets attached to MKG and OPC specimens were provided. They were in line with the findings of the experiment. The empirical link between the interfacial shear bond stress and local slip took into account the impact of the coarse aggregate's size. The interfacial shear bond-slip behavior is adequately predicted by this connection.

References

ACI 440.2R-08 (2008) Guide for the design and construction of externally bonded FRP systems for strengthening existing structures. ACI Comm 440 144

Akçaoğlu T (2017) Determining aggregate size & shape effect on concrete microcracking under compression by means of a degree of reversibility method. Constr Build Mater 143:376–386. https://doi.org/10.1016/j.conbuildmat.2017.03.128

Akçaoğlu T, Tokyay M, Çelik T (2002) Effect of coarse aggregate size on interfacial cracking under uniaxial compression. Mater Lett 57:828–833. https://doi.org/10.1016/S0167-577X(02)00881-9

Akçaoğlu T, Tokyay M, Çelik T (2004) Effect of coarse aggregate size and matrix quality on ITZ and failure behavior of concrete under uniaxial compression. Cem Concr Compos 26:633–638. https://doi.org/10.1016/S0958-9465(03)00092-1

Akçaoğlu T, Tokyay M, Çelik T (2005) Assessing the ITZ microcracking via scanning electron microscope and its effect on the failure behavior of concrete. Cem Concr Res 35:358–363. https://doi.org/10.1016/j.cemconres.2004.05.042

Akcil A, Koldas S (2006) Acid mine drainage (AMD): causes, treatment and case studies. J Clean Prod 14:1139–1145. https://doi.org/10.1016/j.jclepro.2004.09.006

Al-Bayati HKA, Tighe SL, Baaj H (2016) Effect of different treatment methods on the interfacial transition zone microstructure to coarse recycled concrete aggregate. 2016 Transp Assoc Canada's Conf Exhib TAC 2016

Alzeebaree R, Çevik A, Nematollahi B et al (2019) Mechanical properties and durability of unconfined and confined geopolymer concrete with fiber reinforced polymers exposed to sulfuric acid. Constr Build Mater 215:1015–1032. https://doi.org/10.1016/j.conbuildmat.2019.04.165

Ali-Ahmad, Subramaniam MK, Ghosn M (2004) "Fracture analysis of the debonding between FRP and concrete using digital image correlation. In Proc., FRAMCOS-5 Int. Conf. on Fracture of Concrete and Concrete Structures, 787–793. Colorado, CO: FraMCoS-5.

Ali-Ahmad, Subramaniam MK, Ghosn M (2006) Experimental investigation and fracture analysis of debonding between concrete and FRP sheets. J Eng Mech 132 (9):914–923. https://doi.org/10.1061/(ASCE)0733-9399

Athawale P, B.E. (2012) Analysis of factors affecting effective bond length for fiber reinforced polymer composite laminate externally bonded to concrete substrate. MSc thesis, Texas Tech University, May, p 116

ASTM C618 (2019) Standard Specification for Coal Fly Ash and Raw or Calcined Natural Pozzolan for Use in Concrete; ASTM International: West Conshohocken, PA, USA

Bakharev T (2005) Resistance of geopolymer materials to acid attack. Cem Concr Res 35:658–670. https://doi.org/10.1016/j.cemconres.2004.06.005

Bakharev T, Sanjayan JG, Cheng YB (2000) Effect of admixtures on properties of alkali-activated slag concrete. Cem Concr Res 30:1367–1374. https://doi.org/10.1016/S0008-8846(00)00349-5

Ben Ouezdou M, Belarbi A, Bae S-W (2009) Effective bond length of FRP sheets externally bonded to concrete. Int J Concr Struct Mater 3:127–131. https://doi.org/10.4334/ijcsm.2009.3.2.127

Bilotta A, Di LM, Nigro E (2011) FRP-to-concrete interface debonding: experimental calibration of a capacity model. Compos Part B Eng 42:1539–1553. https://doi.org/10.1016/j.compositesb.2011.04.016

Boschetto G, Pellegrino C, Tinazzi D, Modena C (2006) Bond behaviour between FRP sheets and concrete: an experimental study. In Proc., 2nd FIB Cong. Naples, Italy: FIB

BS EN 12390-6 (2009) Testing Hardened Concrete. Tensile Splitting Strength of Test Specimens; British Standard Institution (BSI): Brussels, Belgium

BS EN 12390-2 (2009) Testing Hardened Concrete-Part 2: Making and Curing Specimens for Strength Tests; BSI (British Standard Institution): Brussels, Belgium

BS EN 12390-3 (2009) Testing Hardened Concrete Part 3: Compressive Strength of Test Specimens; BSI (British Standard Institution): London, UK

BSI (British Standard Institution) (2005) Eurocode 8: Design of structures for earthquake resistance. Part 3: strengthening and repair of buildings. BS EN 1998-3. Brussels, Belgium: BSI

CSA (Canadian Standard Association) (2002) Design and construction of building components with fibre-reinforced polymers. CSA S806-02. Rexdale, ON, Canada: CSA

Canadian Standard Association (2012) Design and construction of building components with fibre-reinforced polymers (CAN/CSA S806-02). Can/Csas806-12 177

Carloni C, Subramaniam KV (2013) Investigation of sub-critical fatigue crack growth in FRP/concrete cohesive interface using digital image analysis Composites, Part B 51: 35–43. https://doi.org/10.1016/j.compositesb.2013.02.015

CEB-FIP (1990) Evaluation of the time-dependent behavior of concrete. Bulletin D'information No 199. CEB Comité Euro international du Béton, Laussane, Switzerland

Chen B, Liu J (2004) Effect of aggregate on the fracture behavior of high strength concrete. Constr Build Mater 18:585–590. https://doi.org/10.1016/j.conbuildmat.2004.04.013
Chen JF, Teng JG (2001) Anchorage strength models for FRP and steel plates bonded to concrete. J Struct Eng 127:784–791
Chen W, Hao H, Jong M et al (2017) Quasi-static and dynamic tensile properties of basalt fibre reinforced polymer. Compos Part B Eng 125:123–133. https://doi.org/10.1016/j.compositesb.2017.05.069
Chen W, Pham TM, Sichembe H et al (2018) Experimental study of flexural behaviour of RC beams strengthened by longitudinal and U-shaped basalt FRP sheet. Compos Part B Eng 134:114–126. https://doi.org/10.1016/j.compositesb.2017.09.053
CIDAR (Centre for Infrastructure Diagnosis/Assessment and Rehabilitation) (2006) Design guideline for RC structures retrofitted with FRP and metal plates. Beams and slabs, draft 3. Canberra, Australia: Standards Australia (CIDAR/CBIR)
Ćosić K, Korat L, Ducman V, Netinger I (2015) Influence of aggregate type and size on properties of pervious concrete. Constr Build Mater 78:69–76. https://doi.org/10.1016/j.conbuildmat.2014.12.073
Dai J, Ueda T, Sato Y (2005) Development of the nonlinear bond stress-slip model of fiber reinforced plastics sheet-concrete interfaces with a simple method. J Compos Constr 9:52–62. https://doi.org/10.1061/(asce)1090-0268(2005)9:1(52)
Di Maio A, Giaccio G, Zerbino R (1996) Failure mechanism of concrete, combined effects of coarse aggregates and specimen geometry. Constr Build Mater 10:571–575. https://doi.org/10.1016/S0950-0618(96)00025-6
Diab HM, Farghal OA (2014) Bond strength and effective bond length of FRP sheets/plates bonded to concrete considering the type of adhesive layer. Compos Part B Eng 58:618–624. https://doi.org/10.1016/j.compositesb.2013.10.075
Elarbi AM (2011) Durability performance of frp strenghtened concrete beams and columns exposed to hygrothermal environment, 193
Farhan KZ, Johari MAM, Demirboğa R (2020) Assessment of important parameters involved in the synthesis of geopolymer composites: a review. Constr Build Mater 264. https://doi.org/10.1016/j.conbuildmat.2020.120276
Fawzia S, Zhao XL, Al-Mahaidi R, Rizkalla SH (2020) Preliminary bond-slip model for CFRP sheets bonded to steel plates. Compos Civ Eng CICE 2006:737–740
Fazli H, Mohd Yassin AY, Shafiq N, Teo W (2018a) Effective bond length of CFRP sheets externally bonded to concrete beams under marine environment. Constr Build Mater 167:726–738. https://doi.org/10.1016/j.conbuildmat.2018.02.085
Fazli H, Yassin AYM, Shafiq N, Teo W (2018b) Pull-off testing as an interfacial bond strength assessment of CFRP-concrete interface exposed to a marine environment. Int J Adhes Adhes 84:335–342. https://doi.org/10.1016/j.ijadhadh.2018.04.015
Ferracuti B, Savoia M, Mazzotti C (2007) Interface law for FRP-concrete delamination. Compos Struct 80:523–531. https://doi.org/10.1016/j.compstruct.2006.07.001
FIB (International Federation for Structural Concrete) (2001) Externally bonded FRP reinforcement for RC structures: technical report on the design and use of externally bonded fibre reinforced polymer reinforcement (FRP EBR) for reinforced concrete structures. FIB B14. Lausanne: FIB
Foster S, Khomwan N (2005) Determination of bond stress versus slip for externally bonded FRP from standardized bond strength tests. In Proc., Int. Symp. on Bond Behaviour of FRP in Structures, 85–90. Hong Kong: International Institute for FRP in Construction
Franco A, Royer-Carfagni G (2014) Effective bond length of FRP stiffeners. Int J Non Linear Mech 60:46–57. https://doi.org/10.1016/j.ijnonlinmec.2013.12.003
Freddi F, Sacco E (2014) An interface damage model accounting for in-plane effects. Int J Solids Struct 51:4230–4244. https://doi.org/10.1016/j.ijsolstr.2014.08.010
Ghiassi B, Xavier J, Oliveira DV, Lourenço PB (2013) Application of digital image correlation in investigating the bond between FRP and masonry. Compos Struct 106:340–349

Girgin ZC, Arioglu N, Arioglu E (2007) Evaluation of strength criteria for very-high-strength concretes under triaxial compression. ACI Struct J 104:278–284. https://doi.org/10.14359/18617

Godat A, Neale KW, Labossière P (2007) Numerical modeling of FRP shear-strengthened reinforced concrete beams. J Compos Constr 11:640–649. https://doi.org/10.1061/(asce)1090-0268(2007)11:6(640)

Gonilho Pereira C, Castro-Gomes J, Pereira de Oliveira L (2009) Influence of natural coarse aggregate size, mineralogy and water content on the permeability of structural concrete. Constr Build Mater 23:602–608. https://doi.org/10.1016/j.conbuildmat.2008.04.009

Guades EJ (2016) Experimental investigation of the compressive and tensile strengths of geopolymer mortar: the effect of sand/fly ash (S/FA) ratio. Constr Build Mater 127:484–493. https://doi.org/10.1016/j.conbuildmat.2016.10.030

Guades EJ (2017) Effect of coarse aggregate size on the compressive behaviour of geopolymer concrete. Eur J Environ Civ Eng 23:693–709. https://doi.org/10.1080/19648189.2017.1304276

Hampshire AP (2012) Confinement of Geopolymer Concrete with Carbon Fibre Reinforced Polymer. Canberra, Australia: University of New South Wales

Hansen TC, Nielsen K (1965) Influence of aggregate properties on concrete shrinkage. ACI J Proc 62. https://doi.org/10.14359/7723

Isabella C, Xu H, Luckey GC, van Deventer JSJ (2003) The effect of aggregate particle size on formation of geopolymeric gel. Paper 9:2005

Jiang C, Wu YF, Jiang JF (2017) Effect of aggregate size on stress-strain behavior of concrete confined by fiber composites. Compos Struct 168:851–862. https://doi.org/10.1016/j.compstruct.2017.02.087

Joseph B, Mathew G (2012) Influence of aggregate content on the behavior of fly ash based geopolymer concrete. Sci Iran 19:1188–1194. https://doi.org/10.1016/j.scient.2012.07.006

Junaid MT, Elbana A, Altoubat S (2020) Flexural response of geopolymer and fiber reinforced geopolymer concrete beams reinforced with GFRP bars and strengthened using CFRP sheets. Structures 24:666–677. https://doi.org/10.1016/j.istruc.2020.02.003

Junaid MT, Elbana A, Altoubat S (2019) Predicting flexural capacity of concrete beams reinforced with GFRP bars and strengthened with CFRP sheets. In Proc., 4th World Cong. on Civil, Structural, and Environmental Engineering, 1–7. Italy, Rome: CSEE'19

Júlio ENBS, Branco FAB, Silva VD (2005) Concrete-to-concrete bond strength: Influence of an epoxy-based bonding agent on a roughened substrate surface. Mag Concr Res 57:463–468. https://doi.org/10.1680/macr.2005.57.8.463

Kang THK, Howell J, Kim S, Lee DJ (2012) A state-of-the-art review on debonding failures of FRP laminates externally adhered to concrete. Int J Concr Struct Mater 6:123–134. https://doi.org/10.1007/s40069-012-0012-1

Karaguler ME, Yatagan MS (2018) Effect of aggregate size on the restrained shrinkage of the concrete and mortar. MOJ Civ Eng 4:15–21. https://doi.org/10.15406/mojce.2018.04.00092

Karakoç MB, Türkmen I, Maraş MM et al (2014) Mechanical properties and setting time of ferrochrome slag based geopolymer paste and mortar. Constr Build Mater 72:283–292. https://doi.org/10.1016/j.conbuildmat.2014.09.021

Karbhari VM, Ghosh K (2009) Comparative durability evaluation of ambient temperature cured externally bonded CFRP and GFRP composite systems for repair of bridges. Compos Part A Appl Sci Manuf 40:1353–1363. https://doi.org/10.1016/j.compositesa.2009.01.011

Khalifa A, Gold WJ, Nanni A, Abdel Aziz MI (1998) Contribution of externally bonded frp to shear capacity of RC flexural members. J Compos Constr 2:195–202. https://doi.org/10.1061/(asce)1090-0268(1998)2:4(195)

Kim JK, Yi ST, Park CK, Eo SH (1999) Size effect on compressive strength of plain and spirally reinforced concrete cylinders. ACI Struct J 96:88–94. https://doi.org/10.14359/599

Ko H, Sato Y (2007) Bond stress-slip relationship between FRP sheet and concrete under cyclic load. J Compos Constr 11:419–426. https://doi.org/10.1061/(asce)1090-0268(2007)11:4(419)

Lahoti M, Narang P, Tan KH, Yang EH (2017) Mix design factors and strength prediction of metakaolin-based geopolymer. Ceram Int 43:11433–11441. https://doi.org/10.1016/j.ceramint.2017.06.006
Lokuge W, Karunasena W (2016) Ductility enhancement of geopolymer concrete columns using fibre-reinforced polymer confinement. J Compos Mater 50 (14):1887–1896. https://doi.org/10.1177%2F0021998315597553
Lokuge WP, Setunge S, Sanjayan JG (2011) Stress-strain model for high strength concrete confined by FRP. In: Proceedings of the 21st Australasian Conference on the Mechanics of Structures and Materials: incorporating sustainable practice in mechanics and structures of materials, pp 481–485. https://doi.org/10.1201/b10571-85
Lu X (2004) Study on FRP-concrete interface
Lu XZ, Teng JG, Ye LP, Jiang JJ (2005) Bond-slip models for FRP sheets/plates bonded to concrete. Eng Struct 27:920–937. https://doi.org/10.1016/j.engstruct.2005.01.014
Mahmoud N, Ahmad E (2015) Investigating the effect of the cement paste and transition zone on strength development of concrete containing nanosilica and silica fume. Mater Des 75:174
Mastali M, Kinnunen P, Dalvand A et al (2018) Drying shrinkage in alkali-activated binders–a critical review. Constr Build Mater 190:533–550. https://doi.org/10.1016/j.conbuildmat.2018.09.125
Mazzotti C, Savoia M, Ferracuti B (2008) An experimental study on delamination of FRP plates bonded to concrete. Constr Build Mater 22:1409–1421. https://doi.org/10.1016/j.conbuildmat.2007.04.009
Mihashi H, Nomura N, Niiseki S (1991) Influence of aggregate size on fracture process zone of concrete detected with three dimensional acoustic emission technique. Cem Concr Res 21:737–744. https://doi.org/10.1016/0008-8846(91)90168-H
Mostofinejad D, Shameli SM (2013) Externally bonded reinforcement in grooves (EBRIG) technique to postpone debonding of FRP sheets in strengthened concrete beams. Constr Build Mater 38:751–758. https://doi.org/10.1016/j.conbuildmat.2012.09.030
Nakaba K, Kanakubo T, Furuta T, Yoshizawa H (2001) Bond behavior between fiber-reinforced polymer laminates and concrete. ACI Struct J 98:359–367. https://doi.org/10.14359/10224
Neto P, Alfaiate J, Dias-da-Costa D, Vinagre J (2016) Mixed-mode fracture and load misalignment on the assessment of FRP-concrete bond connections. Compos Struct 135:49–60. https://doi.org/10.1016/j.compstruct.2015.08.139
Neubauer U, Rostasy FS (1997) Design aspects of concrete structures strengthened with externally bonded CFRP-plates. In: Proceedings of the 7th international conference on structural faults and repair, Edinburgh July, pp 109–118
Neupane K (2016) Fly ash and GGBFS based powder-activated geopolymer binders: a viable sustainable alternative of Portland cement in concrete industry. Mech Mater 103:110–122. https://doi.org/10.1016/j.mechmat.2016.09.012
Nikoloutsopoulos N, Sotiropoulou A, Kakali G, Tsivilis S (2021) Physical and mechanical properties of fly ash based geopolymer concrete compared to conventional concrete. Buildings 11. https://doi.org/10.3390/buildings11050178
Ozbakkaloglu T, Xie T (2016) Geopolymer concrete-filled FRP tubes: Behavior of circular and square columns under axial compression. Composites, Part B 96: 215–230. https://doi.org/10.1016/j.compositesb.2016.04.013
Pan J, Leung CK (2007) Effect of concrete composition on FRP/concrete bond capacity. J Compos Constr 11:611–618. https://doi.org/10.1061/(asce)1090-0268(2007)11:6(611)
Part WK, Ramli M, Cheah CB (2017) An overview on the influence of various factors on the properties of geopolymer concrete derived from industrial byproducts. Handb Low Carbon Concr 263–334. https://doi.org/10.1016/B978-0-12-804524-4.00011-7
Peng H, Cui C, Cai CS et al (2019) Microstructure and microhardness property of the interface between a metakaolin/GGBFS-based geopolymer paste and granite aggregate. Constr Build Mater 221:263–273. https://doi.org/10.1016/j.conbuildmat.2019.06.090

Pellegrino C, Tinazzi D, Modena C (2008) Experimental study on bond behavior between concrete and FRP reinforcement. J Compos Constr 12(2):180–189.https://doi.org/10.1061/(asce)1090-0268.

Perera DS, Uchida O, Vance ER, Finnie KS (2007) Influence of curing schedule on the integrity of geopolymers. J Mater Sci 42:3099–3106. https://doi.org/10.1007/s10853-006-0533-6

Perry C, Gillott JE (1977) The influence of mortar-aggregate bond strength on the behaviour of concrete in uniaxial compression. Cem Concr Res 7:553–564. https://doi.org/10.1016/0008-8846(77)90117-X

Pham TM, Hao H (2016) Impact behavior of FRP-strengthened RC beams without stirrups. J Compos Constr 20:04016011. https://doi.org/10.1061/(asce)cc.1943-5614.0000671

Popovics S (1973) A numerical approach to the complete stress-strain curve of concrete. Cem Concr Res 3:583–599. https://doi.org/10.1016/0008-8846(73)90096-3

Rathinam K, Basanth B, Kanagarajan V (2016) Geopolymer concrete beam strengthening with glass fibre reinforced polymer. Gradjevinar 68:11–18

Sarker PK, Borak A., Mukherjee A., Rangan VB (2017) Bond strength of fly ash geopolymer with CFRP fabric after exposure to high temperature. In Proc., 4th Conference of Smart Monitoring, Assessment and Rehabilitation of Civil Structures, Switzerland: SMAR 2017

Sato Y, Vecchio FJ (2003) Tension stiffening and crack formation in reinforced concrete members with fiber-reinforced polymer sheets. J Struct Eng 129:717–724. https://doi.org/10.1061/(asce)0733-9445(2003)129:6(717)

Sato Y, Kimura K, Kobatake Y (1997) Bond behavior between CFRP sheet and concrete (part 1) J Struct Constr Eng 62 (500): 75–82.https://doi.org/10.3130/aijs.62.75_5

Savoia M, Ferracuti B, Vincenzi L (2009) Inverse analysis for the calibration of FRP—concrete interface law. Adv Struct Eng 12:613–625. https://doi.org/10.1260/136943309789867845

Shen D, Deng S, Zhang J et al (2015a) Behavior of reinforced concrete box beam with initial cracks repaired with basalt fiber-reinforced polymer sheet. J Reinf Plast Compos 34:1540–1554. https://doi.org/10.1177/0731684415595469

Shen D, Ji Y, Yin F, Zhang J (2015b) Dynamic bond stress-slip relationship between basalt FRP sheet and concrete under initial static loading. J Compos Constr 19:04015012. https://doi.org/10.1061/(asce)cc.1943-5614.0000568

Shen D, Shi H, Ji Y, Yin F (2015c) Strain rate effect on effective bond length of basalt FRP sheet bonded to concrete. Constr Build Mater 82:206–218. https://doi.org/10.1016/j.conbuildmat.2015.02.016

Shen D, Shi X, Ji Y, Yin F (2015d) Strain rate effect on bond stress-slip relationship between basalt fiber-reinforced polymer sheet and concrete. J Reinf Plast Compos 34:547–563. https://doi.org/10.1177/0731684415574539

Singh B, Ishwarya G, Gupta M, Bhattacharyya SK (2015) Geopolymer concrete: a review of some recent developments. Constr Build Mater 85:78–90. https://doi.org/10.1016/j.conbuildmat.2015.03.036

Subramaniam K, Carloni C, Nobile L (2011) An understanding of the width effect in FRP–concrete debonding. Strain 47(2): 127–137.https://doi.org/10.1111/j.1475-1305.2008.00481.x .

Sucharda O, Mateckova P, Bilek V (2020) Non-linear analysis of an RC beam without shear reinforcement with a sensitivity study of the material properties of concrete. Slovak J Civ Eng 28:33–43. https://doi.org/10.2478/sjce-2020-0005

Tang Z, Li W, Tam VWY, Yan L (2020) Mechanical behaviors of CFRP-confined sustainable geopolymeric recycled aggregate concrete under both static and cyclic compressions. Compos Struct 252:112750. https://doi.org/10.1016/j.compstruct.2020.112750.

Tasdemir C, Tasdemir MA, Lydon FD, Barr BIG (1996) Effects of silica fume and aggregate size on the brittleness of concrete. Cem Concr Res 26:63–68. https://doi.org/10.1016/0008-8846(95)00180-8

Tasong WA, Lynsdale CJ, Cripps JC (1999) Aggregate-cement paste interface: Part I. Influence of aggregate geochemistry. Cem Concr Res 29:1019–1025. https://doi.org/10.1016/S0008-8846(99)00086-1

Teng JG, Chen JF, Smith ST, Lam L (2003) Behaviour and strength of FRP-strengthened RC structures: a state-of-the-art review. Struct & Build 156:51–62. https://doi.org/10.1680/stbu.156.1.51.37880

Torres A, Hu J, Ramos A (2015) The effect of the cementitious paste thickness on the performance of pervious concrete. Constr Build Mater 95:850–859. https://doi.org/10.1016/j.conbuildmat.2015.07.187

van Mier JGM (1998) Failure of concrete under uniaxial compression: an overview. Fract Mech Concr Struct Fram 1169–1182

Wolinski S, Hordijk DA, Reinhardt HW, Cornelissen HAW (1987) Influence of aggregate size on fracture mechanics parameters of concrete. Int J Cem Compos Light Concr 9:95–103. https://doi.org/10.1016/0262-5075(87)90025-X

Wu Y-F, Zhou Y-W (2010) Unified strength model based on Hoek-Brown failure criterion for circular and square concrete columns confined by FRP. J Compos Constr 14:175–184. https://doi.org/10.1061/(asce)cc.1943-5614.0000062

Wu Z, Islam SM, Said H (2009) A three-parameter bond strength model for FRP-concrete interface. J Reinf Plast Compos 28:2309–2323. https://doi.org/10.1177/0731684408091961

Xiang J, Liu L, Cui X et al (2019) Effect of Fuller-fine sand on rheological, drying shrinkage, and microstructural properties of metakaolin-based geopolymer grouting materials. Cem Concr Compos 104:103381. https://doi.org/10.1016/j.cemconcomp.2019.103381

Xu J, Corr DJ, Shah SP (2015) Nanomechanical properties of C-S-H gel/cement grain interface by using nanoindentation and modulus mapping. J Zhejiang Univ Sci A 16:38–46. https://doi.org/10.1631/jzus.A1400166

Yang YX, Yue QR, Hu YC (2001) Experimental study on bond performance between carbon fiber sheets and concrete. Jianzhu Jiegou Xuebao/J Build Struct 22:36

Yang T, Zhu H, Zhang Z (2017) Influence of fly ash on the pore structure and shrinkage characteristics of metakaolin-based geopolymer pastes and mortars. Constr Build Mater 153:284–293. https://doi.org/10.1016/j.conbuildmat.2017.05.067

Yao J, Teng JG, Chen JF (2005) Experimental study on FRP-to-concrete bonded joints. Compos Part B Eng 36:99–113. https://doi.org/10.1016/j.compositesb.2004.06.001

Yu F, Sun D, Wang J, Hu M (2019) Influence of aggregate size on compressive strength of pervious concrete. Constr Build Mater 209:463–475. https://doi.org/10.1016/j.conbuildmat.2019.03.140

Yuan H, Teng JG, Seracino R et al (2004) Full-range behavior of FRP-to-concrete bonded joints. Eng Struct 26:553–565. https://doi.org/10.1016/j.engstruct.2003.11.006

Yuan C, Chen W, Pham TM et al (2019a) Effect of aggregate size on the dynamic interfacial bond behaviour between basalt fiber reinforced polymer sheets and concrete. Constr Build Mater 227. https://doi.org/10.1016/j.conbuildmat.2019.07.310

Yuan C, Chen W, Pham TM, Hao H (2019b) Effect of aggregate size on bond behaviour between basalt fibre reinforced polymer sheets and concrete. Compos Part B Eng 158:459–474. https://doi.org/10.1016/j.compositesb.2018.09.089

Zhang P, Wang K, Li Q et al (2020) Fabrication and engineering properties of concretes based on geopolymers/alkali-activated binders—a review. J Clean Prod 258. https://doi.org/10.1016/j.jclepro.2020.120896

Zhong R, Wille K (2016) Compression response of normal and high strength pervious concrete. Constr Build Mater 109:177–187. https://doi.org/10.1016/j.conbuildmat.2016.01.051

Zuhua Z, Xiao Y, Huajun Z, Yue C (2009) Role of water in the synthesis of calcined kaolin-based geopolymer. Appl Clay Sci 43:218–223. https://doi.org/10.1016/j.clay.2008.09.003

Chapter 9
Reinforcement Bonding of MKG Concrete

Abstract In this chapter, the bond behavior between MKG concrete and steel reinforcements is studied by pull-out experiments. The effects of bond size (bond length and cover thickness) and loading rate on the bond behavior are analyzed. The mechanism which behinds the good bonding performance of reinforcements to MKG concrete is discussed.

9.1 Introduction

In the design of reinforced concrete structure, the bonding behavior of reinforcements in concrete is one of the key problems. The bonding characteristics determine the stress transfer between the reinforcements and concrete. Thus, the necessary bearing capacity of the structure is relied on the good bonding formed on the interface (Xu 1990; Jin and Du 2012). A lot of studies have been conducted to elucidate the bonding between OPC concrete and steel reinforcements (Tepfers 1979; Jiang 1984; Xu et al. 1988a, b; Xu 1990; Zhao and Jin 2002; Jin and Du 2012; Yan and Chen 2012; Torre-Casanova et al. 2013). The constitutive model and calculation formula of the bond performance of reinforcements in OPC concrete have been established. Compared with that, the bond behavior of reinforcements in geopolymer concrete is relatively immature. Geopolymer concrete usually exhibits lower elastic modulus (Hung et al. 2013), higher bonding force (Latella et al. 2006; Ji and Song 2009) and more obvious brittleness (Ueng et al. 2012; Sarker et al. 2013) compared with OPC concrete. These differences in mechanical properties may lead to significant changes in the bonding behaviors (Dai et al. 2014). However, from the previous literatures, the studies of the bonding behavior of reinforcements in geopolymer concrete is relatively limited.

Fernandez-Jimenez et al. (2006) was the first to test the bonding performance of steel reinforcements in fly ash-based geopolymer (FAG) concrete. The research shows that FAG concrete has good bonding performance with steel reinforcement. The reinforcements are usually pull-out from OPC concrete in pull-out experiment. However, the reinforcements may fail in tensile when pull-out from geopolymer concrete under the same test conditions. Subsequently, Sofi et al. (2007) conducted

D. Yan et al., *Metakaolin-Based Geopolymers*,
https://doi.org/10.1007/978-981-97-0652-5_9

tests on the reinforcements bonding performance of more types of FAG concrete. The results showed that sudden splitting failure was the main failure form of FAG concrete in pull-out experiment. This indicates that geopolymer concrete is of significant brittleness. The bonding strength of reinforcement in geopolymer concrete is majorly controlled by the tensile strength of concrete. The study of Sarker (2011) also shows that splitting failure is the main failure mode of the pull-out specimen of geopolymer concrete. The bonding strength of reinforcement in geopolymer concrete is proportional to ratio of concrete cover thickness to the reinforcement diameter. Under the same concrete strength and test condition, the bonding strength of reinforcement in geopolymer concrete is greater than that in OPC concrete. Both Sarker and Fernandez-Jimenez believe that the higher bonding strength of reinforcement in geopolymer may be related to the higher splitting strength of geopolymer concrete. Castel and Foster (2015) also show that geopolymer concrete has a relatively higher bonding strength than OPC concrete. Moreover, the bonding strength is related to the curing condition. A certain degree of heat-curing could improve the bonding strength. On the other hand, the preliminary experiments have shown that the compressive and tensile strengths of geopolymer concrete are sensitive to the loading rate (Luo et al. 2012; Khandelwal et al. 2013;2015a, b). However, the comprehensive investigations on dynamic bonding behavior between geopolymer concrete and reinforcements is very limited.

The above studies show that the bonding strength between geopolymer concrete and reinforcements is significantly different from that of OPC concrete. The correctness of applying the design method and calculation formula of reinforcement bonding in OPC concrete in geopolymer concrete is still uncertain. So, it is urgent to carry out in-depth research to establish the design theory of geopolymer reinforced concrete structure. In this chapter, the pull-out experiments are conducted to evaluate the bonding behavior of reinforcements in MKG concrete. The effects of bond size (concrete cover thickness and bond length) and loading rate are analyzed. The bonding mechanism of steel reinforcements in geopolymer concrete is revealed. It would shed light on the rational design method of reinforced geopolymer concrete.

9.2 Bond Size Effect on Reinforcement Bonding in MKG Concrete

9.2.1 Experimental Method

9.2.1.1 Material and Preparation

The materials utilized in the preparation of geopolymer concrete are as follows:

(1) Metakaolin: It has an average particle size of 5.91 μm, with 90% of particles passing through a sieve size of 13.59 μm. The particle size distribution is illustrated in Fig. 9.1. The chemical composition of metakaolin was determined

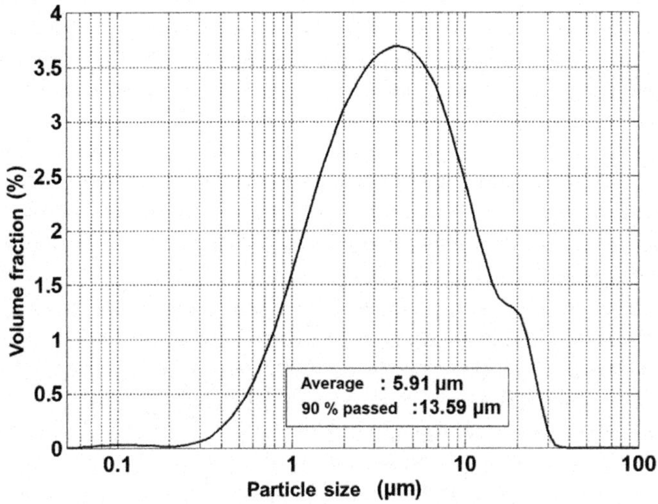

Fig. 9.1 Particle size distribution of metakaolin

using X-ray fluorescence spectroscopy (XRF), and the detailed composition is provided in Table 9.1.

(2) Sodium silicate solution: The solution has a SiO_2/Na_2O mole ratio (modulus of silicate) of 3.0. The content of SiO_2 is 16.4 L %, while the Na_2O content is 5.97%.

(3) Sodium hydroxide: It is in flake form and possesses a purity of 98%.

(4) Fine aggregate: Local river sand is used as the fine aggregate, with a fineness modulus of 1.96.

(5) Coarse aggregate: Local graded gravel is employed as the coarse aggregate, consisting of 33.7% particles ranging between 5 and 10 mm and 66.3% particles between 10 and 20 mm in size.

The proportion design of geopolymer concrete refers to the research work of literature (Zhang 2003; Hung et al. 2013). The geopolymer is prepared using metakaolin as aluminosilicate source and modified sodium silicate as an activator. In the preparation of geopolymer slurry, the content of Na_2O in the activator is 8.6%, the content of SiO is 16.8%, and the modulus *Ms* (the mole ratio of SiO to Na_2O) of the activator is about 2.0. After mixing the activator with metakaolin at the mass ratio of 2.08:1, the content of Na_2O in the slurry is about 17.8% of the mass of metakaolin, the ratio of silicon to Aluminum and the ratio of sodium to Aluminum are 5.0 and 1.0, respectively. The water content of the slurry is 45.6%. In the design of geopolymer concrete,

Table 9.1 Chemical composition of metakaolin

Chemical composition	SiO_2	Al_2O_3	TiO_2	Fe_2O_3	Na_2O	K_2O	CaO	MgO	LOI
Specific content (%)	59.72	32.6	1.13	3.97	0.07	0.14	0.17	0.28	1.83

Table 9.2 Composition of geopolymer concrete

Mixtures	MK	Activator	Coarse aggregate	Fine aggregate
Specific content (kg/m^3)	200	415	1200	600

the mass ratio of aggregate to geopolymer slurry is about 3.0, and the sand percentage of aggregate is about 33.3%. The raw material ratio of geopolymer concrete is shown in Table 9.2. The measured cubic compressive strength of geopolymer concrete at 28 d is 42 ± 3 MPa.

The preparation process of geopolymer concrete refers to the work of reference (Hardjito et al. 2005). Firstly, flake sodium hydroxide was added into sodium silicate according to the design ratio. After it was fully dissolved, it was sealed and placed in room temperature (20 + 5 °C) for 24 h. When mixing, the static activator was first added into the mixer, and then the metakaolin powder in accordance with the design ratio was added. After 5–10 min of mixing, the slurry reached a uniform state, and then added fine sand on the basis of the design ratio, and continued to stir for 5 min. After the mortar was evenly mixed, the graded gravel was added as required by the design and kept stirring for 10 min. The mixed concrete was loaded into the cylindrical test mold with fixed steel bar in twice. After each loading, it was placed on the shaking table to vibrate for 1 min to eliminate bubbles. When all the fillings were completed, the open end of the mold was closed with plastic film and rubber band, and placed in a curing condition of 20 ± 3 °C and humidity greater than 90%.

After 7 days, the specimen was removed and placed under the aforementioned curing conditions to continue curing. After 21 days, the specimen was taken out and the concrete loading surface of the specimen was leveled with fine mortar. After 2 more days, the reinforcement pulling test was carried out.

9.2.1.2 Specimen and Test

The bond strength between geopolymer concrete and steel bar was tested by the center pull out test of steel bar. Rebar center pull out test is the most common test method in the research of rebar bond performance because of its simple device and clear force (RILEM/CEB/FIP 1983; Du et al. 2006; Fernández-Jiménez et al. 2006; Sofi et al. 2007). The specimen used in this paper is shown in Fig. 9.2. The outer layer of geopolymer concrete is cylindrical, and the diameter and height are determined according to the thickness of concrete protective layer (c) and the bonding length of reinforcement (l_b) for the same value. See Table 9.3 for detailed size settings. The drawn steel bars are arranged through the axis of the cylindrical concrete specimen, and the steel bars are HRB400 deformed steel bars with diameters (d_s) of 12 mm.

To minimize the impact of end effects on the bonding interface, a specific length of an unbonded section was introduced at both the loading and free ends of the concrete specimen. This was achieved by isolating the steel bar and concrete within this designated section using plastic casing. The purpose of this arrangement was to

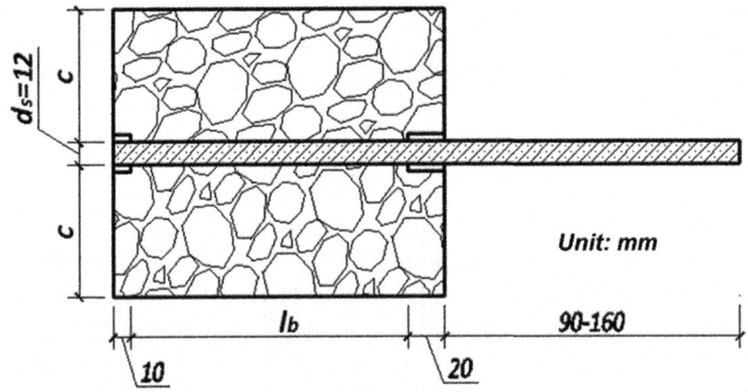

Fig. 9.2 Illustration of pull-out test specimen

Table 9.3 Dimension settings for pull-out specimens

Series	Cylinder diameter d_c (mm)	Column height h_c (mm)	Thickness of concrete protective layer c (mm)	Bonding length of reinforcement l_b (mm)	Size parameters	
					$\frac{c}{d_s}$	$\frac{l_b}{d_s}$
C1	45	180	16.5	150	1.38	12.5
C2	68	180	28	150	2.33	12.5
C3	105	180	46.5	150	3.88	12.5
C4	152	180	70	150	5.83	12.5
L1	105	54	46.5	24	3.88	2
L2	105	90	46.5	60	3.88	5
L3	105	114	46.5	84	3.88	7
L4	105	174	46.5	144	3.88	12

Note d_s is the diameter of reinforcement bar, which is 12 mm in this test

ensure that any potential influence arising from stress concentration or discontinuities near the specimen ends did not affect the behavior and performance of the bonded region.

Studies have shown that the thickness of concrete protective layer and the bonding length of reinforcement are two key factors affecting the bonding performance of reinforcement in concrete (Tepfers 1979; Xu et al. 1988a, b, 1994). Based on this, this paper designed two groups of tests: group C kept the bonding length of reinforcement unchanged, and the thickness of concrete protective layer changed from 1.38 times the diameter of reinforcement (16.5 mm) to 5.83 times the diameter of reinforcement (70 mm); In group L, the thickness of concrete protective layer was kept unchanged, and the bonding length of reinforcement was changed from 2 times the diameter of reinforcement (24 mm) to 12 times the diameter of reinforcement (144 mm). See Table 9.3 for detailed specimen size Settings.

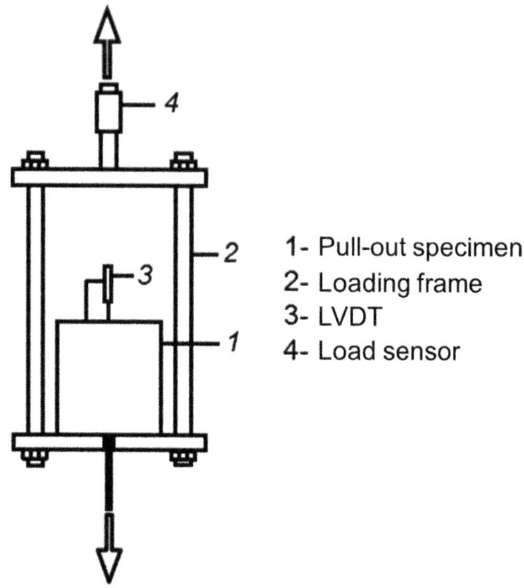

Fig. 9.3 Loading and measuring device

The reinforcement pull-out test was conducted using a loading device, as illustrated in Fig. 9.3. The test was performed on a 250 kN hydraulic servo test machine. The test specimen was connected to the testing machine's beam through a steel frame, allowing the pulling load to be applied to the extended section of the steel bar via the actuating chuck at the lower end. To monitor the relative slip between the steel bar and the concrete, an electronic displacement sensor (LVDT) was positioned at the free end of the pulled specimen. Concurrently, the load sensor fixed on the testing machine's beam and connected to the steel frame monitored the pulling force exerted on the specimen. Data from these sensors were continuously and simultaneously collected by an integrated measurement system. The loading rate was controlled based on displacement and could be categorized into static and dynamic loading modes. For the static pull-out test, the loading rate was set at 0.4 mm/min, equivalent to a strain rate of approximately 10^{-6}/s (quasi-static loading). In contrast, the dynamic pull-out test had a loading rate of 400 mm/min, corresponding to a strain rate of approximately 10^{-4}/s (seismic action) (Bischoff and Perry 1991).

9.2.2 Results and Discussion

There are usually three failure modes of ordinary concrete center pull-out specimens (Azizinamini et al. 1995; Reinhardt and Balázs 1995):

(1) The concrete splits along the direction parallel to the reinforcement;
(2) The concrete on the surface of the steel bar is shear damaged and the steel bar is pulled out;
(3) When the reinforcement reaches the ultimate tensile strength, it breaks and is not pulled out from the concrete specimen.

In the existing geopolymer concrete center pull out test (Azizinamini et al. 1995; Fernández-Jiménez et al. 2006; Sofi et al. 2007), the failure mode of concrete shear failure and steel bar pull out is not common, and most of the specimens have concrete split failure or steel bar can not pull out, which is consistent with the test results in this paper. There is no pull-out failure mode of reinforcement in geopolymer concrete either in static or dynamic specimens. This shows that geopolymer concrete has strong shear resistance and its anchoring performance to deformed steel bar is mainly controlled by the splitting resistance of concrete. When the thickness of concrete protective layer is less than 3 times the diameter of steel bar or the bonding length of steel bar is less than 7 times the diameter of steel bar, the geopolymer concrete layer cracks along the axial direction of the specimen. However, when the thickness of concrete protective layer is greater than 3 times the diameter of steel bar and the bonding length of steel bar is greater than 7 times the diameter of steel bar, the steel bar of pulled specimen reaches the ultimate tensile strength and is pulled apart outside the bonding zone. Therefore, ensuring sufficient thickness of concrete protective layer and reinforcing bar bonding length is the key to exerting excellent anchoring performance of geopolymer concrete, and is also the key point in anchoring design of reinforced one-ground polymer concrete.

Figure 9.4 illustrates the failure zones at the interface under two different conditions. Examination of Fig. 9.4 reveals that near the loading end, concrete deposition in the shape of debris is predominantly observed on the surface of the reinforcement, with transverse cracks forming at the rib tips. The rib marks on the concrete surface become indistinct after the peeling of the reinforcement. Moving further away from the loading end, the accumulation of concrete debris and transverse cracks decreases significantly, and more concrete debris adheres to the reinforcement surface. Moreover, the rib marks on the concrete surface become clearer following reinforcement removal. This observation indicates that stress distribution at the interface of reinforced polymer concrete remains uneven until concrete cracking occurs, with higher bond stress near the loading end. In this region, concrete primarily provides the main bonding force for the reinforcement, as well as being the site where transverse and splitting cracks initially develop. Consequently, particular attention should be given to reinforcing the concrete protective layer at the stress concentration areas when designing reinforced polymer concrete structures.

Another noteworthy observation is the presence of concrete residue commonly adhering to the reinforcement surface, indicating a significant chemical bonding force and friction between the geopolymer concrete and the reinforcement. This finding is consistent with the conclusions drawn by Latella et al. (2006).

Figure 9.5 shows the failure mode of the pulled specimen. As can be seen from Fig. 9.5, from the perspective of bonding strength, the bonding strength of dynamic

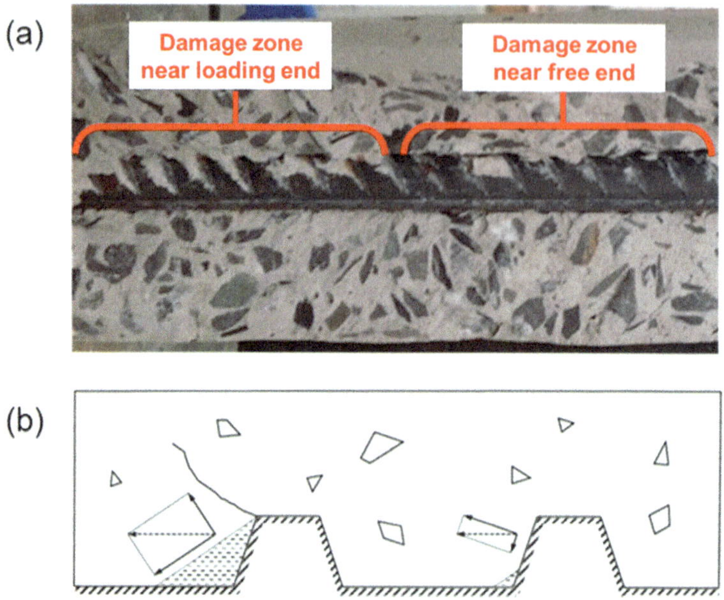

Fig. 9.4 The state of interfacial failure zones: **a** splitting section; **b** stress state of failure zones

pull-out specimen is significantly higher than that of static pull-out specimen. This phenomenon reflects that geopolymer concrete has the same loading rate sensitivity as ordinary concrete although it is significantly different from ordinary concrete in microstructure (Li and Xu 2009; Xu et al. 2009; Yan et al. 2014). At the same time, the loading rate also affects the failure mode of specimens. Most dynamic specimens have three main splitting cracks, while most static specimens have two main splitting cracks. The increase of the number of splitting cracks increases the energy consumption and the bonding strength of geopolymer concrete. This law is similar to the rate dependence of reinforcing bar bonding properties in ordinary concrete (Yan and Chen 2012).

The transfer of force between steel bars and concrete relies on chemical bonding, friction, and mechanical anchoring force. In the case of deformed steel bars, the primary mechanism for bearing the tensile force is the mechanical anchoring force generated through mutual extrusion between the rib front and concrete, as depicted in Fig. 9.6a. Due to the perpendicular nature of the extrusion pressure on the action surface, both radial stress and circumferential stress are inevitably induced in the concrete protective layer. The latter is the main contributor to the splitting failure of the concrete protective layer. While the chemical bond force and friction force between the steel bar and reinforced concrete play a limited role in carrying the tensile force of the steel bar, they provide an additional path for stress transfer along the tangential contact surface. This direct stress transfer pathway helps reduce the radial and circumferential stresses within the concrete cover, as illustrated in Fig. 9.6b.

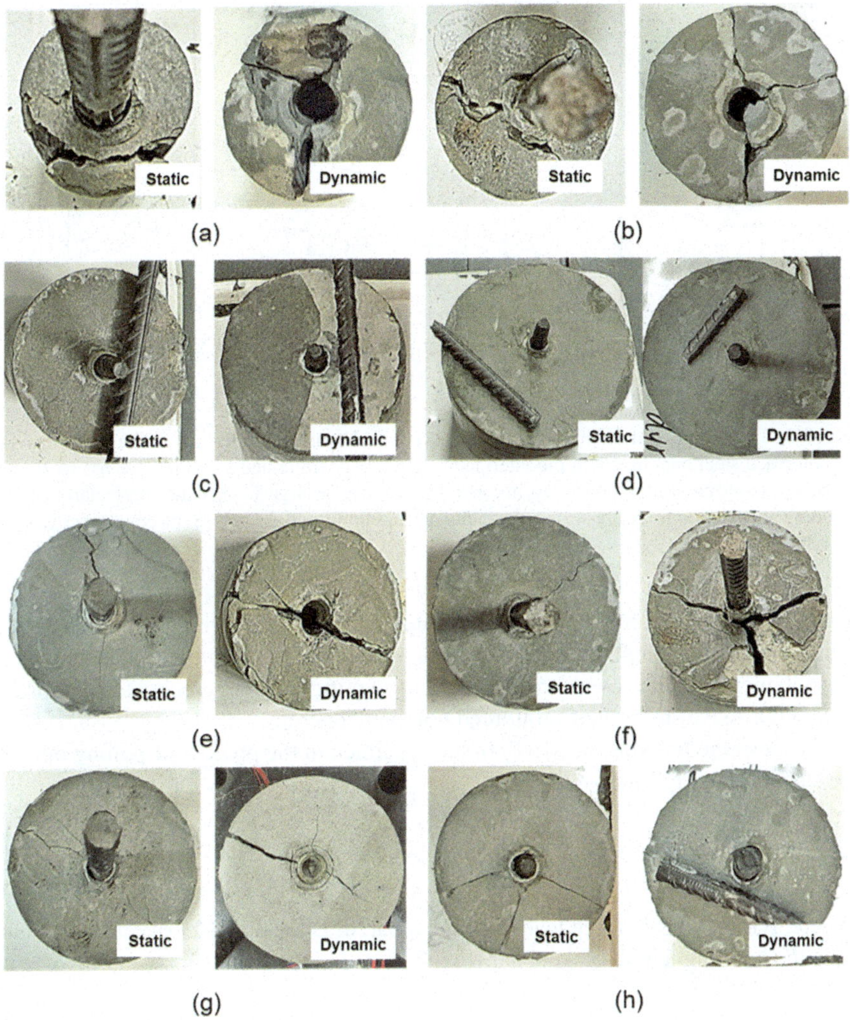

Fig. 9.5 Failure modes of pulled-out specimens: **a** C1 group; **b** C2 group; **c** C3 group; **d** C4 group; **e** L1 group; **f** L2 group; **g** L3 group; **h** L4 group

Under constant tensile force conditions, Fig. 9.6a highlights that smaller chemical bonding and friction forces lead to higher radial and circumferential stresses within the concrete protective layer. This results in earlier cracking and reduced bonding strength between the reinforcement and the concrete cover. Conversely, Fig. 9.6b demonstrates that larger chemical bonding and friction forces can mitigate the radial and circumferential stresses in the concrete protective layer, reducing cracking and enhancing the bonding strength of the steel bar. Previous studies have shown that polymer concrete exhibits higher chemical bonding force and friction force with steel

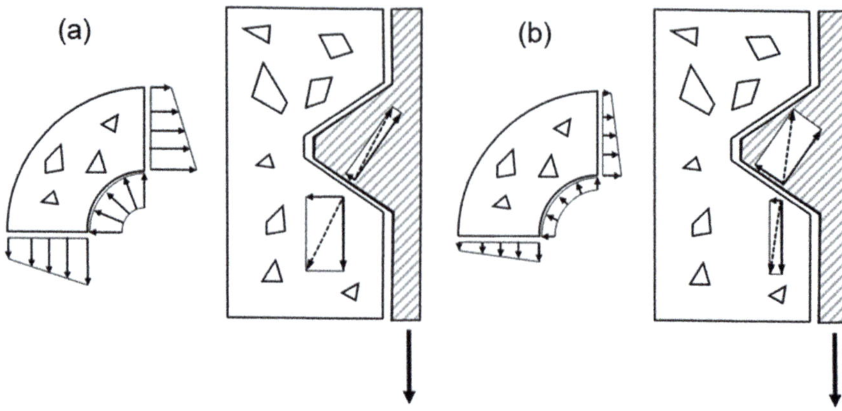

Fig. 9.6 The influence mechanism of interfacial chemical bonding and friction on the bond behavior of deformed steel bars. **a** When the interfacial chemical bonding and friction force are weak, the radial stress and circumferential stress in concrete cover will be large; **b** when the interfacial chemical bonding and friction force are strong, the radial and circumferential stress in the concrete cover will be reduced

when compared to ordinary concrete (Latella et al. 2006; Ueng et al. 2012). As a result, under equivalent loading conditions, the radial and circumferential stresses in the polymer concrete cover decrease, helping to mitigate cover cracking and improve the compressive strength of reinforced polymer concrete.

The average bonding stress ($\overline{\tau}$) of the specimen in the process of pulling out can be calculated as follows:

$$\overline{\tau} = \frac{F}{\pi \cdot d_s \cdot l_b} \tag{9.1}$$

where, F is the tensile force loaded on the rebar; d_s is the diameter of reinforcement; l_b is the bonding length of steel bar (Fig. 9.7).

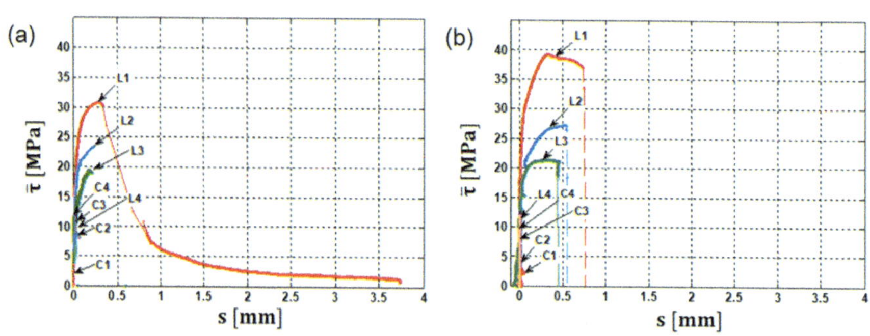

Fig. 9.7 $\overline{\tau}$ – **s** curves: **a** static loading; **b** dynamic loading

In specimens with a bond length less than 7 times the diameter of the steel bar, the average bond stress-slip curves differ between dynamic and static loading. Under dynamic loading, the slippage at the peak of average bonding stress and upon sudden unloading is significantly higher compared to static loading. The average bond stress-slip curve under dynamic loading shows a plateau before sudden unloading. However, when the bonding length reaches or exceeds 12 times the diameter of the steel bar, the average bond stress-slip curves for both static and dynamic loading become similar. Regardless of whether the specimen fails due to tensile fracture of the steel bar or splitting of the concrete, the average bond stress-slip curve appears as an almost vertical straight line. The only difference lies in the height of the average bond stress peak (bond strength) among the curves.

Figure 9.8 illustrates the impact of concrete protective layer thickness on the average bonding strength between reinforced concrete and geopolymer concrete. As depicted in Fig. 9.8, there is a significant increase in the average bonding strength of the steel bar and geopolymer concrete as the protective layer thickness increases. When the thickness of the protective layer reaches 3.9 times the diameter of the steel bar, the steel bar cannot be pulled out of the concrete specimen due to its ultimate tensile strength. The effect of loading rate was not evident in the test group with different protective layer thicknesses. This is primarily because the bonding length of reinforcement in the group with varying protective layer thickness exceeded 12 times the diameter of the reinforcement. As the bonding length increases, the influence of loading rate on bonding strength gradually diminishes.

The relationship between the bonding strength of steel bar and geopolymer concrete and the thickness of protective layer can be estimated by the following formula:

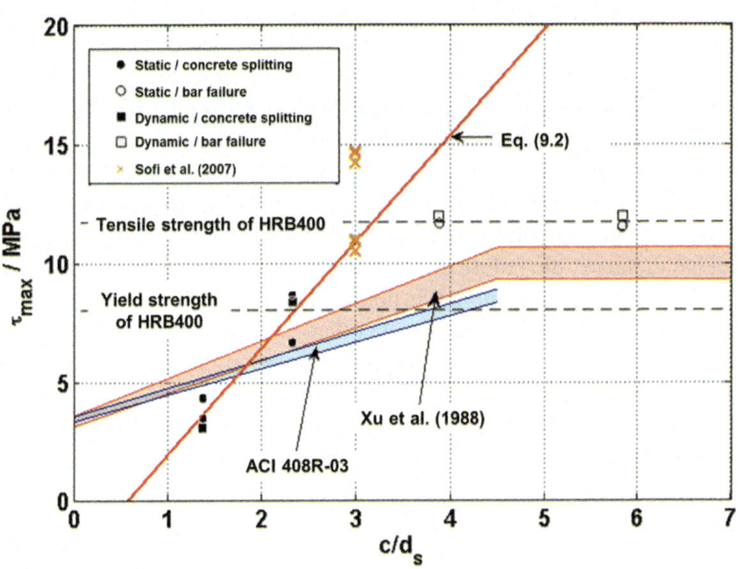

Fig. 9.8 The relationship between the thickness of concrete cover and bond strength

$$\overline{\tau}_{\max} = 4.467\left(\frac{c}{d_s}\right) - 2.537 \quad (9.2)$$

where: $\overline{\tau}_{\max}$ is the bonding strength of reinforcement in geopolymer concrete, MPa; c is the protective layer thickness of geopolymer concrete, mm; d_s is the diameter of steel bar, mm.

According to the empirical formula, the recommended thickness of the geopolymer concrete protective layer is approximately 2.4 times the diameter of the reinforcement when the reinforcement reaches its yield strength. When the reinforcement reaches its ultimate tensile strength, the suggested thickness of the geopolymer concrete protective layer increases to about 3.2 times the diameter of the reinforcement.

The dark and light areas in Fig. 9.8 show the bonding strength range of deformed steel bars in ordinary concrete. The dark area is the empirical formula given by Xu et al. when the tensile strength of concrete is 2.2–2.5 MPa (Xu et al. 1988a, b), and the light area is the empirical formula given by ACI Standards Committee when the compressive strength of concrete is 35–45 MPa (ACI Committee 408 R-03 2003). When the thickness of the protective layer is larger than 4.5 times of steel bar, the failure mode of ordinary concrete will change from the splitting failure of the protective layer to the pulling out failure of the reinforcement. At this time, the thickness of the protective layer of concrete has no influence on the bonding strength of the reinforcement.

When the thickness of the concrete protective layer reaches 12.5 times the diameter of the steel bar, empirical formulas suggest that the bond strength of ordinary concrete falls within the range of 8.3–10.6 MPa. However, in contrast, the geopolymer concrete specimens tested in this study did not experience pull-out failure and exhibited a bonding strength of 11.7 MPa or higher. This higher bonding strength of the steel bar in geopolymer concrete may be attributed to the increased chemical bonding force and friction between the geopolymer concrete and the steel bar.

Figure 9.9 illustrates the impact of bonding length on the bonding strength of reinforced geopolymer concrete. As depicted in Fig. 9.9, both under static loading (thick dashed line) and dynamic loading (thick solid line), cracking of the specimen occurs when the bonding length is equal to or less than 7 times the steel bar diameter (indicated by solid data points). However, when the bonding length exceeds 7 times the steel bar diameter, the failure mode shifts from concrete splitting to steel pulling (indicated by hollow data points). It is noteworthy that the bond strength of the reinforcement under dynamic loading is higher than that under static loading for varying bond lengths. Nonetheless, the increase in bond strength gradually diminishes with an increase in bond length.

The bond strength of the specimen with different bond lengths can be estimated by the following formula when the specimen has split failure:

$$\overline{\tau}_{\max} = \frac{246.9}{\frac{l_b}{d_s} + 4.278},\ \dot{\varepsilon} = 10^{-6}/\mathrm{s}, \quad (9.3a)$$

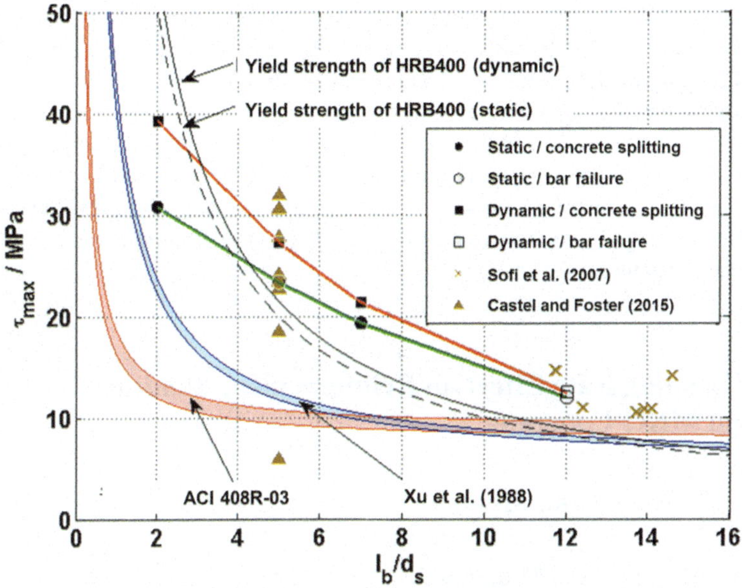

Fig. 9.9 The relationship between bond length and bond strength

$$\overline{\tau}_{\max} = \frac{270.5}{\frac{l_b}{d_s} + 6.766}, \dot{\varepsilon} = 10^{-3}/\text{s}, \tag{9.3b}$$

where, $\overline{\tau}_{\max}$ is the bond strength of reinforcement in geopolymer concrete, in MPa; Is the bonding length of reinforcement, l_b is the diameter of reinforcement, both in mm; The strain rates are 10^{-6}/s (static loading) and 10^{-3}/s (dynamic loading), respectively.

When the steel bar reaches the yield strength, the average bond stress T in the pulled specimen and the yield strength and bond length of the steel bar should satisfy the following equation:

$$\overline{\tau}_y = \frac{f_y A_s}{\pi d_s l_b} = \frac{f_y d_s}{4 l_b} \tag{9.4}$$

where, f_y is the yield strength of reinforcement; A_s is the cross-sectional area of reinforcement; d_s is the diameter of reinforcement; l_b is the bonding length of reinforcement.

Based on Eq. (9.4), the thin dotted black line and the thin solid black line in Fig. 9.9 represent the average bonding stress in the pulled specimen when the yield strength of HRB400 steel bar is reached under static and dynamic loading, respectively. By utilizing Eqs. (9.3a, 9.3b) and (9.4), it is possible to determine the critical bonding length of HRB400 reinforcement in geopolymer concrete under central bonding conditions. The critical bonding length is approximately 3.8 times the diameter of the steel bar under static loading, and 3.1 times the diameter of the steel bar under dynamic loading. In contrast, the dark and light colored areas in the figure depict the estimated range of bonding strengths for steel bars in ordinary concrete with grades ranging from C35 to C45. The critical bonding length of HRB400 steel bars in ordinary concrete falls within the range of 11–14 times the diameter of the steel bars. Based on the findings of this study, it can be concluded that the critical bonding length of the steel bar in geopolymer concrete is significantly smaller compared to that in ordinary concrete. Consequently, it is deemed safe to design the bond of steel bars in geopolymer concrete using the design methods employed for ordinary reinforced concrete.

9.3 Loading Rate Effect on Reinforcement Bonding in MKG Concrete

9.3.1 Experimental Method

9.3.1.1 Experimental Materials

Metakaolin powder was used as the aluminosilicate raw material for making geopolymer. The chemical composition of the powder was tested by X-ray fluorescence (XRF) method. The main results are shown in Table 9.4. The silica and aluminum oxides accounted for about 97% of the total mass. The microscopic morphology of the metakaolin powder can be observed by electron microscopy as shown in Fig. 9.10, and its average particle size is about 6 μm.

Activator—The activator used to activate the metakaolin was a mixture of liquid sodium silicate and sodium hydroxide. The liquid sodium silicate contained 17.39 wt% SiO_2 and 5.97 wt% Na_2O. The sodium hydroxide was 98% of purity. In order to keep the molar ratio of SiO_2/Na_2O at about 1.5, the sodium hydroxide needed to be dissolved in the liquid sodium silicate. Previous studies have shown that this enables the preparation of geopolymers with better properties (Yan et al. 2016a, b). The prepared solution needs to be stored at room temperature (20 ± 3 °C) for 24 h to stabilize its chemicals.

Coarse and fine aggregates—Crushed stone with specific gravity of 2.65 was used as coarse aggregate and river sand with specific gravity of 2.62 was used as fine aggregate. The water absorption of crushed stone was 0.8%. The fineness modulus of river sand is 1.96. The cumulative sieving curves of both are shown in Fig. 9.11.

Table 9.4 Chemical composition of metakaolin powder

Component	SiO_2	Al_2O_3	TiO_2	Fe_2O_3	Na_2O	K_2O	CaO	LOI
Mass content, %	57.26	39.68	1.78	0.43	0.27	0.21	0.04	0.33

Reproduced from [Rate-Dependent Bonding of Steel Reinforcement in Geopolymer Concrete] by [Dongming Yan] with permission from [ACI Materials Journal]

Fig. 9.10 Typical micromorphology of as-received metakaolin powder. Reproduced from [Rate-Dependent Bonding of Steel Reinforcement in Geopolymer Concrete] by [Dongming Yan] with permission from [ACI Materials Journal]

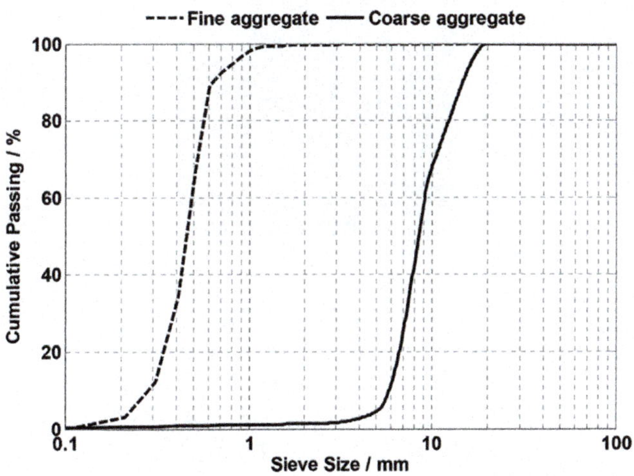

Fig. 9.11 Cumulative sieving curves of coarse and fine aggregates. Reproduced from [Rate-Dependent Bonding of Steel Reinforcement in Geopolymer Concrete] by [Dongming Yan] with permission from [ACI Materials Journal]

Steel bars—Both plain and deformed bars were utilized in accordance with the National Codes of China GB1499.1-2008 and GB1499.2-2007. The nominal diameter (d_s) of both types of bars is 12 mm. The deformed bar has a lateral rib height (h_{r1}) of 1.2 mm and a rib spacing (s_{r0}) of 8 mm. A comprehensive depiction of the bar geometries can be found in Table 9.5 and Fig. 9.12. The nominal yield strengths for the plain and deformed steel bars are 300 MPa and 400 MPa, respectively (Table 9.6).

Table 9.5 Geometry dimensions of reinforcements

Geometry parameter	Plain rebar	Deformed rebar
Nominal diameter d_s, mm	12	12
Core diameter d_i, mm	12	11.6
Lateral rib height h_{r1}, mm	–	1.2
Axial rib height h_{r2}, mm	–	1.6
Lateral rib spacing s_{r0}, mm	–	8
Lateral rib front angle γ, degrees	–	45
Lateral rib inclined angle θ, degrees	–	55

(a)

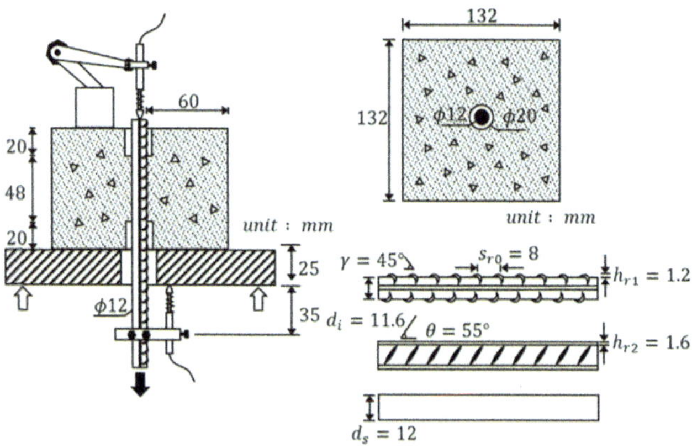

(b)

Fig. 9.12 **a** Pullout experiment; and **b** test arrangement and specimen dimensions. Reproduced from [Rate-Dependent Bonding of Steel Reinforcement in Geopolymer Concrete] by [Dongming Yan] with permission from [ACI Materials Journal]

Table 9.6 Mixture proportion of geopolymer concrete

Component	Metakaolin powder	Alkaline activator	Coarse aggregate	Fine aggregate
Mass content, kg/m^3	183	417	1200	600

Reproduced from [Rate-Dependent Bonding of Steel Reinforcement in Geopolymer Concrete] by [Dongming Yan] with permission from [ACI Materials Journal]

9.3.1.2 Mixture Proportion and Synthesis Program

In order to achieve the desired workability and mechanical strength, the mixture of MKG was made with 2.0 Si/Al ratio and 1.0 Na/Al ratio. This specific composition was determined in previous research conducted by Yan et al. (2016a, b). The mixing ratios for the MKG concrete are shown in Table 9.6. The metakaolin was first mixed with the activator in a concrete mixer for about 5 min. Subsequently, fine aggregate was added to the mixed slurry and continued mixing for 5 min. Finally, coarse aggregate was added to the mixed mortar and continued mixing for about 10 min.

The prepared MKG concrete was poured into molds. The steel bars were inserted in the center of the molds. Additionally, cubic molds were used for making standard specimens for strength test according to the code GB/T50081-2010. The casting process involved two steps, with each step filling 1/2 mold and shaking the concrete for 2 min. Once the molds were full of concrete, they were sealed in plastic films and placed in curing room maintained at 20 ± 1 °C and 90% RH. After 7 days of initial curing, the specimens were removed from the molds. After curing for 49 days, the specimens were tested. This elongated curing process reduce uncertainties in their strengths (Hardjito et al. 2004).

9.3.1.3 Specimen Preparation and Test Methods

The specimen with short embedment length ($l_b/d_s < 5$) was used. It is illustrated in Fig. 9.12. The concrete cover was of 60 mm thickness ($c/d_s = 5$). The steel bars were in 48 mm bond length ($l_b/d_s = 4$). Two 20 mm debond regions were set at each end of the specimen to eliminate the end effect.

An electro-hydraulic servo testing machine was used for the pull-out test. As shown in Fig. 9.12, the specimen is restrained against the upper beam of the testing machine through a steel frame. The pullout load is applied to the rebar through the lower loading head. The loading head speed was controlled from 0.4 to 400 mm/min (corresponding to bond stress rates on the order of 0.1–100 MPa/s). This loading rate simulated working conditions ranging from quasi-static to seismic action (Bischoff and Perry 1991). The test loads were measured by load cells attached to the upper beam of the testing machine. A linear variable differential transformer (LVDT) was added at each end of the specimen for measuring the bond slip of the reinforcement with respect to the concrete. Three specimens were tested for each loading rate experimental group. The specimens were labeled according to the following rules: the plain bar specimens were labeled as PB, and the deformed bar specimens were labeled as DB; the four pullout loading rates were labeled as V1, V2, V3, and V4, which corresponded to 0.4, 4, 40, and 400 mm/min. The dynamic compressive and split tensile strengths of MKG concrete were tested according to the standard concrete strength test method (GB/T50081-2010). The dimensions of the cubic specimens were 100 × 100 × 100 mm. The loading rates for the strength tests were controlled to range from 0.2 to 200 mm/min (for the compression test, the corresponding stress rates in the concrete ranged from 0.1 to 100 MPa/s; and for the splitting test, they ranged from 0.01 to 10 MPa/s). Four specimens were tested for each loading rate test set. The specimens in the compression test were labeled C and the samples in the splitting test were labeled T. In the strength tests, the four loading rates were labeled V1, V2, V3, and V4, corresponding to 0.2, 2, 20, and 200 mm/min.

9.3.2 Results and Discussion

9.3.2.1 Stress-Slip Relation

Assuming that the bond stress is uniformly distributed, the nominal bond stress ($\overline{\tau}$) can be calculated using the Eq. 9.5 according to Reinhardt and Balázs (1995):

$$\overline{\tau} = \frac{F}{\pi d_s l_b} \tag{9.5}$$

where F is the loading force; d_s is the bar diameter, which is equal to 12 mm; and l_b is the bond length, which is equal to 48 mm.

The free end slip (s_f) and the loading end slip (s_l) were measured from the LVDTs set at both ends of the specimen (shown in Fig. 9.12). Considering the deformation of the reinforcement between the LVDT at the loading end and the front edge of the bond zone, it is necessary to subtract the length change of this section of reinforcement, and the final nominal slip ($\overline{s}$) is calculated as follows:

$$\overline{s} = \frac{1}{2}(s_f + s_l - \frac{4Fl_a}{\pi d_s^2 E_s}) \tag{9.6}$$

where l_a is the length between the LVDT and the bond region, which is equal to 80 mm; and E_s is the Young's modulus of the steel bar, which is equal to 206 GPa.

PB specimens usually undergo rebar pullout damage, while DB specimens usually fail by cracking of the concrete cover. The bond stress-slip curves for the PB specimens and DB specimens are shown in Figs. 9.13 and 9.14, respectively. Figure 9.15 shows typical bond stress-slip curves for PB (PB-V2-3) and DB (DB-V2-1) specimens. There is a clear difference between the bond stress-slip curves of the two rebar types. Different bonding mechanisms and failure modes are the main reasons for the differences. However, in general, the bond stress-slip curves of both PB and DB specimens can be expressed as a three-stage evolutionary pattern, as shown in Fig. 9.15.

Initial linear stage (OA): In this stage, except for a short nonlinear segment (slip less than 0.05 mm), the bond stress increases linearly with slip. The linear relationship between bonding stress and slip indicates that the bonding interface is elastic under small slip. The nonlinear segment is mainly caused by plastic deformation of defects present at the interface, such as the opening of microcracks or free sliding of debonding areas (Tassios and Vintzeleou 1987; Gambarova and Rosati 1996). The bonding interface defects between steel bars and geopolymer concrete may be caused by the shrinkage of geopolymer concrete during the curing process.

Pre-peak non-linear stage (AB): When the bond stress approaches its peak (bond strength), the stress–strain slip behavior exhibits strong nonlinearity. As shown in Figs. 9.13 and 9.14, the transition points (defined as linear limits) from linear stress–strain relationships to nonlinear stress–strain relationships for PB and DB specimens typically occur at 65% to 97% and 70% to 100% of peak strength, respectively. The appearance of transition points indicates that damage and deformation begin to occur at the steel concrete interface (Mains 1951).

Post-peak stage (BC): After the peak bonding stress is reached, the bonding stress rapidly decreases with the increase of slip. For PB specimens, when the loading speed reaches 400 mm/min, the stress–strain slip curve shows severe oscillation, as shown in Fig. 9.13d. For DB specimens, the concrete cover of geopolymer concrete cracks, leading to a sudden decrease in stress. When the loading rate reaches 400 mm/min,

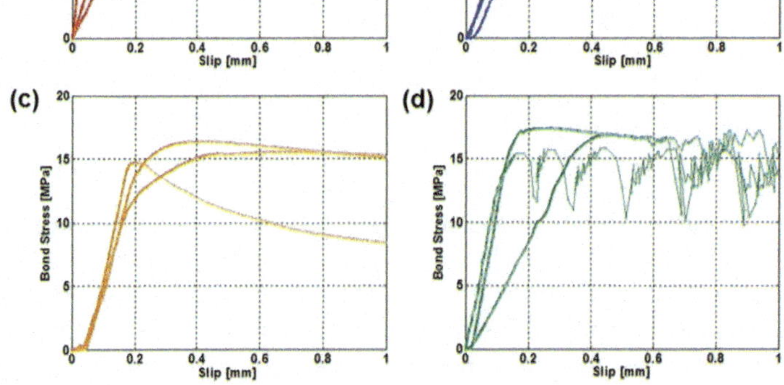

Fig. 9.13 Bond stress-slip curves of PB specimens: **a** PB-V1; **b** PB-V2; **c** PB-V3; and **d** PB-V4. Reproduced from [Rate-Dependent Bonding of Steel Reinforcement in Geopolymer Concrete] by [Dongming Yan] with permission from [ACI Materials Journal]

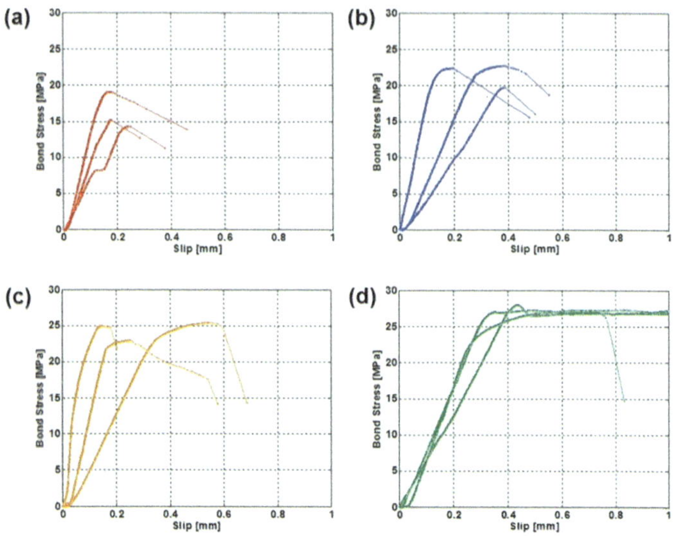

Fig. 9.14 Bond stress-slip curves of DB specimens: **a** DB-V1; **b** DB-V2; **c** DB-V3; and **d** DB-V4. Reproduced from [Rate-Dependent Bonding of Steel Reinforcement in Geopolymer Concrete] by [Dongming Yan] with permission from [ACI Materials Journal]

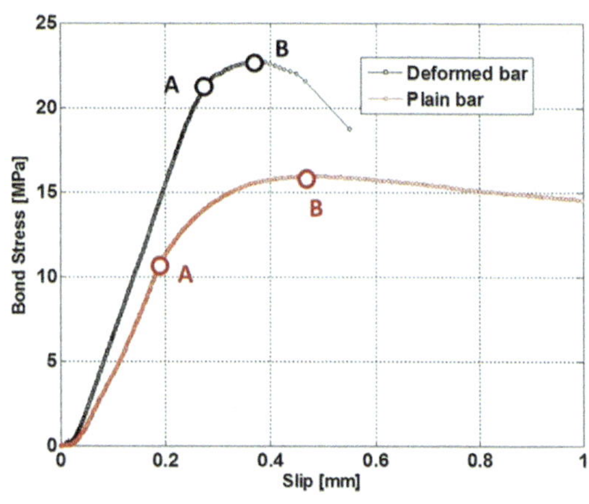

Fig. 9.15 Typical bond stress-slip curves of PB (red) and DB (black) specimens. Reproduced from [Rate-Dependent Bonding of Steel Reinforcement in Geopolymer Concrete] by [Dongming Yan] with permission from [ACI Materials Journal]

the stress-slip curve exhibits a bond stress plateau (Fig. 9.14d), mainly due to plastic deformation of the steel bars reaching their yield strength. This result is very different from the behavior of steel bars in OPC concrete, where the yield of steel bars typically requires a relatively long bond length (approximately $10d_s$ to $14d_s$) (Xu et al. 1988a, b). This also indicates that the bond between geopolymer concrete and deformed steel bars is stronger than that between OPC concrete and steel bars.

9.3.2.2 Bonding Behavior of Plain Bar

The bonding behavior of PB specimen was investigated by fitting the results by power function. The dynamic increasing factor (DIF), as proposed by Malvar (1998), was employed to quantify the observed rate dependency. Furthermore, the enhanced interfacial adhesion between MKG and steel was examined by analyzing the failure surfaces, establishing a connection between the rate behavior of the bond strength.

The results of the pull-out test are shown in Table 9.7. Calculate the loading rate by taking the time average of the linear segment corresponding to the stress slip curve. Previous studies have shown that the effect of loading rate on the bond strength of plain steel bars in OPC concrete can be ignored (Vos and Reinhardt 1982; Li et al. 2015). However, for MKG concrete, the linear limit and bond strength of plain steel bars significantly increase with the increase of loading rate, as shown in Fig. 9.13 and Table 9.7. Using power functions (Vos and Reinhardt 1982) to fit experimental data can obtain a rate dependent equation for bond strength.

$$\overline{f}_b = \overline{f}_{b,ref}\left(\frac{\dot{\sigma}}{\dot{\sigma}_{ref}}\right)^{\eta_b} \tag{9.7}$$

where, $\dot{\sigma}$ denotes the loading rate and $\overline{f}_b$ is the corresponding bond strength. The reference loading rate $\dot{\sigma}_{ref}$ was 0.1 MPa/s according to the Comité Euro-International Du Béton (CEB) recommendation (Comite Euro-International Du Beton (CEB) 1988). The reference bond strength $\overline{f}_{b,ref}$ and the power index η_b were calculated by fitting. Here, $\overline{f}_{b,ref}$ is 11.67 MPa, and η_b is 0.0274.

To quantify and compare the rate dependence of bond strength, DIF was obtained by normalizing dynamic bond strength $\overline{f}_b$ with static bond strength $\overline{f}_{b,ref}$ at different loading rates

$$\mathrm{DIF} = \frac{\overline{f_b}}{\overline{f}_{b,ref}} \tag{9.8}$$

The DIFs for plain bars in OPC concrete in literatures (Li et al. 2015) were plotted for comparison. In Fig. 9.16a, the DIF of bond strength of PB specimens increase by approximately 16% when the loading rate increases. However, the DIF in OPC concrete is almost a constant.

Upon inspection of the damaged section, it can be seen that there are many concrete powders adhered to the surface of the steel bars (as shown in Fig. 9.17). Collecting and weighing these powders reveals an increase in the amount of powder adhering to the steel bars as the loading rate increases. When the loading rate increases from 0.1 to 100 MPa/s, the mass of the adhesive powder increases from approximately 0.037 to 0.115 g. At a loading rate of 400 mm/min, the thin layer between plain steel bars and concrete continues to fail and slide under dynamic tensile forces, leading to the occurrence of "stick slip" behavior (Berman et al. 1996; Tambe and Bhushan 2005). This is also the reason for the oscillation of the bond stress of the plain round steel

Table 9.7 Test results of pullout experiments

Bar type	Test group	Specimen	Loading head speed, mm/min	Loading rate, MPa/s	Linear limit, MPa	Bond strength, MPa
Deformed	DB-V1	DB-V1-1	0.4	0.117	13.83	19.09
		DB-V1-2	0.4	0.094	7.924	14.34
		DB-V1-3	0.4	0.099	11.75	15.16
	DB-V2	DB-V2-1	4	0.83	19.57	22.72
		DB-V2-2	4	0.83	19.41	19.70
		DB-V2-3	4	1.04	17.56	22.40
	DB-V3	DB-V3-1	40	7.0	21.48	25.42
		DB-V3-2	40	8.1	21.65	23.00
		DB-V3-3	40	6.1	11.47	24.97
	DB-V4	DB-V4-1	400	64	23.31	27.22
		DB-V4-2	400	57	27.38	27.99
		DB-V4-3	400	68	21.73	27.25
Plain	PB-V1	PB-V1-1	0.4	0.086	11.42	13.32
		PB-V1-2	0.4	0.098	12.21	12.67
		PB-V1-3	0.4	0.106	11.68	14.74
	PB-V2	PB-V2-1	4	0.96	11.85	14.45
		PB-V2-2	4	0.96	14.01	15.33
		PB-V2-3	4	1.02	11.07	15.92
	PB-V3	PB-V3-1	40	8.6	14.55	14.72
		PB-V3-2	40	8.5	10.18	15.56
		PB-V3-3	40	8.5	13.02	16.39
	PB-V4	PB-V4-1	400	61	15.05	16.89
		PB-V4-2	400	73	14.01	17.39
		PB-V4-3	400	82	12.55	15.36

Reproduced from [Rate-Dependent Bonding of Steel Reinforcement in Geopolymer Concrete] by [Dongming Yan] with permission from [ACI Materials Journal]

bars, as shown in Fig. 9.13d. As the loading rate increases, the formation of the powder layer also explains to some extent the observed rate dependence. Detailed observation shows that the failure interface transfers from the steel reinforced concrete interface to adjacent MKG concrete as the loading rate increases.

9.3.2.3 Bonding Behavior of Deformed Bar

The bond between deformed bars and concrete is formed through mechanical interlocking, along with improved interfacial adhesion resulting from the work of Wu

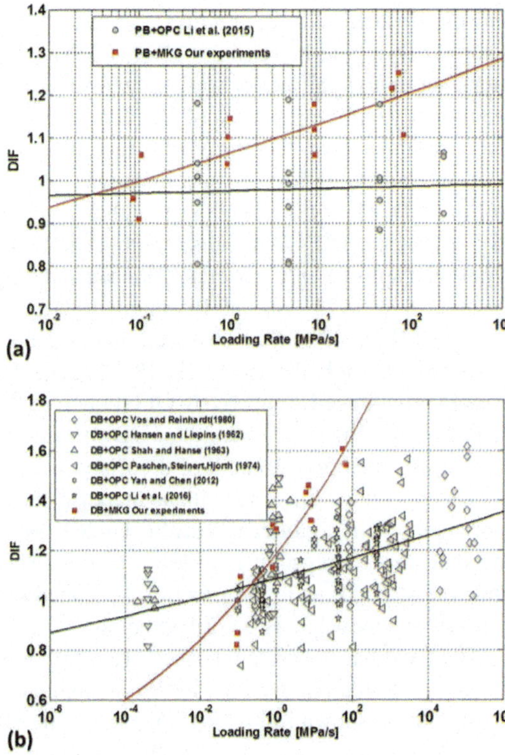

Fig. 9.16 DIF of bond strength for: **a** plain bars; and **b** deformed bars in MKG concrete (red) and OPC concrete (gray). Reproduced from [Rate-Dependent Bonding of Steel Reinforcement in Geopolymer Concrete] by [Dongming Yan] with permission from [ACI Materials Journal]

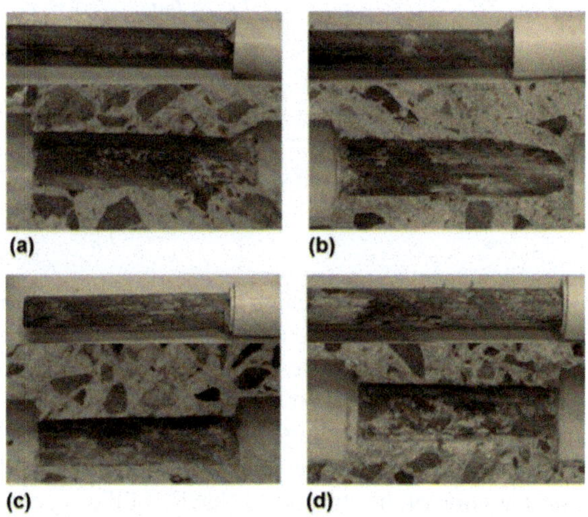

Fig. 9.17 Failed interface between plain steel bars and MKG concrete under loading rate of: **a** 0.4 mm/min; **b** 4 mm/min; **c** 40 mm/min; and **d** 400 mm/min. Reproduced from [Rate-Dependent Bonding of Steel Reinforcement in Geopolymer Concrete] by [Dongming Yan] with permission from [ACI Materials Journal]

and Chen (2015). In this section, the overall rate dependency is elucidated using a similar Dynamic Increasing Factor (DIF) analysis conducted previously. Furthermore, an analytical model was utilized to differentiate between the effect arising from the dynamic properties of MKG confinement and the effect originating from the chemical adhesion at the MKG-steel interface (Table 9.8).

The bond strength of DB specimen was fitted using the power function, as shown in Eq. (9.7). The fitting results show that $\overline{f}_{b,ref} = 17.35$ MPa and $\eta_b = 0.0749$. In Fig. 9.16b, the DIFs of bond strength of DB specimen were plotted together with reported results of OPC concrete (Hansen and Liepins 1962; Shah and Hansen 1963; Paschen et al. 1974; Vos and Reinhardt 1981; Yan and Chen 2012; Li et al. 2016).

In Fig. 9.16b, it can be observed that an increased loading rate results in higher bond strength for both MKG and OPC concrete. But, the rate effect in MKG concrete is more pronounced compared to OPC concrete. As the loading rate changes from 0.1 to 100 MPa/s, the bond strength of DB specimen shows an increase of around 55%. In contrast, the bond strength of deformed bars in OPC concrete experiences a less dependency in loading rate, with an increase of about 15% under similar loading rate increments. This significant rate effect in MKG concrete suggests a different bonding mechanism may present. In order to distinguish the effects of dynamic increase in concrete strength and rate effect of the concrete-steel interface, a bond mechanics analysis is conducted as described below.

The local bar-concrete interface properties are considered to analyze the pullout results, a unified bond model by Wu and Chen (2015) was referred. For specimen configuration shown in Fig. 9.18, the bond strength is dependent as

$$\frac{\overline{f}_b}{f_c} = \frac{h_{r1}}{s_{r0}}(1 + \kappa \cot \beta) \tag{9.9}$$

$$\frac{\overline{p}_n}{f_c} = \frac{h_{r1}}{s_{r0}}(\cot \beta - \kappa) \tag{9.10}$$

$$\kappa = \frac{\overline{f}_s}{\overline{f}_n} \tag{9.11}$$

where $\overline{f}_b$ is the bond strength and $\overline{p}_n$ is the confining pressure; f_c is the concrete compressive strength; h_{r1} is the lateral rib height and s_{r0} is the rib spacing; β is the effective bearing angle; κ is the interfacial bonding coefficient at the rib front, which represents the ratio between the shear stress ($\overline{f}_s$) to the normal stress ($\overline{f}_n$). It describes the local bonding ability of the bar-concrete interface.

The $\overline{p}_n/f_c$ ratio can be calculated by the thick-walled cylinder model with the consideration of the softening behavior of the concrete (Wang and Liu 2003):

$$\frac{\overline{p}_n}{f_c} = \frac{\overline{p}_n}{f_t}\frac{f_t}{f_c} = \Psi(\frac{\mathrm{c}}{\mathrm{d_s}})\frac{f_t}{f_c} \tag{9.12}$$

Table 9.8 Extract results of interfacial bonding coefficient

Data source	Material	s_{r0}/h_{r1}	c/d$_s$	Ψ(c/d$_s$)	$\dot{\sigma}$[a]	f_t/f_c	$\overline{f}_b/f_c$	κ
Our experiments	MKG	8.667	5.0	6.532	0.117	0.081	0.425	0.523
					0.094	0.082	0.322	0.360
					0.099	0.082	0.340	0.388
					0.83	0.080	0.468	0.597
					0.83	0.080	0.406	0.501
					1.04	0.080	0.458	0.582
					7.0	0.078	0.482	0.627
					8.1	0.078	0.433	0.553
					6.1	0.078	0.476	0.617
					64	0.077	0.472	0.623
					57	0.077	0.488	0.647
					68	0.077	0.472	0.623
Yan and Chen (2012)	OPC	10.462	5.9	7.541	0.158	0.098	0.681	0.724
					0.158	0.098	0.710	0.757
					0.158	0.098	0.740	0.792
					1.58	0.099	0.687	0.724
					1.58	0.099	0.743	0.787
					1.58	0.099	0.754	0.799
					15.8	0.100	0.748	0.787
					15.8	0.100	0.763	0.803
					15.8	0.100	0.791	0.834
					62	0.101	0.737	0.770
					62	0.101	0.755	0.790
					62	0.101	0.785	0.824
Morris (2015)	OPC	10.134	4.5	5.962	0.031	0.089	0.451	0.597
					0.034	0.089	0.498	0.669
					0.035	0.089	0.510	0.687
					0.80	0.091	0.538	0.719
					0.80	0.091	0.538	0.719
					43	0.092	0.525	0.690
					45	0.092	0.545	0.718
					46	0.092	0.555	0.732
Li et al. (2016)	OPC	8.981	2.9	7.374	0.44	0.093	0.412	0.444
					0.44	0.093	0.366	0.385
					0.44	0.093	0.320	0.323
					44.4	0.095	0.334	0.335
					44.4	0.095	0.299	0.289

(continued)

Table 9.8 (continued)

Data source	Material	s_{r0}/h_{r1}	c/d$_s$	Ψ(c/d$_s$)	$\dot{\sigma}$[a]	f_t/f_c	$\overline{f}_b/f_c$	κ
					44.4	0.095	0.344	0.348
					444	0.096	0.377	0.387
					444	0.096	0.310	0.300
					444	0.096	0.354	0.357

Reproduced from [Rate-Dependent Bonding of Steel Reinforcement in Geopolymer Concrete] by [Dongming Yan] with permission from [ACI Materials Journal]

[a]The unit of loading rate ($\dot{\sigma}$) is MPa/s (ksi./s)

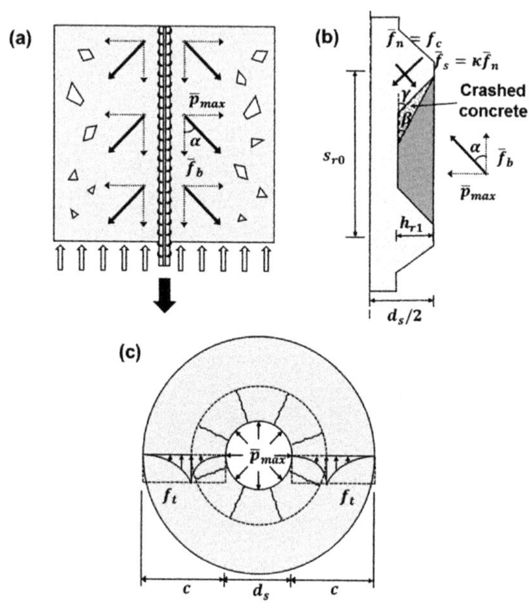

Fig. 9.18 Schematic representation of: **a** axial and radial stress; **b** near-rib stress analysis; and **c** hydraulically pressured thick-walled cylinder model with concrete softening behavior. Reproduced from [Rate-Dependent Bonding of Steel Reinforcement in Geopolymer Concrete] by [Dongming Yan] with permission from [ACI Materials Journal]

where, $\Psi(c/d_s)$ is a parameter related to the concrete cover thickness c and bar diameter d_s. It can be derived by Wang and Liu (2003).

Combining Eqs. (9.9), (9.10), and (9.12), the interfacial bonding coefficient κ can be determined from the bond strength and the concrete strength as follow:

$$\kappa = \frac{1}{2}\left(\sqrt{c_p^2 + 4c_b - 4} - c_p\right) \tag{9.13}$$

$$\cot\beta = \kappa + c_p \tag{9.14}$$

where,

$$\begin{cases} c_p = \Psi\left(\frac{c}{d_s}\right)\frac{s_{r0}}{h_{r1}}\frac{f_t}{f_c} \\ c_b = \frac{s_{r0}}{h_{r1}}\frac{\overline{f}_b}{f_c} \end{cases}$$

When the bar geometry and confinement configuration (h_{r1}, s_{r0}, c and d_s are known a priori) are given, the effective bearing angle β and the interfacial bonding coefficient κ could be calculated basing on the ratios $\overline{f}_b/f_c$ and f_t/f_c. The latter ones can be obtained from pull-out experiments and the dynamic properties of concrete which were extracted as detailed in following.

The compressive strength f_c and tensile strength f_t of MKG concrete are tested using the standard compression and splitting test. The results are listed in Table 9.9. They are further fitted by the power function similar to Eq. (9.7)

$$f_c = f_{c,ref}\left(\frac{\dot{\sigma}}{\dot{\sigma}_{c,ref}}\right)^{\eta_c} \tag{9.15}$$

$$f_t = f_{t,ref}\left(\frac{\dot{\sigma}}{\dot{\sigma}_{t,ref}}\right)^{\eta_t} \tag{9.16}$$

where, $\dot{\sigma}_{c,ref}$ is the reference loading rates for the compressive strength test (1 MPa/s) and $\dot{\sigma}_{t,ref}$ are that for the tensile strength test (0.1 MPa/s). The fitted reference compressive strength $f_{c,ref}$ is 48.88 MPa and reference tensile strength $f_{t,ref}$ is 3.64 MPa. The power index η_c and η_t are 0.0396 and 0.0299, respectively. The Eqs. (9.15) and (9.16) and the bond strength test results presented in Table 9.7 can be used to calculate the ratio $\overline{f}_b/f_c$ and f_t/f_c as shown in Fig. 9.19.

The ratio $\overline{f}_b/f_c$ is found to increase with the loading rate. At the same time, ratio f_t/f_c decreases. This is different from the expected positive correlation, so there must be a more potential mechanism to change the dependence of bonding strength on concrete strength, and the interface chemical bonding strength that depends on the rate is a possible reason.

The DIFs of f_c and f_t were calculated by Eq. (9.8) and plotted in Fig. 9.20. Results from Feng et al. (2015a, b) show a similar trend with this research, which is obtained from the test of fly-ash-based geopolymer concrete with compressive strengths ranging from 25.6 to 68.3 MPa. The CEB-recommended rate-dependent strength model (Comite Euro-International Du Beton (CEB) 1988) for OPC concrete with compressive strength of 45 MPa is also compared in Fig. 9.20.

Upon comparison, it is found that the rate effect of compressive strength show both in MKG concrete and fly-ash-based geopolymer concrete. But the rate effect of OPC concrete is significantly lower compared with the formers. The rate effect of tensile strength of MKG concrete is lower than that in fly-ash-based geopolymer concrete. At the same time, it is similar to that in OPC concrete. The variation in the DIF of MKG concrete strength does not exceed 25% when loading rate change from 0.1 to 100 MPa/s. This increase is approximately half of the observed enhancement (55%)

Table 9.9 Concrete strength

Test type	Test group	Specimen	Loading head speed, mm/min	Loading rate, MPa/s	Strength, MPa
Compressive	C-V1	C-V1-1	0.2	0.597	37.85
		C-V1-2	0.2	0.672	48.53
		C-V1-3	0.2	0.609	45.79
		C-V1-4	0.2	0.621	46.40
	C-V2	C-V2-1	2	6.54	54.55
		C-V2-2	2	6.34	57.34
		C-V2-3	2	5.88	49.25
		C-V2-4	2	5.56	56.20
	C-V3	C-V3-1	20	69.1	58.67
		C-V3-2	20	66.8	53.52
		C-V3-3	20	70.3	55.44
		C-V3-4	20	68.9	58.26
	C-V4	C-V4-1	200	656	69.52
		C-V4-2	200	670	64.76
		C-V4-3	200	681	59.60
		C-V4-4	200	675	66.70
Tensile	T-V1	T-V1-1	0.2	0.0140	3.49
		T-V1-2	0.2	0.0142	3.69
		T-V1-3	0.2	0.0138	3.26
		T-V1-4	0.2	0.0135	3.89
	T-V2	T-V2-1	2	0.210	3.12
		T-V2-2	2	0.186	3.45
		T-V2-3	2	0.223	3.90
		T-V2-4	2	0.191	3.87
	T-V3	T-V3-1	20	2.22	4.00
		T-V3-2	20	2.37	3.15
		T-V3-3	20	2.05	3.54
		T-V3-4	20	2.23	4.52
	T-V4	T-V4-1	200	20.8	3.36
		T-V4-2	200	21.1	4.01
		T-V4-3	200	20.1	5.40
		T-V4-4	200	20.2	4.96

Reproduced from [Rate-Dependent Bonding of Steel Reinforcement in Geopolymer Concrete] by [Dongming Yan] with permission from [ACI Materials Journal]

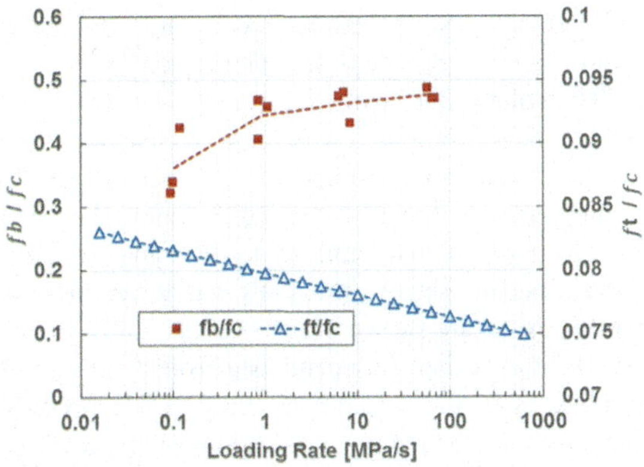

Fig. 9.19 fb/fc and ft/fc of MKG concrete. Reproduced from [Rate-Dependent Bonding of Steel Reinforcement in Geopolymer Concrete] by [Dongming Yan] with permission from [ACI Materials Journal]

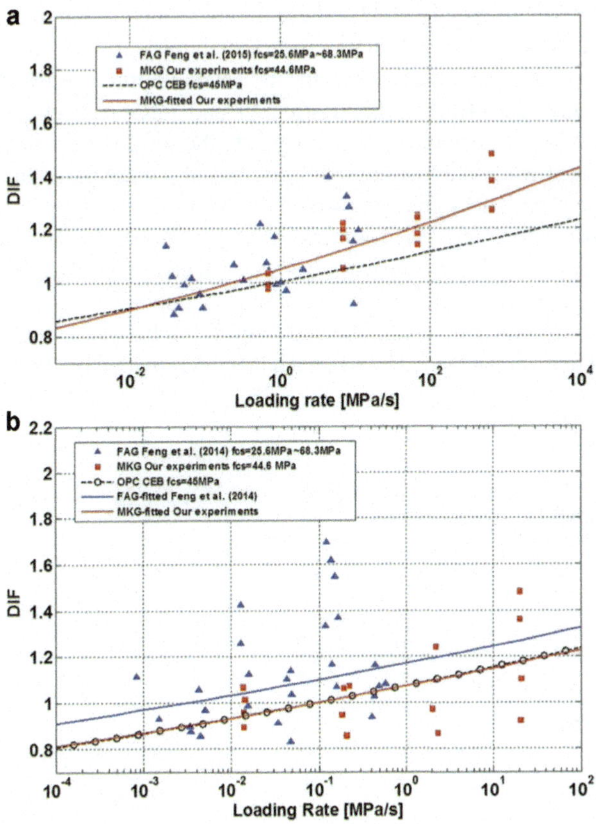

Fig. 9.20 DIFs of: **a** compressive strength; and **b** splitting tensile strength of MKG concrete. Reproduced from [Rate-Dependent Bonding of Steel Reinforcement in Geopolymer Concrete] by [Dongming Yan] with permission from [ACI Materials Journal]

in bond strength for DB specimens. Thus, the rate effect of the MKG concrete-steel bond is closely related to the interface.

The values of κ, extracted from the DB result of MKG concrete, are presented in Table 9.8 and compared with the κ values obtained from the reported results for OPC concrete interface (Yan and Chen 2012; Morris 2015; Li et al. 2016). To facilitate comparison in terms of rate dependency, the Dynamic Increasing Factors (DIFs) of κ were calculated similarly to the bond strength using Eq. (9.8). The DIFs were determined relative to the static value obtained at the lowest tested loading rate. These results are shown in Fig. 9.21. Notably, the κ values for the DB-OPC concrete interface exhibited independence from loading rates. In contrast, within a similar loading rates, the κ values for the DB-MKG concrete interface show a significant increase with the loading rate. Additionally, when considering the normalized κ with respect to that under 1 MPa/s, it follows a similar increasing trend as the DIF of bond strength for PB specimen (Fig. 9.22). These findings suggest that the primary factor contributing to the rate effect of interface chemical bond.

The observed bar appearance after pull-out test could further support the connection between the rate effect of bond strength and interfacial chemical bond. In Fig. 9.23, it can be seen that the scratch damage to the surrounding concrete becomes more pronounced when the loading rate is increased. Under high loading rates, the MKG concrete experiences greater crushing due to compression by the deformed bar rib. The enhanced crushing zone signifies an increase in bond adhesion attributed to the enhanced value of κ. This leads to a lower effective bearing angle β, resulting in higher shear stress along the interface. Similar phenomena were also observed in the bond behavior between enamel-coated reinforcing bars and OPC concrete (Wu et al. 2012), where the enhanced interfacial adhesion provided by the enamel coating

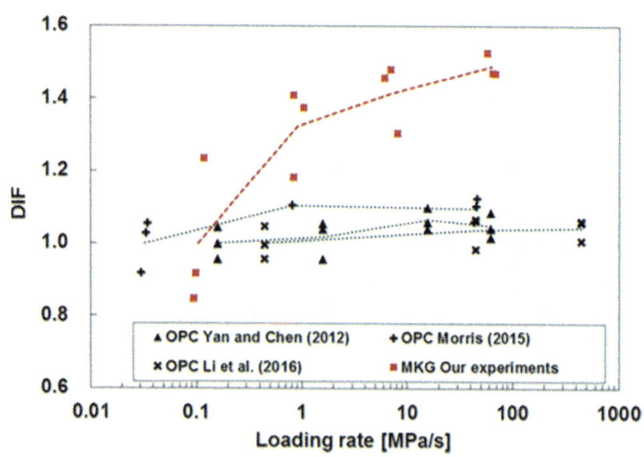

Fig. 9.21 DIF of interfacial bonding coefficient κ. Reproduced from [Rate-Dependent Bonding of Steel Reinforcement in Geopolymer Concrete] by [Dongming Yan] with permission from [ACI Materials Journal]

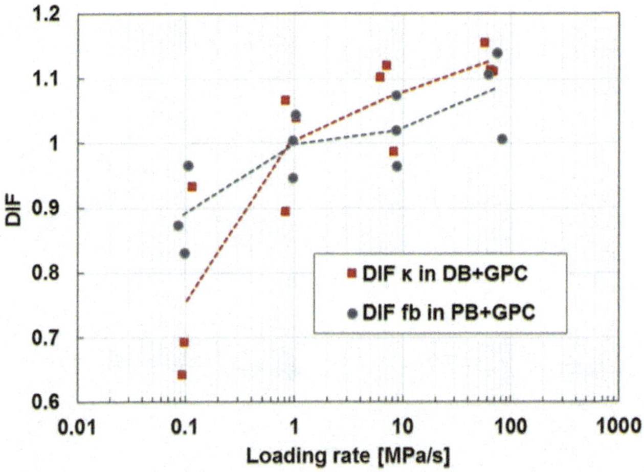

Fig. 9.22 Comparison of DIFs for κ and that of bond strength of plain bars in MKG concrete. Dashed lines are average values. Reproduced from [Rate-Dependent Bonding of Steel Reinforcement in Geopolymer Concrete] by [Dongming Yan] with permission from [ACI Materials Journal]

resulted in a higher κ and a significant decrease in angle β, accompanied by a higher bond strength compared to that of uncoated reinforcing bars.

9.3.2.4 Morphology of Bar-Concrete Interface in MKG Concrete

The experiments and analyses have provided evidence that the interfacial chemical bond between MKG concrete and steel bar has significant rate effect. To further investigate the underlying cause of this rate dependency, additional microstructural evidence is presented in this section.

Morphology images of the bar-concrete interface reveal the presence of a transition layer with a thickness ranging from 10 to 15 μm. It is between the steel bar and the surrounding MKG, as depicted in Fig. 9.24a. Within this layer, the porous gel in MKG gradually densifies as approaching the bar surface. It is within approximately 2 μm, as shown in Fig. 9.24b. This densification result to an increase in chemical bond between MKG and steel. In fact, Yong et al. (2007) show that iron aluminate oxides and Fe–O–Si bonds form at the interface between MKG and steel, resulting in strong chemical adhesion. As the chemical adhesion increases, bond failure is more likely to occur within the matrix surrounding the steel bar rather than along the interface. This phenomenon contributes to the observed rate dependency of the bond strength between MKG and steel.

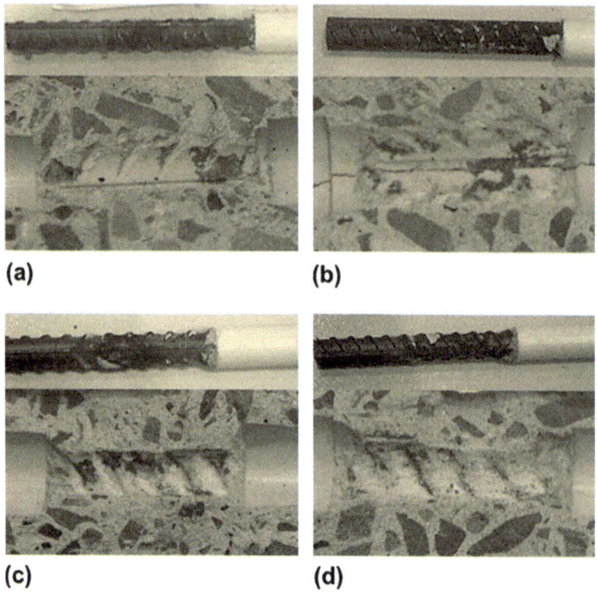

Fig. 9.23 Failed interface between deformed steel bars and MKG concrete under loading rate of: **a** 0.4 mm/min; **b** 4 mm/min; **c** 40 mm/min; and **d** 400 mm/min. Reproduced from [Rate-Dependent Bonding of Steel Reinforcement in Geopolymer Concrete] by [Dongming Yan] with permission from [ACI Materials Journal]

9.4 Conclusions

Based on the comparison and analysis of the bonding properties between steel bars in geopolymer concrete and ordinary concrete, the following conclusions and suggestions can be made:

(1) Geopolymer concrete exhibits strong shear resistance, allowing for efficient stress transfer from the steel bars. The anchorage performance of steel bars is primarily governed by the splitting resistance of the concrete.
(2) The anchorage performance of reinforcements in geopolymer concrete is sensitive to loading rate. Under dynamic loading conditions, bond slip and bond strength are higher compared to static loading, and there is an increase in failure energy consumption.
(3) The influence of loading rate on bond strength diminishes as the bonding length increases. When the bonding length exceeds 12 times the diameter of the reinforcement, the impact of loading rate on bond strength becomes negligible.
(4) Geopolymer concrete requires less protective layer thickness and bonding length compared to ordinary concrete to achieve the same bond strength for reinforcing bars. The critical protective thickness for HRB400 reinforcement in geopolymer concrete is approximately 2.4 times the diameter of the reinforcement, while the critical bonding length is about 3.8 times the diameter of the reinforcement

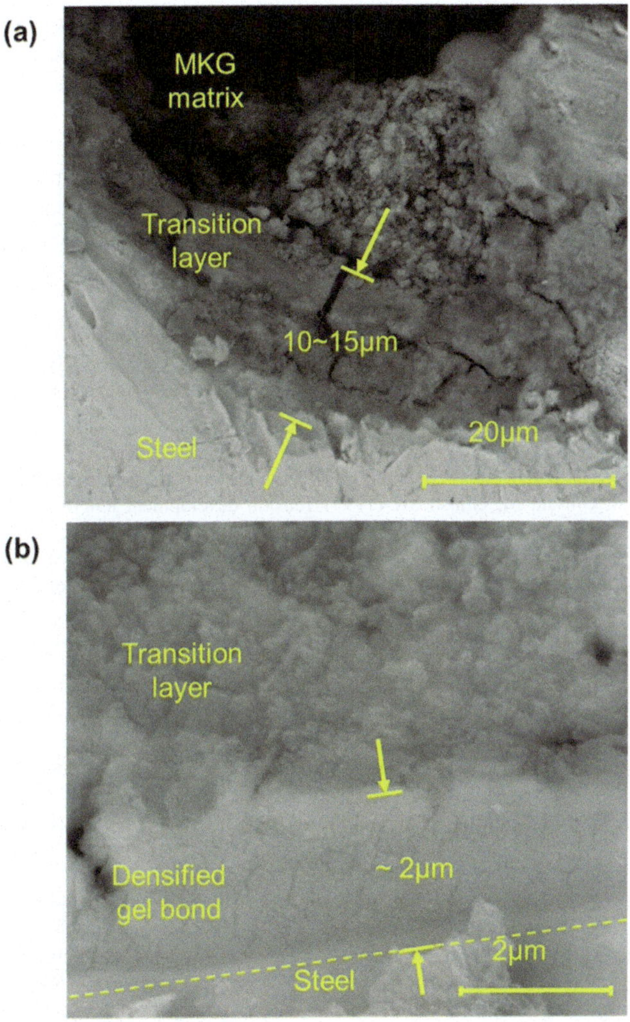

Fig. 9.24 SEM morphology of: **a** transition layer; and **b** local densified gel at MKG-steel interface. Reproduced from [Rate-Dependent Bonding of Steel Reinforcement in Geopolymer Concrete] by [Dongming Yan] with permission from [ACI Materials Journal]

for static loading and 3.1 times the diameter of the reinforcement for dynamic loading.

(5) Compared with OPC concrete, the bonding strength of MKG concrete exhibits a higher rate dependence. The smooth and deformed steel bars in MKG concrete exhibit a power law relationship with the loading rate. The change in Dynamic Increasing Factor (DIF) indicates that with the increase of loading rate, the chemical bonding between MKG and the steel bar interface is greatly enhanced.

(6) The analysis of bond stress indicates that the dynamic increase in strength of MKG concrete cannot fully explain the significant increase in bond strength with loading rate. The main reason for this enhancement is the change in bonding rate between MKG and the steel bar interface.
(7) Microscopic characterization shows that the rate dependence of bond strength of steel bars in geopolymer concrete mainly comes from the chemical adhesion enhancement at the interface. This chemical adhesion induced bond enhancement can reduce the splitting stress of steel bars under high-speed loading, thereby improving dynamic bond strength.

References

ACI Committee 408 R-03 (2003) Bond and development of straight reinforcing bars in tension

Azizinamini A, Chisala M, Ghosh SK (1995) Tension development length of reinforcing bars embedded in high-strength concrete. Eng Struct 17:512–522. https://doi.org/10.1016/0141-0296(95)00096-P

Berman AD, Ducker WA, Israelachvili JN (1996) Origin and characterization of different stick-slip friction mechanisms. Langmuir 12:4559–4562

Bischoff PH, Perry SH (1991) Compressive behaviour of concrete at high strain rates. Mater Struct 24:425–450

Castel A, Foster SJ (2015) Bond strength between blended slag and class F fly ash geopolymer concrete with steel reinforcement. Cem Concr Res 72:48–53. https://doi.org/10.1016/j.cemconres.2015.02.016

Comite Euro-International Du Beton (CEB) (1988) Concrete structures under impact and Impulsive loading—synthesis report. Lausanne, Switzerland

Dai JG, Munir S, Ding Z (2014) Comparative study of different cement-based inorganic pastes towards the development of FRIP strengthening technology. J Compos Constr 18.https://doi.org/10.1061/(ASCE)CC.1943-5614.0000420

Du F, Xiao J, Gao X (2006) An overview of test methods on the bond behavior between concrete and steel-bars. Struct Eng 22:93–97

Feng KN, Ruan D, Pan Z et al (2015a) Discussion: effect of strain rate on splitting tensile strength of geopolymer concrete. Mag Concr Res 67:906–907. https://doi.org/10.1680/macr.15.00089

Feng KN, Ruan D, Pan Z et al (2015b) Mechanical behavior of geopolymer concrete subjected to high strain rate compressive loadings. Mater Struct Constr 48:671–681. https://doi.org/10.1617/s11527-014-0322-7

Fernández-Jiménez AM, Palomo A, López-Hombrados C (2006) Engineering properties of alkali-activated fly ash concrete. ACI Mater J 103:106–112

Gambarova PG, Rosati GP (1996) Bond and splitting in reinforced concrete: test results on bar pull-out. Mater Struct Constr 29:267–276

GB/T50081-2010 (2010) Standard for test method of mechanical properties on ordinary concrete

Hansen RJ, Liepins AA (1962) Behavior of bond under dynamic loading. ACI J Proc pp 563–584

Hardjito D, Wallah SE, Sumajouw DMJ, Rangan BV (2004) On the development of fly ash-based geopolymer concrete. ACI Mater J 101:467–472. https://doi.org/10.14359/13485

Hardjito D, Wallah SE, Sumajouw DM, Rangan BV (2005) Fly ash-based geopolymer concrete. Aust J Struct Eng 6:77–86

Hung TC, Huang JS, Wang YW, Fan YC (2013) Microstructure and properties of metakaolin-based inorganic polymer foams. J Mater Sci 48:7446–7455. https://doi.org/10.1007/s10853-013-7559-3

Ji X, Song Y (2009) Mechanic analysis on the failure of bond behavior between concrete and steel bar when suffered from frost injury. J Hydraul Eng 40:1495–1499

Jiang D (1984) Bonding and slip of reinforcement and concrete. J Tongji Univ 8–17

Jin L, Du X (2012) Meso numerical simulation of reinforced concrete members. J Hydraul Eng 43:1230–1236

Khandelwal M, Ranjith PG, Pan Z, Sanjayan JG (2013) Effect of strain rate on strength properties of low-calcium fly-ash-based geopolymer mortar under dry condition. Arab J Geosci 6:2383–2389. https://doi.org/10.1007/s12517-011-0507-0

Latella BA, Perera DS, Escott TR, Cassidy DJ (2006) Adhesion of glass to steel using a geopolymer. J Mater Sci 41:1261–1264. https://doi.org/10.1007/s10853-005-4234-3

Li W, Xu J (2009) Mechanical properties of basalt fiber reinforced geopolymeric concrete under impact loading. Mater Sci Eng A 505:178–186. https://doi.org/10.1016/j.msea.2008.11.063

Li X, Wu Z, Zheng J, Dong W (2015) Effect of loading rate on the bond behavior of plain round bars in concrete under lateral pressure. Constr Build Mater 94:826–836. https://doi.org/10.1016/j.conbuildmat.2015.07.085

Li X, Wu Z, Zheng J, Alahdal A (2016) Effect of loading rate on bond behavior of deformed reinforcing bars in concrete under biaxial lateral pressures. J Struct Eng 142:04016027. https://doi.org/10.1061/(asce)st.1943-541x.0001479

Luo X, Xu J, Bai E, Li W (2012) Systematic study on the basic characteristics of alkali-activated slag-fly ash cementitious material system. Constr Build Mater 29:482–486. https://doi.org/10.1016/j.conbuildmat.2011.09.021

Mains RM (1951) Measurement of the distribution of tensile and bond stresses along reinforcing bars. ACI J Proc pp 225–252

Malvar LJ (1998) Review of static and dynamic properties of steel reinforcing bars. ACI Mater J 95:609–616

Morris GJ (2015) Experimental evaluation of local bond behaviour of deformed reinforcing bars in concrete structures. University of Canterbury

Paschen H, Steinert J, Hjorth O (1974) Untersuchung über das Verbundverhalten von Betonstählen bei Kurzzeitbeanspruchung: Bericht über den Forschungsauftrag des Bundesministers für Raumordnung. Bauwes Und Städtebau B II:9–81

Reinhardt HW, Balázs GL (1995) Steel-concrete interfaces: experimental aspects. Stud Appl Mech 42:255–279. https://doi.org/10.1016/S0922-5382(06)80014-0

RILEM/CEB/FIP (1983) Bond test for reinforcement steel: 2. Pull-out test (Revised Edition). CEB Manual on Concrete Reinforcement Technology

Sarker PK (2011) Bond strength of reinforcing steel embedded in fly ash-based geopolymer concrete. Mater Struct Constr 44:1021–1030. https://doi.org/10.1617/s11527-010-9683-8

Sarker PK, Haque R, Ramgolam KV (2013) Fracture behaviour of heat cured fly ash based geopolymer concrete. Mater Des 44:580–586. https://doi.org/10.1016/j.matdes.2012.08.005

Shah IK, Hansen RJ (1963) Behavior of bond under dynamic loading

Sofi M, Van Deventer JSJ, Mendis PA, Lukey GC (2007) Bond performance of reinforcing bars in inorganic polymer concrete (IPC). J Mater Sci 42:3107–3116. https://doi.org/10.1007/s10853-006-0534-5

Tambe NS, Bhushan B (2005) Friction model for the velocity dependence of nanoscale friction. Nanotechnology 16:2309–2324. https://doi.org/10.1088/0957-4484/16/10/054

Tassios TP, Vintzeleou EN (1987) Concrete-to-concrete friction. J Struct Eng (United States) 113:832–849. https://doi.org/10.1061/(ASCE)0733-9445(1987)113:4(832)

Tepfers R (1979) Cracking of concrete cover along anchored deformed reinforcing bars. Mag Concr Res 31:3–12. https://doi.org/10.1680/macr.1979.31.106.3

Torre-Casanova A, Jason L, Davenne L, Pinelli X (2013) Confinement effects on the steel-concrete bond strength and pull-out failure. Eng Fract Mech 97:92–104. https://doi.org/10.1016/j.engfracmech.2012.10.013

Ueng TH, Lyu SJ, Chu HW et al (2012) Adhesion at interface of geopolymer and cement mortar under compression: an experimental study. Constr Build Mater 35:204–210. https://doi.org/10.1016/j.conbuildmat.2012.03.008
Vos E, Reinhardt HW (1981) Bond resistance of deformed bars, plain bars and strands under impact loading. Delft, The Netherlands
Vos E, Reinhardt HW (1982) Influence of loading rate on bond behaviour of reinforcing steel and prestressing strands. Matériaux Constr 15:3–10. https://doi.org/10.1007/BF02473553
Wang X, Liu X (2003) A strain-softening model for steel-concrete bond. Cem Concr Res 33:1669–1673. https://doi.org/10.1016/S0008-8846(03)00137-6
Wu C, Chen G (2015) Unified model of local bond between deformed steel rebar and concrete: indentation analogy theory and validation. J Eng Mech 141. https://doi.org/10.1061/(ASCE)EM.1943-7889.0000945
Wu C, Chen G, Volz JS et al (2012) Local bond strength of vitreous enamel coated rebar to concrete. Constr Build Mater 35:428–439. https://doi.org/10.1016/j.conbuildmat.2012.04.067
Xu Y (1990) Experimental study on bond anchorage performance of deformed reinforcement concrete. Tsinghua University
Xu Y, Shao Z, Shen W (1988a) Bond strength between reinforcing bars and concrete. Build Sci 4:8–14
Xu Y, Shao Z, Shen W (1988b) Bond and anchorage strength between reinforcement and concrete. Build Sci 8–14
Xu Y, Shen W, Wang H (1994) Experimental study on bonding and anchoring performance of reinforced concrete. J Build Struct 26–37
Xu J, Li W, Fan F, Bai E (2009) Study on impact mechanical properties of geopolymer concrete. J Vib Shock 28:46–50
Yan D, Chen G (2012) Effect of loading rate on the bonding strength between rebar and concrete. Adv Mater Res 450–451:122–125. https://doi.org/10.4028/www.scientific.net/AMR.450-451.122
Yan D, Li H, Liu J, Zheng H (2014) Dynamic properties of concrete specimens with different size. J Hydraul Eng 45:95–99
Yan D, Chen S, Xu S (2016a) Bonding behavior between rebar and geopolymer concrete. J Hydraul Eng 47:1167–1176
Yan D, Chen S, Zeng Q et al (2016b) Correlating the elastic properties of metakaolin-based geopolymer with its composition. Mater Des 95:306–318. https://doi.org/10.1016/j.matdes.2016.01.107
Yong SL, Feng DW, Lukey GC, van Deventer JSJ (2007) Chemical characterisation of the steel-geopolymeric gel interface. Colloids Surfaces A Physicochem Eng Asp 302:411–423. https://doi.org/10.1016/j.colsurfa.2007.03.004
Zhang Y (2003) Study on formation mechanism and performance of high performance geopolymer concrete structure. Southeast University
Zhao Y, Jin W (2002) Test study on bond stress-slip relationship of concrete and steel bar. J Build Struct 23:32–37

SPRINGER NATURE

GPSR Compliance

The European Union's (EU) General Product Safety Regulation (GPSR) is a set of rules that requires consumer products to be safe and our obligations to ensure this.

If you have any concerns about our products, you can contact us on ProductSafety@springernature.com

In case Publisher is established outside the EU, the EU authorized representative is:

Springer Nature Customer Service Center GmbH
Europaplatz 3
69115 Heidelberg, Germany

The manufacturer's authorised representative in the EU is Springer Nature Customer Service Centre GmbH, Europaplatz 3, 69115 Heidelberg, Germany. If you have any concerns regarding our products, please contact ProductSafety@springernature.com

Printed and bound by CPI Group (UK) Ltd, Croydon, CR0 4YY
01/10/2026
02225870-0007